Learn LLVM 17

A beginner's guide to learning LLVM compiler tools
and core libraries with C++

Kai Nacke

Amy Kwan

BIRMINGHAM—MUMBAI

Learn LLVM 17

Group Product Manager: Kunal Sawant

Publishing Product Manager: Teny Thomas

Book Project Manager: Prajakta Naik

Senior Editor: Ruvika Rao and Nithya Sadanandan

Technical Editor: Jubit Pincy

Copy Editor: Safis Editing

Indexer: Pratik Shirodkar

Production Designer: Vijay Kamble

DevRel Marketing Coordinator: Shrinidhi Manoharan

Business Development Executive: Kriti Sharma

First published: April 2021

Second published: January 2024

Production reference: 1271223

Published by Packt Publishing Ltd.

Grosvenor House

11 St Paul's Square

Birmingham

B3 1R.

ISBN 978-1-83763-134-6

www.packtpub.com

Writing a book takes time and energy. Without the support and understanding of my wife, Tanya, and my daughter Polina, this book would not have been possible. Thank you both for always encouraging me!

Because of some personal challenges, this project was at risk, and I am grateful to Amy for joining me as an author. Without her, the book would not be as good as it is now.

Once again, the team at Packt not only provided guidance on my writing but also showed an understanding of my slow writing, and always motivated me to carry on. I owe them a great thank you.

- Kai Nacke

2023 has been a very transformative year for me, and contributing my knowledge of LLVM to this book has been one of the reasons why this year has been so significant. I never would have thought that I would be approached by Kai to embark on this exciting journey to share LLVM 17 with you all! Thank you to Kai, for his technical mentorship and guidance, the team at Packt, and, of course, to my family and close loved ones for providing me with the support and motivation in writing this book.

- Amy Kwan

Contributors

About the authors

Kai Nacke is a professional IT architect currently residing in Toronto, Canada. He holds a diploma in computer science from the Technical University of Dortmund, Germany. and his diploma thesis on universal hash functions was recognized as the best of the semester.

With over 20 years of experience in the IT industry, Kai has extensive expertise in the development and architecture of business and enterprise applications. In his current role, he evolves an LLVM/clang-based compiler.

For several years, Kai served as the maintainer of LDC, the LLVM-based D compiler. He is the author of *D Web Development* and *Learn LLVM 12*, both published by Packt. In the past, he was a speaker in the LLVM developer room at the **Free and Open Source Software Developers' European Meeting (FOSDEM)**.

Amy Kwan is a compiler developer currently residing in Toronto, Canada. Originally, from the Canadian prairies, Amy holds a Bachelor of Science in Computer Science from the University of Saskatchewan. In her current role, she leverages LLVM technology as a backend compiler developer. Previously, Amy has been a speaker at the **LLVM Developer Conference** in 2022 alongside Kai Nacke.

About the reviewers

Akash Kothari is a Research Assistant at the Illinois LLVM Compiler Research Lab. He earned his Ph.D. in Computer Science from the University of Illinois at Urbana-Champaign. Specializing in performance engineering, program synthesis, and formal semantics and verification, Akash's interests extend to exploring the history of computing and programming systems.

Shuo Niu, a Master of Engineering in computer engineering, is a dynamic force in the realm of compiler technology. With five prolific years at Intel PSG specializing in FPGA HLD compilers, he led innovations in the compiler middle-end optimizer. His expertise in developing cutting-edge features has empowered users to achieve remarkable performance enhancements on FPGA boards.

Table of Contents

Part 2: From Source to Machine Code Generation

3

Turning the Source File into an Abstract Syntax Tree   51

4

Basics of IR Code Generation   83

Part 3: Taking LLVM to the Next Level

8

9

10

Part 4: Roll Your Own Backend

11

The Target Description   283

12

Instruction Selection   317

13

Beyond Instruction Selection   353

Preface

Constructing a compiler is a complex and fascinating task. The LLVM project provides reusable components for your compiler and the LLVM core libraries implement a world-class optimizing code generator, which translates a source language-independent intermediate representation of machine code for all popular CPU architectures. The compilers for many programming languages already take advantage of LLVM technology.

This book teaches you how to implement your own compiler and how to use LLVM to achieve it. You will learn how the frontend of a compiler turns source code into an abstract syntax tree, and how to generate **Intermediate Representation** (**IR**) from it. Furthermore, you will also explore adding an optimization pipeline to your compiler, which allows you to compile the IR to performant machine code.

The LLVM framework can be extended in several ways, and you will learn how to add new passes, and even a completely new backend to LLVM. Advanced topics such as compiling for a different CPU architecture and extending clang and the clang static analyzer with your own plugins and checkers are also covered. This book follows a practical approach and is packed with example source code, which makes it easy to apply the gained knowledge within your own projects.

What's new in this edition

Learn LLVM 17 now features a new chapter dedicated to introducing the concept and syntax of the TableGen language used within LLVM, in which readers can leverage to define classes, records, and an entire LLVM backend. Furthermore, this book also presents an emphasis on backend development, which discusses various new backend concepts that can be implemented for an LLVM backend, such as implementing the GlobalISel instruction framework and developing machine function passes.

Who this book is for

This book is for compiler developers, enthusiasts, and engineers who are interested in learning about the LLVM framework. It is also useful for C++ software engineers looking to use compiler-based tools for code analysis and improvement, as well as casual users of LLVM libraries who want to gain more knowledge of LLVM essentials. Intermediate-level experience with C++ programming is mandatory to understand the concepts covered in this book more effectively.

What this book covers

Chapter 1, Installing LLVM, explains how to set up and use your development environment. At the end of the chapter, you will have compiled the LLVM libraries and learned how to customize the build process.

Chapter 2, The Structure of a Compiler, gives you an overview of the components of a compiler. At the end of the chapter, you will have implemented your first compiler producing LLVM IR.

Chapter 3, Turning the Source File into an Abstract Syntax Tree, teaches you in detail how to implement the frontend of a compiler. You will create your own frontend for a small programming language, ending with the construction of an abstract syntax tree.

Chapter 4, Basics of IR Code Generation, shows you how to generate LLVM IR from an abstract syntax tree. At the end of the chapter, you will have implemented a compiler for the example language, emitting assembly text or object code files as a result.

Chapter 5, IR Generation for High-Level Language Constructs, illustrates how you translate source language features commonly found in high-level programming languages to LLVM IR. You will learn about the translation of aggregate data types, the various options to implement class inheritance and virtual functions, and how to comply with the application binary interface of your system.

Chapter 6, Advanced IR Generation, shows you how to generate LLVM IR for exception-handling statements in the source language. You will also learn how to add metadata for type-based alias analysis, and how to add debug information to the generated LLVM IR, and you will extend your compiler-generated metadata.

Chapter 7, Optimizing IR, explains the LLVM pass manager. You will implement your own pass, both as part of LLVM and as a plugin, and you will learn how to add your new pass to the optimizing pass pipeline.

Chapter 8, The TableGen Language, introduces LLVM's own domain-specific language called TableGen. This language is used to reduce the coding effort of the developer, and you will learn about the different ways you can define data in the TableGen language, and how it can be leveraged in the backend.

Chapter 9, JIT Compilation, discusses how you can use LLVM to implement a **just-in-time** (**JIT**) compiler. By the end of the chapter, you will have implemented your own JIT compiler for LLVM IR in two different ways.

Chapter 10, Debugging Using LLVM Tools, explores the details of various libraries and components of LLVM, which helps you to identify bugs in your application. You will use the sanitizers to identify buffer overflows and other bugs. With the libFuzzer library, you will test functions with random data as input, and XRay will help you find performance bottlenecks. You will use the clang static analyzer to identify bugs at the source level, and you will learn that you can add your own checker to the analyzer. You will also learn how to extend clang with your own plugin.

Chapter 11, The Target Description, explains how you can add support for a new CPU architecture. This chapter discusses the necessary and optional steps like defining registers and instructions, developing instruction selection, and supporting the assembler and disassembler.

Chapter 12, Instruction Selection, demonstrates two different approaches to instruction selection, specifically explaining how SelectionDAG and GlobalISel work and showing how to implement these functionalities in a target, based on the example from the previous chapter. In addition, you will learn how to debug and test instruction selection.

Chapter 13, Beyond Instruction Selection, explains how you complete the backend implementation by exploring concepts beyond instruction selection. This includes adding new machine passes to implement target-specific tasks and points you to advanced topics that are not necessary for a simple backend but may be interesting for highly optimizing backends, such as cross-compilation to another CPU architecture.

To get the most out of this book

You need a computer running Linux, Windows, Mac OS X, or FreeBSD, with the development toolchain installed for the operating system. Please see the table for the required tools. All tools should be in the search path of your shell.

Software/Hardware covered in the book	OS Requirements
A C/C++ compiler: gcc 7.1.0 or later, clang 5.0 or later, Apple clang 10.0 or later, Visual Studio 2019 16.7 or later	Linux (any), Windows, Mac OS X, or FreeBSD
CMake 3.20.0 or later	
Ninja 1.11.1	
Python 3.6 or later	
Git 2.39.1 or later	

To create the flame graph in *Chapter 10, Debugging Using LLVM Tools*, you need to install the scripts from https://github.com/brendangregg/FlameGraph. To run the script, you also need a recent version of Perl installed, and to view the graph you need a web browser capable of displaying SVG files, which all modern browsers do. To see the Chrome Trace Viewer visualization in the same chapter, you need to have the Chrome browser installed.

If you are using the digital version of this book, we advise you to type the code yourself or access the code via the GitHub repository (link available in the next section). Doing so will help you avoid any potential errors related to the copying and pasting of code.

Download the example code files

You can download the example code files for this book from GitHub at https://github.com/PacktPublishing/Learn-LLVM-17. In case there's an update to the code, it will be updated on the existing GitHub repository.

We also have other code bundles from our rich catalog of books and videos available at https://github.com/PacktPublishing/. Check them out!

Conventions used

There are a number of text conventions used throughout this book.

Code in text: Indicates code words in text, database table names, folder names, filenames, file extensions, pathnames, dummy URLs, user input, and Twitter handles. Here is an example: "You can observe in the code that a quantum circuit operation is being defined and a variable called numOnes."

A block of code is set as follows:

```
#include "llvm/IR/IRPrintingPasses.h"
#include "llvm/IR/LegacyPassManager.h"
#include "llvm/Support/ToolOutputFile.h"
```

When we wish to draw your attention to a particular part of a code block, the relevant lines or items are set in bold:

```
  switch (Kind) {
// Many more cases
  case m88k:            return "m88k";
  }
```

Bold: Indicates a new term, an important word, or words that you see onscreen. For example, words in menus or dialog boxes appear in the text like this. Here is an example: "For development on OS X, it is best to install **Xcode** from the Apple store."

Tips or important notes
Appear like this.

Get in touch

Feedback from our readers is always welcome.

General feedback: If you have questions about any aspect of this book, mention the book title in the subject of your message and email us at customercare@packtpub.com.

Errata: Although we have taken every care to ensure the accuracy of our content, mistakes do happen. If you have found a mistake in this book, we would be grateful if you would report this to us. Please visit www.packtpub.com/support/errata, selecting your book, clicking on the Errata Submission Form link, and entering the details.

Piracy: If you come across any illegal copies of our works in any form on the Internet, we would be grateful if you would provide us with the location address or website name. Please contact us at copyright@packt.com with a link to the material.

If you are interested in becoming an author: If there is a topic that you have expertise in and you are interested in either writing or contributing to a book, please visit authors.packtpub.com.

Share Your Thoughts

Once you've read *Learn LLVM 17*, we'd love to hear your thoughts! Scan the QR code below to go straight to the Amazon review page for this book and share your feedback.

https://packt.link/r/1837631344

Your review is important to us and the tech community and will help us make sure we're delivering excellent quality content.

Download a free PDF copy of this book

Thanks for purchasing this book!

Do you like to read on the go but are unable to carry your print books everywhere?

Is your eBook purchase not compatible with the device of your choice?

Don't worry, now with every Packt book you get a DRM-free PDF version of that book at no cost.

Read anywhere, any place, on any device. Search, copy, and paste code from your favorite technical books directly into your application.

The perks don't stop there, you can get exclusive access to discounts, newsletters, and great free content in your inbox daily

Follow these simple steps to get the benefits:

1. Scan the QR code or visit the link below

https://packt.link/free-ebook/9781837631346

2. Submit your proof of purchase

3. That's it! We'll send your free PDF and other benefits to your email directly

Part 1:
The Basics of Compiler Construction with LLVM

In this section, you will learn how to compile LLVM by yourself and tailor the build to your needs. You will understand how LLVM projects are organized, and you will create your first project utilizing LLVM. Finally, you will explore the overall structure of a compiler, while creating a small compiler yourself.

This section comprises the following chapters:

1

Installing LLVM

In order to learn how to work with LLVM, it is best to begin by compiling LLVM from source. LLVM is an umbrella project and the GitHub repository contains the source of all projects belonging to LLVM. Each LLVM project is in a top-level directory of the repository. Besides cloning the repository, your system must also have all tools installed that are required by the build system. In this chapter, you will learn about the following topics:

- Getting the prerequisites ready, which will show you how to set up your build system

- Cloning the repository and building from source, which will cover how to get the LLVM source code, and how to compile and install the LLVM core libraries and clang with CMake and Ninja

- Customizing the build process, which will talk about the various possibilities for influencing the build process

Compiling LLVM versus installing binaries

You can install LLVM binaries from various sources. If you are using Linux, then your distribution contains the LLVM libraries. Why bother compiling LLVM yourself?

First, not all install packages contain all the files required for developing with LLVM. Compiling and installing LLVM yourself prevents this problem. Another reason stems from the fact that LLVM is highly customizable. With building LLVM, you learn how you can customize LLVM, and this will enable you to diagnose problems that may arise if you bring your LLVM application to another platform. And last, in the third part of this book, you will extend LLVM itself, and for this, you need the skill of building LLVM yourself.

However, it is perfectly fine to avoid compiling LLVM for the first steps. If you want to go on this route, then you only need to install the prerequisites as described in the next section.

> **Note**
>
> Many Linux distributions split LLVM into several packages. Please make sure that you install the development package. For example, on Ubuntu, you need to install the `llvm-dev` package. Please also make sure that you install LLVM 17. For other versions, the examples in this book may require changes.

Getting the prerequisites ready

To work with LLVM, your development system should run a common operating system such as Linux, FreeBSD, macOS, or Windows. You can build LLVM and clang in different modes. A build with debug symbols enabled can take up to 30 GB of space. The required disk space depends heavily on the chosen build options. For example, building only the LLVM core libraries in release mode, targeting only one platform, requires about 2 GB of free disk space, which is the bare minimum needed.

To reduce compile times, a fast CPU (such as a quad-core CPU with a 2.5 GHz clock speed) and a fast SSD are also helpful. It is even possible to build LLVM on a small device such as a Raspberry Pi – it only takes a lot of time. The examples within this book were developed on a laptop with an Intel quad-core CPU running at a 2.7 GHz clock speed, with 40 GB RAM and 2.5 TB SSD disk space. This system is well suited for the development task.

Your development system must have some prerequisite software installed. Let's review the minimal required version of these software packages.

To check out the source from GitHub, you need **Git** (`https://git-scm.com/`). There is no requirement for a specific version. The GitHub help pages recommend using at least version 1.17.10. Due to known security issues found in the past, it is recommended to use the latest available version, which is 2.39.1 at the time of writing.

The LLVM project uses **CMake** (`https://cmake.org/`) as the build file generator. At least the 3.20.0 version is required. CMake can generate build files for various build systems. In this book, **Ninja** (`https://ninja-build.org/`) is used because it is fast and available on all platforms. The latest version, 1.11.1, is recommended.

Obviously, you also need a **C/C++ compiler**. The LLVM projects are written in modern C++, based on the C++17 standard. A conforming compiler and standard library are required. The following compilers are known to work with LLVM 17:

- gcc 7.1.0 or later
- clang 5.0 or later
- Apple clang 10.0 or later
- Visual Studio 2019 16.7 or later

> Tip
>
> Please be aware that with further development of the LLVM project, the requirements for the compiler are most likely to change. In general, you should use the latest compiler version available for your system.

Python (`https://python.org/`) is used during the generation of the build files and for running the test suite. It should be at least the 3.8 version.

Although not covered in this book, there can be reasons that you need to use Make instead of Ninja. In this case, you need to use **GNU Make** (`https://www.gnu.org/software/make/`) version 3.79 or later. The usage of both build tools is very similar. It is sufficient to replace `ninja` in each command with `make` for the scenarios described below.

LLVM also depends on the `zlib` library (`https://www.zlib.net/`). You should have at least version 1.2.3.4 installed. As usual, we recommend using the latest version, 1.2.13.

To install the prerequisite software, the easiest way is to use the package manager from your operating system. In the following sections, the commands required to install the software are shown for the most popular operating systems.

Ubuntu

Ubuntu 22.04 uses the `apt` package manager. Most of the basic utilities are already installed; only the development tools are missing. To install all packages at once, you type the following:

```
$ sudo apt -y install gcc g++ git cmake ninja-build zlib1g-dev
```

Fedora and RedHat

The package manager of Fedora 37 and RedHat Enterprise Linux 9 is called `dnf`. Like Ubuntu, most of the basic utilities are already installed. To install all packages at once, you type the following:

```
$ sudo dnf -y install gcc gcc-c++ git cmake ninja-build \
    zlib-devel
```

FreeBSD

On FreeBSD 13 or later, you have to use the `pkg` package manager. FreeBSD differs from Linux-based systems in that the clang compiler is already installed. To install all other packages at once, you type the following:

```
$ sudo pkg install -y git cmake ninja zlib-ng
```

OS X

For development on OS X, it is best to install **Xcode** from the Apple store. While the Xcode IDE is not used in this book, it comes with the required C/C++ compilers and supporting utilities. For installation of the other tools, the package manager **Homebrew** (https://brew.sh/) can be used. To install all packages at once, you type the following:

```
$ brew install git cmake ninja zlib
```

Windows

Like OS X, Windows does not come with a package manager. For the C/C++ compiler, you need to download **Visual Studio Community 2022** (https://visualstudio.microsoft.com/vs/community/), which is free for personal use. Please make sure that you install the workload named **Desktop Development with C++**. You can use the package manager **Scoop** (https://scoop.sh/) to install the other packages. After installing Scoop as described on the website, you open **x64 Native Tools Command Prompt for VS 2022** from your Windows menu. To install the required packages, you type the following:

```
$ scoop install git cmake ninja python gzip bzip2 coreutils
$ scoop bucket add extras
$ scoop install zlib
```

Please watch the output from Scoop closely. For the Python and zlib packages, it advises adding some registry keys. These entries are needed so that other software can find these packages. To add the registry keys, you'd best copy and paste the output from Scoop, which looks like the following:

```
$ %HOMEPATH%\scoop\apps\python\current\install-pep-514.reg
$ %HOMEPATH%\scoop\apps\zlib\current\register.reg
```

After each command, a message window from the registry editor will pop up asking whether you really want to import those registry keys. You need to click on **Yes** to finish the import. Now all prerequisites are installed.

For all examples in this book, you must use the **x64 Native Tools Command Prompt** for VS 2022. Using this command prompt, the compiler is automatically added to the search path.

> Tip
>
> The LLVM code base is very large. To comfortably navigate the source, we recommend using an IDE that allows you to jump to the definition of classes and search through the source. We find **Visual Studio Code** (https://code.visualstudio.com/download), which is an extensible cross-platform IDE, very comfortable to use. However, this is no requirement for following the examples in this book.

Cloning the repository and building from source

With the build tools ready, you can now check out all LLVM projects from GitHub and build LLVM. This process is essentially the same on all platforms:

1. Configure Git.
2. Clone the repository.
3. Create the build directory.
4. Generate the build system files.
5. Finally, build and install LLVM.

Let's begin with configuring Git.

Configuring Git

The LLVM project uses Git for version control. If you have not used Git before, then you should do some basic configuration of Git first before continuing: to set the username and email address. Both pieces of information are used if you commit changes.

One can check whether they previously had an email and username already configured in Git with the following commands:

```
$ git config user.email
$ git config user.name
```

The preceding commands will output the respective email and username that you already have set when using Git. However, in the event that you are setting the username and email for the first time, the following commands can be entered for first-time configuration. In the following commands, you can simply replace Jane with your name and jane@email.org with your email:

```
$ git config --global user.email "jane@email.org"
$ git config --global user.name "Jane"
```

These commands change the global Git configuration. Inside a Git repository, you can locally overwrite those values by not specifying the --global option.

By default, Git uses the **vi** editor for commit messages. If you prefer another editor, then you can change the configuration in a similar way. To use the **nano** editor, you type the following:

```
$ git config --global core.editor nano
```

For more information about Git, please see the *Git Version Control Cookbook* (https://www.packtpub.com/product/git-version-control-cookbook-second-edition/9781789137545).

Now you are ready to clone the LLVM repository from GitHub.

Cloning the repository

The command to clone the repository is essentially the same on all platforms. Only on Windows, it is recommended to turn off the auto-translation of line endings.

On all non-Windows platforms, you type the following command to clone the repository:

```
$ git clone https://github.com/llvm/llvm-project.git
```

Only on Windows, add the option to disable auto-translation of line endings. Here, you type the following:

```
$ git clone --config core.autocrlf=false \
  https://github.com/llvm/llvm-project.git
```

This Git command clones the latest source code from GitHub into a local directory named `llvm-project`. Now change the current directory into the new `llvm-project` directory with the following command:

```
$ cd llvm-project
```

Inside the directory are all LLVM projects, each one in its own directory. Most notably, the LLVM core libraries are in the `llvm` subdirectory. The LLVM project uses branches for subsequent release development ("release/17.x") and tags ("llvmorg-17.0.1") to mark a certain release. With the preceding clone command, you get the current development state. This book uses LLVM 17. To check out the first release of LLVM 17 into a branch called `llvm-17`, you type the following:

```
$ git checkout -b llvm-17 llvmorg-17.0.1
```

With the previous steps, you cloned the whole repository and created a branch from a tag. This is the most flexible approach.

Git also allows you to clone only a branch or a tag (including history). With `git clone --branch release/17.x https://github.com/llvm/llvm-project`, you only clone the `release/17.x` branch and its history. You then have the latest state of the LLVM 17 release branch, so you only need to create a branch from the release tag like before if you need the exact release version. With the additional `--depth=1` option, which is known as a **shallow clone with Git**, you prevent the cloning of the history, too. This saves time and space but obviously limits what you can do locally, including checking out a branch based on the release tags.

Creating a build directory

Unlike many other projects, LLVM does not support inline builds and requires a separate build directory. Most easily, this is created inside the llvm-project directory, which is your current directory. Let us name the build directory, build, for simplicity. Here, the commands for Unix and Windows systems differ. On a Unix-like system, you use the following:

```
$ mkdir build
```

And on Windows, use the following:

```
$ md build
```

Now you are ready to create the build system files with the CMake tool inside this directory.

Generating the build system files

In order to generate build system files to compile LLVM and clang using Ninja, you run the following:

```
$ cmake -G Ninja -DCMAKE_BUILD_TYPE=Release \
    -DLLVM_ENABLE_PROJECTS=clang -B build -S llvm
```

The -G option tells CMake for which system to generate build files. Often-used values for that option are as follows:

- Ninja – for the Ninja build system
- Unix Makefiles – for GNU Make
- Visual Studio 17 VS2022 – for Visual Studio and MS Build
- Xcode – for Xcode projects

With the -B option, you tell CMake the path of the build directory. Similarly, you specify the source directory with the -S option. The generation process can be influenced by setting various variables with the -D option. Usually, they are prefixed with CMAKE_ (if defined by CMake) or LLVM_ (if defined by LLVM).

As mentioned previously, we are also interested in compiling clang alongside LLVM. With the LLVM_ENABLE_PROJECTS=clang variable setting, this allows CMake to generate the build files for clang in addition to LLVM. Furthermore, the CMAKE_BUILD_TYPE=Release variable tells CMake that it should generate build files for a release build.

The default value for the −G option depends on your platform, and the default value for the build type depends on the toolchain. However, you can define your own preference with environment variables. The CMAKE_GENERATOR variable controls the generator, and the CMAKE_BUILD_TYPE variable specifies the build type. If you use **bash** or a similar shell, then you can set the variables with the following:

```
$ export CMAKE_GENERATOR=Ninja
$ export CMAKE_BUILD_TYPE=Release
```

If you are using the Windows command prompt instead, then you set the variables with the following:

```
$ set CMAKE_GENERATOR=Ninja
$ set CMAKE_BUILD_TYPE=Release
```

With these settings, the command to create the build system files becomes the following, which is easier to type:

```
$ cmake -DLLVM_ENABLE_PROJECTS=clang -B build -S llvm
```

You will find more about CMake variables in the *Customizing the build process* section.

Compiling and installing LLVM

After the build files are generated, LLVM and clang can be compiled with the following:

```
$ cmake --build build
```

This command runs Ninja under the hood because we told CMake to generate Ninja files in the configuration step. However, if you generate build files for a system such as Visual Studio, which supports multiple build configurations, then you need to specify the configuration to use for the build with the --config option. Depending on the hardware resources, this command runs for between 15 minutes (server with lots of CPU cores, memory, and fast storage) and several hours (dual-core Windows notebook with limited memory).

By default, Ninja utilizes all available CPU cores. This is good for the speed of compilation but may prevent other tasks from running; for example, on a Windows-based notebook, it is almost impossible to surf the internet while Ninja is running. Fortunately, you can limit the resource usage with the −j option.

Let's assume you have four CPU cores available and Ninja should only use two (because you have parallel tasks to run); you then use this command for compilation:

```
$ cmake --build build -j2
```

After compilation is finished, a best practice is to run the test suite to check whether everything works as expected:

```
$ cmake --build build --target check-all
```

Again, the runtime of this command varies widely with the available hardware resources. The check-all Ninja target runs all test cases. Targets are generated for each directory containing test cases. Using check-llvm instead of check-all runs the LLVM tests but not the clang tests; check-llvm-codegen runs only the tests in the CodeGen directory from LLVM (that is, the llvm/test/CodeGen directory).

You can also do a quick manual check. One of the LLVM applications is **llc**, the LLVM compiler. If you run it with the -version option, it shows the LLVM version, the host CPU, and all supported architectures:

```
$ build/bin/llc --version
```

If you have trouble getting LLVM compiled, then you should consult the *Common Problems* section of the *Getting Started with the LLVM System* documentation https://releases.llvm.org/17.0.1/docs/GettingStarted.html#common-problems) for solutions to typical problems.

As the last step, you can install the binaries:

```
$ cmake --install build
```

On a Unix-like system, the install directory is /usr/local. On Windows, C:\Program Files\LLVM is used. This can be changed, of course. The next section explains how.

Customizing the build process

The CMake system uses a project description in the CMakeLists.txt file. The top-level file is in the llvm directory, llvm/CMakeLists.txt. Other directories also have CMakeLists.txt files, which are recursively included during the generation process.

Based on the information provided in the project description, CMake checks which compilers are installed, detects libraries and symbols, and creates the build system files, for example, build.ninja or Makefile (depending on the chosen generator). It is also possible to define reusable modules, for example, a function to detect whether LLVM is installed. These scripts are placed in the special cmake directory (llvm/cmake), which is searched automatically during the generation process.

The build process can be customized with the definition of CMake variables. The command-line option -D is used to set a variable to a value. The variables are used in the CMake scripts. Variables defined by CMake itself are almost always prefixed with CMAKE_ and these variables can be used in all projects. Variables defined by LLVM are prefixed with LLVM_ but they can only be used if the project definition includes the use of LLVM.

Variables defined by CMake

Some variables are initialized with the value of environment variables. Most notable are CC and CXX, which define the C and C++ compilers to be used for building. CMake tries to locate a C and a C++ compiler automatically, using the current shell search path. It picks the first compiler found. If you have several compilers installed, for example, gcc and clang or different versions of clang, then this might not be the compiler you want for building LLVM.

Suppose you like to use clang17 as a C compiler and clang++17 as a C++ compiler. Then, you can invoke CMake in a Unix shell in the following way:

```
$ CC=clang17 CXX=clang++17 cmake -B build -S llvm
```

This sets the value of the environment variables only for the invocation of cmake. If necessary, you can specify an absolute path for the compiler executables.

CC is the default value of the CMAKE_C_COMPILER CMake variable, and CXX is the default value of the CMAKE_CXX_COMPILER CMake variable. Instead of using the environment variables, you can set the CMake variables directly. This is equivalent to the preceding call:

```
$ cmake -DCMAKE_C_COMPILER=clang17 \
  -DCMAKE_CXX_COMPILER=clang++17 -B build -S llvm
```

Other useful variables defined by CMake are as follows:

Variable name	Purpose
CMAKE_INSTALL_PREFIX	This is a path prefix that is prepended to every path during installation. The default is /usr/local on Unix and C:\ Program Files\<Project> on Windows. To install LLVM in the /opt/llvm directory, you specify -DCMAKE_ INSTALL_PREFIX=/opt/llvm. The binaries are copied to / opt/llvm/bin, library files to /opt/llvm/lib, and so on.

Variable name	Purpose
CMAKE_BUILD_TYPE	Different types of build require different settings. For example, a debug build needs to specify options to generate debug symbols and usually link against debug versions of system libraries. In contrast, a release build uses optimization flags and links against production versions of libraries. This variable is only used for build systems that can only handle one build type, for example, Ninja or Make. For IDE build systems, all variants are generated and you have to use the mechanism of the IDE to switch between build types. Possible values are as follows: DEBUG: build with debug symbols RELEASE: build with optimization for speed RELWITHDEBINFO: release build with debug symbols MINSIZEREL: build with optimization for size The default build type is taken from the CMAKE_BUILD_TYPE environment variable. If this variable is not set, then the default depends on the used toolchain and is often empty. In order to generate build files for a release build, you specify -DCMAKE_BUILD_TYPE=RELEASE.
CMAKE_C_FLAGS CMAKE_CXX_FLAGS	These are extra flags used when compiling C and C++ source files. The initial values are taken from the CFLAGS and CXXFLAGS environment variables, which can be used as an alternative.
CMAKE_MODULE_PATH	This specifies additional directories that are searched for CMake modules. The specified directories are searched before the default ones. The value is a semicolon-separated list of directories.
PYTHON_EXECUTABLE	If the Python interpreter is not found or if the wrong one is picked in case you have installed multiple versions, you can set this variable to the path of the Python binary. This variable only has an effect if the Python module of CMake is included (which is the case for LLVM).

Table 1.1 - Additional useful variables provided by CMake

CMake provides built-in help for variables. The --help-variable var option prints help for the var variable. For instance, you can type the following to get help for CMAKE_BUILD_TYPE:

```
$ cmake --help-variable CMAKE_BUILD_TYPE
```

You can also list all variables with the following command:

```
$ cmake --help-variable-list
```

This list is very long. You may want to pipe the output to more or a similar program.

Using LLVM-defined build configuration variables

The build configuration variables defined by LLVM work in the same way as those defined by CMake except that there is no built-in help. The most useful variables are found in the following tables, where they are divided into variables that are useful for first-time users installing LLVM, and also variables for more advanced LLVM users.

Variables useful for first-time users installing LLVM

Variable name	Purpose
LLVM_TARGETS_TO_BUILD	LLVM supports code generation for different CPU architectures. By default, all these targets are built. Use this variable to specify the list of targets to build, separated by semicolons. The current targets are AArch64, AMDGPU, ARM, AVR, BPF, Hexagon, Lanai, LoongArch, Mips, MSP430, NVPTX, PowerPC, RISCV, Sparc, SystemZ, VE, WebAssembly, X86, and XCore. all can be used as shorthand for all targets. The names are case-sensitive. To only enable the PowerPC and the System Z target, you specify -DLLVM_TARGETS_TO_BUILD="PowerPC;SystemZ".
LLVM_EXPERIMENTAL_TARGETS_TO_BUILD	In addition to the official targets, the LLVM source tree also contains experimental targets. These targets are under development and often do not yet support the full functionality of a backend. The current list of experimental targets is ARC, CSKY, DirectX, M68k, SPIRV, and Xtensa. To build the M68k target, you specify -D LLVM_EXPERIMENTAL_TARGETS_TO_BUILD=M68k.

Variable name	Purpose
LLVM_ENABLE_PROJECTS	This is a list of the projects you want to build, separated by semicolons. The source for the projects must be on the same level as the llvm directory (side-by-side layout). The current list is bolt, clang, clang-tools-extra, compiler-rt, cross-project-tests, libc, libclc, lld, lldb, mlir, openmp, polly, and pstl. all can be used as shorthand for all projects in this list. Additionally, you can specify the flang project here. Due to some special build requirements, it is not yet part of the all list. To build clang and bolt together with LLVM, you specify -DLLVM_ENABLE_PROJECT="clang;bolt".

Table 1.2 - Useful variables for first-time LLVM users

Variables for advanced users of LLVM

LLVM_ENABLE_ASSERTIONS	If set to ON, then assertion checks are enabled. These checks help to find errors and are very useful during development. The default value is ON for a DEBUG build and otherwise OFF. To turn assertion checks on (e.g. for a RELEASE build), you specify –DLLVM_ENABLE_ASSERTIONS=ON.
LLVM_ENABLE_EXPENSIVE_CHECKS	This enables some expensive checks that can really slow down compilation speed or consume large amounts of memory. The default value is OFF. To turn these checks on, you specify -DLLVM_ENABLE_EXPENSIVE_CHECKS=ON.
LLVM_APPEND_VC_REV	LLVM tools such as llc display the LLVM version they are based on besides other information if the –version command-line option is given. This version information is based on the LLVM_REVISION C macro. By default, not only the LLVM version but also the current Git hash is part of the version information. This is handy in case you are following the development of the master branch because it makes clear on which Git commit the tool is based. If not needed, then this can be turned off with –DLLVM_APPEND_VC_REV=OFF.

`LLVM_ENABLE_THREADS`	LLVM automatically includes thread support if a threading library is detected (usually the `pthreads` library). Further, LLVM assumes in this case that the compiler supports **TLS (thread-local storage)**. If you don't want thread support or your compiler does not support TLS, then you can turn it off with `-DLLVM_ENABLE_THREADS=OFF`.
`LLVM_ENABLE_EH`	The LLVM projects do not use C++ exception handling and therefore turn exception support off by default. This setting can be incompatible with other libraries your project is linking with. If needed, you can enable exception support by specifying `-DLLVM_ENABLE_EH=ON`.
`LLVM_ENABLE_RTTI`	LLVM uses a lightweight, self-build system for runtime type information. The generation of C++ RTTI is turned off by default. Like the exception handling support, this may be incompatible with other libraries. To turn generation of C++ RTTI on, you specify `-DLLVM_ENABLE_RTTI=ON`.
`LLVM_ENABLE_WARNINGS`	Compiling LLVM should generate no warning messages if possible. The option to print warning messages is therefore turned on by default. To turn it off, you specify `-DLLVM_ENABLE_WARNINGS=OFF`.
`LLVM_ENABLE_PEDANTIC`	The LLVM source should be C/C++ language standard-conforming; hence, pedantic checking of the source is enabled by default. If possible, compiler-specific extensions are also disabled. To reverse this setting, you specify `-DLLVM_ENABLE_PEDANTIC=OFF`.
`LLVM_ENABLE_WERROR`	If set to `ON`, then all warnings are treated as errors – the compilation aborts as soon as warnings are found. It helps to find all remaining warnings in the source. By default, it is turned off. To turn it on, you specify `-DLLVM_ENABLE_WERROR=ON`.

LLVM_OPTIMIZED_TABLEGEN	Usually, the tablegen tool is built with the same options as all other parts of LLVM. At the same time, tablegen is used to generate large parts of the code generator. As a result, tablegen is much slower in a debug build, increasing the compile time noticeably. If this option is set to ON, then tablegen is compiled with optimization turned on even for a debug build, possibly reducing compile time. The default is OFF. To turn it on, you specify –DLLVM_OPTIMIZED_TABLEGEN=ON.
LLVM_USE_SPLIT_DWARF	If the build compiler is gcc or clang, then turning on this option will instruct the compiler to generate the DWARF debug information in a separate file. The reduced size of the object files reduces the link time of debug builds significantly. The default is OFF. To turn it on, you specify –LLVM_USE_SPLIT_DWARF=ON.

Table 1.3 - Useful variables for advanced LLVM users

> **Note**
>
> LLVM defines many more CMake variables. You can find the complete list in the LLVM documentation about CMake https://releases.llvm.org/17.0.1/docs/CMake.html#llvm-specific-variables. The preceding list contains only the ones you are most likely to need.

Summary

In this chapter, you prepared your development machine to compile LLVM. You cloned the GitHub repository and compiled your own version of LLVM and clang. The build process can be customized with CMake variables. You learned about useful variables and how to change them. Equipped with this knowledge, you can tweak LLVM to your needs.

In the next section, we will be taking a closer look at the structure of a compiler. We will be exploring the different components found inside the compiler, as well as different types of analyses that occur in it – specifically, the lexical, syntactical, and semantic analyses. Finally, we will also briefly touch on interfacing with an LLVM backend for code generation.

2

The Structure of a Compiler

Compiler technology is a well-studied field of computer science. The high-level task is to translate a source language into machine code. Typically, this task is divided into three parts, the **frontend**, the **middle end**, and the **backend**. The frontend deals mainly with the source language, while the middle end performs transformation to improve the code and the backend is responsible for the generation of machine code. Since the LLVM core libraries provide the middle end and the backend, we will focus on the frontend within this chapter.

In this chapter, you will cover the following sections and topics:

- *Building blocks of a compiler*, in which you will learn about the components typically found in a compiler

- *An arithmetic expression language*, which will introduce you to an example language and show how grammar is used to define a language

- *Lexical analysis*, which discusses how to implement a lexer for the language

- *Syntactical analysis*, which covers the construction of a parser from the grammar

- *Semantic analysis*, in which you will learn how a semantic check can be implemented

- *Code generation with the LLVM backend*, which discusses how to interface with the LLVM backend and glue all the preceding phases together to create a complete compiler

Building blocks of a compiler

Since computers became available, thousands of programming languages have been developed. It turns out that all compilers must solve the same tasks and that the implementation of a compiler is best structured according to these tasks. At a high level, there are three components. The frontend turns the source code into an **intermediate representation** (**IR**). Then the middle end performs transformations on the IR, with the goal of either improving performance or reducing the size of the code. Finally, the backend produces machine code from the IR. The LLVM core libraries provide a middle end consisting of very sophisticated transformations and backends for all popular platforms.

Furthermore, the LLVM core libraries also defines an intermediate representation used as input for the middle end and the backend. This design has the advantage that you only need to care about the frontend for the programming language you want to implement.

The input for the frontend is the source code, usually a text file. To make sense of it, the frontend first identifies the words of the language, such as numbers and identifiers, which are usually called tokens. This step is performed by the **lexer**. Next, the syntactical structure formed by the tokens is analyzed. The so-called **parser** performs this step, and the result is the **abstract syntax tree** (**AST**). Last, the frontend needs to check that the rules of the programming language are obeyed, which is done by the **semantic analyzer**. If no errors were detected, then the AST is transformed into IR and handed over to the middle end.

In the following sections, we will construct a compiler for an expression language, which produces LLVM IR from its input. The LLVM `llc` static compiler, representing the backend, can then be used to compile the IR into object code. It all begins with defining the language. Keep in mind that all of the C++ implementation of the files within this chapter will be contained within a directory called `src/`.

An arithmetic expression language

Arithmetic expressions are a part of every programming language. Here is an example of an arithmetic expression calculation language called **calc**. The calc expressions are compiled into an application that evaluates the following expression:

```
with a, b: a * (4 + b)
```

The used variables in the expression must be declared with the keyword, `with`. This program is compiled into an application that asks the user for the values of the a and b variables and prints the result.

Examples are always welcome but, as a compiler writer, you need a more thorough specification than this for implementation and testing. The vehicle for the syntax of the programming language is the grammar.

Formalism for specifying the syntax of a programming language

The elements of a language, for example, keywords, identifiers, strings, numbers, and operators, are called **tokens**. In this sense, a program is a sequence of tokens, and the grammar specifies which sequences are valid.

Usually, grammar is written in the **extended Backus-Naur form** (**EBNF**). A rule in grammar has a left and a right side. The left side is just a single symbol called **non-terminal**. The right side of a rule consists of non-terminals, tokens, and meta-symbols for alternatives and repetitions. Let's have a look at the grammar of the calc language:

```
calc : ("with" ident ("," ident)* ":")? expr ;
expr : term (( "+" | "-" ) term)* ;
```

```
term : factor (( "*" | "/") factor)* ;
factor : ident | number | "(" expr ")" ;
ident : ([a-zAZ])+ ;
number : ([0-9])+ ;
```

In the first line, `calc` is a non-terminal. If not otherwise stated, then the first non-terminal of a grammar is the start symbol. The colon (`:`) is the separator between the left and the right side of the rule. Here, `"with"`, `","` and `":"` are tokens that represent this string. Parentheses are used for grouping. A group can be optional or repeated. A question mark (`?`) after the closing parenthesis denotes an optional group. A star `*` denotes zero or more repetitions and a plus `+` denotes one or more repetitions. `Ident` and `expr` are non-terminals. For each of them, another rule exists. The semicolon (`;`) marks the end of a rule. The pipe `|`, in the second line, denotes an alternative. And last, the brackets `[ ]`, in the last two lines, denote a character class. The valid characters are written inside the brackets. For example, the character class `[a-zA-Z]` matches an upper- or lower-case letter, and `([a-zA-Z])+` matches one or more of these letters. This corresponds to a regular expression.

How does grammar help the compiler writer?

Such grammar may look like a theoretical toy, but it is of value to the compiler writer. First, all the tokens are defined, which is needed to create the lexical analyzer. The rules of the grammar can be translated into the parser. And of course, if questions arise about whether the parser works correctly, then the grammar serves as a good specification.

However, grammar does not define all aspects of a programming language. The meaning – the semantics – of the syntax must also be defined. Formalisms for this purpose were developed, too, but very often, they are specified in plain text, as they were usually drawn up at the initial introduction of the language.

Equipped with this knowledge, the next two sections show how the lexical analysis turns the input into a sequence of tokens and how the grammar is coded in C++ for the syntactical analysis.

Lexical analysis

As already seen in the example in the previous section, a programming language consists of many elements such as keywords, identifiers, numbers, operators, and so on. The task of the lexical analyzer is to take the textual input and create a sequence of tokens from it. The calc language consists of the tokens `with`, `:`, `+`, `-`, `*`, `/`, `(`, `)`, and regular expressions `([a-zA-Z])+` (an identifier) and `([0-9])+` (a number). We assign a unique number to each token to make the handling of tokens easier.

A hand-written lexer

The implementation of a lexical analyzer is often called `Lexer`. Let's create a header file called `Lexer.h` and get started with the definition of `Token`. It begins with the usual header guard and the inclusion of the required headers:

```
#ifndef LEXER_H
#define LEXER_H

#include "llvm/ADT/StringRef.h"
#include "llvm/Support/MemoryBuffer.h"
```

The `llvm::MemoryBuffer` class provides read-only access to a block of memory, filled with the content of a file. On request, a trailing zero character (`'\x00'`) is added to the end of the buffer. We use this feature to read through the buffer without checking the length of the buffer at each access. The `llvm::StringRef` class encapsulates a pointer to a C string and its length. Because the length is stored, the string need not be terminated with a zero character (`'\x00'`) like normal C strings. This allows an instance of `StringRef` to point to the memory managed by `MemoryBuffer`.

With this in mind, we begin by implementing the `Lexer` class:

1. First, the `Token` class contains the definition of the enumeration for the unique token numbers mentioned previously:

   ```
   class Lexer;

   class Token {
     friend class Lexer;

   public:
     enum TokenKind : unsigned short {
       eoi, unknown, ident, number, comma, colon, plus,
       minus, star, slash, l_paren, r_paren, KW_with
     };
   ```

   Besides defining a member for each token, we added two additional values: `eoi` and `unknown`. `eoi` stands for *end of input* and is returned when all characters of the input are processed. `unknown` is used in the event of an error at the lexical level, e.g., # is no token of the language and would therefore be mapped to `unknown`.

2. In addition to the enumeration, the class has a `Text` member, which points to the start of the text of the token. It uses the `StringRef` class mentioned previously:

   ```
   private:
     TokenKind Kind;
   ```

```
   llvm::StringRef Text;

public:
   TokenKind getKind() const { return Kind; }
   llvm::StringRef getText() const { return Text; }
```

This is useful for semantic processing, e.g., for an identifier, it is useful to know the name.

3.   The `is()` and `isOneOf()` methods are used to test whether the token is of a certain kind. The `isOneOf()` method uses a variadic template, allowing a variable number of arguments:

```
   bool is(TokenKind K) const { return Kind == K; }
   bool isOneOf(TokenKind K1, TokenKind K2) const {
     return is(K1) || is(K2);
   }
   template <typename... Ts>
   bool isOneOf(TokenKind K1, TokenKind K2, Ts... Ks) const {
     return is(K1) || isOneOf(K2, Ks...);
   }
};
```

4.   The `Lexer` class itself has a similar simple interface and comes next in the header file:

```
class Lexer {
   const char *BufferStart;
   const char *BufferPtr;

public:
   Lexer(const llvm::StringRef &Buffer) {
     BufferStart = Buffer.begin();
     BufferPtr = BufferStart;
   }

   void next(Token &token);

private:
   void formToken(Token &Result, const char *TokEnd,
                  Token::TokenKind Kind);
};
#endif
```

Except for the constructor, the public interface has only the `next()` method, which returns the next token. The method acts like an iterator, always advancing to the next available token. The only members of the class are pointers to the beginning of the input and the next unprocessed character. It is assumed that the buffer ends with a terminating 0 (just like a C string).

5. Let's implement the `Lexer` class in the `Lexer.cpp` file. It begins with some helper functions to classify characters:

```cpp
#include "Lexer.h"

namespace charinfo {
LLVM_READNONE inline bool isWhitespace(char c) {
  return c == ' ' || c == '\t' || c == '\f' ||
         c == '\v' ||
c == '\r' || c == '\n';
}


LLVM_READNONE inline bool isDigit(char c) {
  return c >= '0' && c <= '9';
}


LLVM_READNONE inline bool isLetter(char c) {
  return (c >= 'a' && c <= 'z') ||
         (c >= 'A' && c <= 'Z');
}
}
```

These functions are used to make conditions more readable.

> **Note**
>
> We are not using the functions provided by the `<cctype>` standard library header for two reasons. First, these functions change behavior based on the locale defined in the environment. For example, if the locale is a German-language area, then German umlauts can be classified as letters. This is usually not wanted in a compiler. Second, since the functions have `int` as a parameter type, a conversion from the `char` type is required. The result of this conversion depends on whether `char` is treated as a signed or unsigned type, causing portability problems.

6. From the grammar in the previous section, we know all the tokens of the language. But the grammar does not define the characters that should be ignored. For example, a space or newline character adds only whitespace and are often ignored. The `next()` method begins with ignoring these characters:

```cpp
void Lexer::next(Token &token) {
  while (*BufferPtr &&
         charinfo::isWhitespace(*BufferPtr)) {
    ++BufferPtr;
  }
```

7. Next, make sure that there are still characters left to process:

```
if (!*BufferPtr) {
    token.Kind = Token::eoi;
    return;
}
```

There is at least one character to process.

8. We first check whether the character is lowercase or uppercase. In this case, the token is either an identifier or the `with` keyword, because the regular expression for the identifier also matches the keyword. The most common solution here is to collect the characters matched by the regular expression and check whether the string happens to be the keyword:

```
if (charinfo::isLetter(*BufferPtr)) {
    const char *end = BufferPtr + 1;
    while (charinfo::isLetter(*end))
        ++end;
    llvm::StringRef Name(BufferPtr, end - BufferPtr);
    Token::TokenKind kind =
        Name == "with" ? Token::KW_with : Token::ident;
    formToken(token, end, kind);
    return;
}
```

The `formToken()` private method is used to populate the token.

9. Next, we check for a number. The code for this is very similar to the preceding snippet:

```
else if (charinfo::isDigit(*BufferPtr)) {
    const char *end = BufferPtr + 1;
    while (charinfo::isDigit(*end))
        ++end;
    formToken(token, end, Token::number);
    return;
}
```

Now only the tokens defined by fixed strings are left.

10. This is done easily with a `switch`. As all of these tokens have only one character, the `CASE` preprocessor macro is used to reduce typing:

```
else {
    switch (*BufferPtr) {
#define CASE(ch, tok) \
case ch: formToken(token, BufferPtr + 1, tok); break
CASE('+', Token::plus);
CASE('-', Token::minus);
```

```
CASE('*', Token::star);
CASE('/', Token::slash);
CASE('(', Token::Token::l_paren);
CASE(')', Token::Token::r_paren);
CASE(':', Token::Token::colon);
CASE(',', Token::Token::comma);
#undef CASE
```

11. Last, we need to check for unexpected characters:

```
    default:
      formToken(token, BufferPtr + 1, Token::unknown);
    }
    return;
  }
}
```

Only the `formToken()` private helper method is still missing.

12. It populates the members of the `Token` instance and updates the pointer to the next unprocessed character:

```
void Lexer::formToken(Token &Tok, const char *TokEnd,
                      Token::TokenKind Kind) {
  Tok.Kind = Kind;
  Tok.Text = llvm::StringRef(BufferPtr,
                             TokEnd - BufferPtr);
  BufferPtr = TokEnd;
}
```

In the next section, we have a look at how to construct a parser for syntactical analysis.

Syntactical analysis

The syntactical analysis is done by the parser, which we will implement next. The base of this is the grammar and the lexer from the previous sections. The result of the parsing process is a dynamic data structure called an **abstract syntax tree** (**AST**). The AST is a very condensed representation of the input and is well-suited for semantic analysis.

First, we will implement the parser, and after that, we will have a look at the parsing process within the AST.

A hand-written parser

The interface of the parser is defined in the header file, `Parser.h`. It begins with some include declarations:

```
#ifndef PARSER_H
#define PARSER_H

#include "AST.h"
#include "Lexer.h"
#include "llvm/Support/raw_ostream.h"
```

The `AST.h` header file declares the interface for the AST and is shown later. The coding guidelines from LLVM forbid the use of the `<iostream>` library, therefore, the header of the equivalent LLVM functionality is included. It is needed to emit an error message:

1. The `Parser` class first declares some private members:

   ```
   class Parser {
     Lexer &Lex;
     Token Tok;
     bool HasError;
   ```

   `Lex` and `Tok` are instances of the classes from the previous section. `Tok` stores the next token (the look-ahead) and `Lex` is used to retrieve the next token from the input. The `HasError` flag indicates whether an error was detected.

2. A couple of methods deal with the token:

   ```
   void error() {
     llvm::errs() << "Unexpected: " << Tok.getText()
                  << "\n";
     HasError = true;
   }

   void advance() { Lex.next(Tok); }

   bool expect(Token::TokenKind Kind) {
     if (Tok.getKind() != Kind) {
       error();
       return true;
     }
     return false;
   }

   bool consume(Token::TokenKind Kind) {
   ```

```
    if (expect(Kind))
        return true;
    advance();
    return false;
}
```

`advance()` retrieves the next token from the lexer. `expect()` tests whether the look-ahead has the expected kind and emits an error message if not. Finally, `consume()` retrieves the next token if the look-ahead has the expected kind. If an error message is emitted, the `HasError` flag is set to true.

3. For each non-terminal of the grammar, a method to parse the rule is declared:

```
AST *parseCalc();
Expr *parseExpr();
Expr *parseTerm();
Expr *parseFactor();
```

> **Note:**
> There are no methods for `ident` and `number`. Those rules only return the token and are replaced by the corresponding token.

4. The public interface follows. The constructor initializes all members and retrieves the first token from the lexer:

```
public:
    Parser(Lexer &Lex)  : Lex(Lex), HasError(false) {
        advance();
    }
```

5. A function is required to get the value of the error flag:

```
bool hasError() { return HasError; }
```

6. And finally, the `parse()` method is the main entry point into parsing:

```
AST *parse();
};

#endif
```

Parser implementation

Let's dive into the implementation of the parser!

1. Our implementation in the `Parser.cpp` file and begins with the `parse()` method:

    ```
    #include "Parser.h"

    AST *Parser::parse() {
      AST *Res = parseCalc();
      expect(Token::eoi);
      return Res;
    }
    ```

    The main point of the `parse()` method is that the whole input has been consumed. Do you remember that the parsing example in the first section added a special symbol to denote the end of the input? We check it here.

2. The `parseCalc()` method implements the corresponding rule. It's worth having a closer look at this method as the other parsing methods follow the same patterns. Let's recall the rule from the first section:

    ```
    calc : ("with" ident ("," ident)* ":")? expr ;
    ```

3. The method begins with declaring some local variables:

    ```
    AST *Parser::parseCalc() {
      Expr *E;
      llvm::SmallVector<llvm::StringRef, 8> Vars;
    ```

4. The first decision to be made is whether the optional group must be parsed or not. The group begins with the `with` token, so we compare the token to this value:

    ```
    if (Tok.is(Token::KW_with)) {
      advance();
    ```

5. Next, we expect an identifier:

    ```
        if (expect(Token::ident))
          goto _error;
        Vars.push_back(Tok.getText());
        advance();
    ```

    If there is an identifier, then we save it in the `Vars` vector. Otherwise, it is a syntax error, which is handled separately.

6. Next in the grammar follows a repeating group, which parses more identifiers, separated with commas:

```
while (Tok.is(Token::comma)) {
    advance();
    if (expect(Token::ident))
      goto _error;
    Vars.push_back(Tok.getText());
    advance();
}
```

By now, this should not be surprising. The repetition group begins with the token (,). The test for the token becomes the condition of the while loop, implementing zero or more repetition. The identifier inside the loop is treated as before.

7. Finally, the optional group requires a colon at the end:

```
    if (consume(Token::colon))
      goto _error;
}
```

8. Last, the rule for expr must be parsed:

```
    E = parseExpr();
```

9. With this call, the parsing of the rule is finished successfully. The collected information is now used to create the AST node for this rule:

```
    if (Vars.empty()) return E;
    else return new WithDecl(Vars, E);
```

Now only the error handling is missing. Detecting a syntax error is easy but recovering from it is surprisingly complicated. Here, a simple approach called **panic mode** is used.

In panic mode, tokens are deleted from the token stream until one is found that the parser can use to continue its work. Most programming languages have symbols that denote an end, e.g., in C++, a ; (end of a statement) or a } (end of a block). Such tokens are good candidates to look for.

On the other hand, the error can be that the symbol we are looking for is missing. In this case, probably a lot of tokens are deleted before the parser can continue. This is not as bad as it sounds. Today, it is more important that a compiler is fast. In the event of an error, the developer looks at the first error message, fixes it, and restarts the compiler. This is quite different from using punch cards, where it was important to get as many error messages as possible, as the next run of the compiler would possibly be only on the next day.

Error handling

Instead of using some arbitrary tokens to look for, another set of tokens is used here. For each non-terminal, there is a set of tokens that can follow this non-terminal in a rule:

1. In the case of `calc`, only the end of input follows this non-terminal. The implementation is trivial:

   ```
   _error:
     while (!Tok.is(Token::eoi))
       advance();
     return nullptr;
   }
   ```

2. The other parsing methods are similarly constructed. `parseExpr()` is the translation of the rule for `expr`:

   ```
   Expr *Parser ::parseExpr() {
     Expr *Left = parseTerm() ;
     while (Tok.isOneOf(Token::plus, Token::minus)) {
       BinaryOp::Operator Op =
           Tok.is(Token::plus) ? BinaryOp::Plus :
                                 BinaryOp::Minus;
       advance();
       Expr *Right = parseTerm();
       Left = new BinaryOp(Op, Left, Right);
     }
     return Left;
   }
   ```

   The repeated group inside the rule is translated as a `while` loop. Note how the use of the `isOneOf()` method simplifies the check for several tokens.

3. The coding of the `term` rule looks the same:

   ```
   Expr *Parser::parseTerm() {
     Expr *Left = parseFactor();
     while (Tok.isOneOf(Token::star, Token::slash)) {
       BinaryOp::Operator Op =
           Tok.is(Token::star) ? BinaryOp::Mul :
                                 BinaryOp::Div;
       advance();
       Expr *Right = parseFactor();
       Left = new BinaryOp(Op, Left, Right);
     }
     return Left;
   }
   ```

This method is strikingly similar to `parseExpr()`, and you may be tempted to combine them into one. In a grammar, it is possible to have one rule dealing with multiplicative and additive operators. The advantage of using two rules instead is that then the precedence of the operators fits well with the mathematical order of evaluation. If you combine both rules, then you need to figure out the evaluation order somewhere else.

4. Last, you need to implement the rule for `factor`:

```
Expr *Parser::parseFactor() {
  Expr *Res = nullptr;
  switch (Tok.getKind()) {
  case Token::number:
    Res = new Factor(Factor::Number, Tok.getText());
    advance(); break;
```

Instead of using a chain of `if` and `else if` statements, a `switch` statement seems more suitable here, because each alternative begins with just one token. In general, you should think about which translation patterns you like to use. If you later need to change the parsing methods, then it is an advantage if not every method has a different way of implementing a grammar rule.

5. If you use a `switch` statement, then error handling happens in the `default` case:

```
case Token::ident:
  Res = new Factor(Factor::Ident, Tok.getText());
  advance(); break;
case Token::l_paren:
  advance();
  Res = parseExpr();
  if (!consume(Token::r_paren)) break;
default:
  if (!Res) error();
```

We guard emitting the error message here because of the fall-through.

6. If there was a syntax error in the parenthesis's expression, then an error message was already emitted. The guard prevents a second error message:

```
    while (!Tok.isOneOf(Token::r_paren, Token::star,
                        Token::plus, Token::minus,
                        Token::slash, Token::eoi))
      advance();
  }
  return Res;
}
```

That was easy, wasn't it? Once you have memorized the patterns used, it is almost tedious work to code the parser based on the grammar rules. This type of parser is called a **recursive descent parser**.

> **A recursive descent parser can't be constructed from every grammar**
>
> A grammar must satisfy certain conditions to be suitable for the construction of a recursive descent parser. This class of grammar is called LL(1). In fact, most grammar that you can find on the internet does not belong to this class of grammar. Most books about the theory of compiler constructions explain the reason for this. The classic book on this topic is the so-called *dragon book*, *Compilers: Principles, Techniques, and Tools* by Aho, Lam, Sethi, and Ullman.

The abstract syntax tree

The result of the parsing process is the AST. The AST is another compact representation of the input program. It captures the essential information. Many programming languages have symbols that are needed as separators but do not carry further meaning. For example, in C++, a semicolon ; denotes the end of a single statement. Of course, this information is important for the parser. As soon as we turn the statement into an in-memory representation, the semicolon is not important anymore and can be dropped.

If you look at the first rule of the example expression language, then it is clear that the `with` keyword, the comma (`,`), and the colon (`:`) are not important for the meaning of a program. What is important is the list of declared variables, which could be used in the expression. The result is that only a couple of classes are required to record the information: `Factor` holds a number or an identifier, `BinaryOp` holds the arithmetic operator and the left and right sides of an expression, and `WithDecl` stores the list of declared variables and the expression. `AST` and `Expr` are only used to create a common class hierarchy.

In addition to the information from the parsed input, tree traversal using the **visitor pattern** is also supported. It's all in `AST.h` header file:

1. It begins with the visitor interface:

    ```cpp
    #ifndef AST_H
    #define AST_H

    #include "llvm/ADT/SmallVector.h"
    #include "llvm/ADT/StringRef.h"

    class AST;
    class Expr;
    class Factor;
    class BinaryOp;
    class WithDecl;

    class ASTVisitor {
    public:
      virtual void visit(AST &) {};
    ```

```
  virtual void visit(Expr &){};
  virtual void visit(Factor &) = 0;
  virtual void visit(BinaryOp &) = 0;
  virtual void visit(WithDecl &) = 0;
};
```

The visitor pattern needs to know each class to visit. Because each class also refers to the visitor, we declare all classes at the top of the file. Please note that the `visit()` methods for AST and Expr have a default implementation, which does nothing.

2. The AST class is the root of the hierarchy:

```
class AST {
public:
  virtual ~AST() {}
  virtual void accept(ASTVisitor &V) = 0;
};
```

3. Similarly, Expr is the root for AST classes related to expressions:

```
class Expr : public AST {
public:
  Expr() {}
};
```

4. The Factor class stores a number or the name of a variable:

```
class Factor : public Expr {
public:
  enum ValueKind { Ident, Number };

private:
  ValueKind Kind;
  llvm::StringRef Val;

public:
  Factor(ValueKind Kind, llvm::StringRef Val)
      : Kind(Kind), Val(Val) {}
  ValueKind getKind() { return Kind; }
  llvm::StringRef getVal() { return Val; }
  virtual void accept(ASTVisitor &V) override {
    V.visit(*this);
  }
};
```

In this example, numbers and variables are treated almost identically, therefore, we decided to create only one AST node class to represent them. The `Kind` member tells us which of both cases the instances represent. In more complex languages, you usually want to have different AST classes, such as a `NumberLiteral` class for numbers and a `VariableAccess` class for a reference to a variable.

5. The `BinaryOp` class holds the data needed for evaluating an expression:

```cpp
class BinaryOp : public Expr {
public:
  enum Operator { Plus, Minus, Mul, Div };

private:
  Expr *Left;
  Expr *Right;
  Operator Op;

public:
  BinaryOp(Operator Op, Expr *L, Expr *R)
      : Op(Op), Left(L), Right(R) {}
  Expr *getLeft() { return Left; }
  Expr *getRight() { return Right; }
  Operator getOperator() { return Op; }
  virtual void accept(ASTVisitor &V) override {
    V.visit(*this);
  }
};
```

In contrast to the parser, the `BinaryOp` class makes no distinction between multiplicative and additive operators. The precedence of the operators is implicitly available in the tree structure.

6. And last, the `WithDecl` class stores the declared variables and the expression:

```cpp
class WithDecl : public AST {
  using VarVector =
                  llvm::SmallVector<llvm::StringRef, 8>;
  VarVector Vars;
  Expr *E;

public:
  WithDecl(llvm::SmallVector<llvm::StringRef, 8> Vars,
          Expr *E)
      : Vars(Vars), E(E) {}
  VarVector::const_iterator begin()
                                { return Vars.begin(); }
  VarVector::const_iterator end() { return Vars.end(); }
```

```
      Expr *getExpr() { return E; }
      virtual void accept(ASTVisitor &V) override {
        V.visit(*this);
      }
  };
  #endif
```

The AST is constructed during parsing. The semantic analysis checks that the tree adheres to the meaning of the language (e.g., that used variables are declared) and possibly augments the tree. After that, the tree is used for code generation.

Semantic analysis

The semantic analyzer walks the AST and checks various semantic rules of the language, e.g. a variable must be declared before use or types of variables must be compatible in an expression. The semantic analyzer can also print out warnings if it finds a situation that can be improved. For the example expression language, the semantic analyzer must check that each used variable is declared because that is what the language requires. A possible extension (which is not implemented here) is to print a warning if a declared variable is not used.

The semantic analyzer is implemented in the Sema class, which is performed by the semantic() method. Here is the complete Sema.h header file:

```
#ifndef SEMA_H
#define SEMA_H

#include "AST.h"
#include "Lexer.h"

class Sema {
public:
  bool semantic(AST *Tree);
};

#endif
```

The implementation is in the Sema.cpp file. The interesting part is the semantic analysis, which is implemented using a visitor. The basic idea is that the name of each declared variable is stored in a set. During the creation of the set, each name can be checked for uniqueness, and later it can be checked that the given name is in the set:

```
#include "Sema.h"
#include "llvm/ADT/StringSet.h"
```

```
namespace {
class DeclCheck : public ASTVisitor {
  llvm::StringSet<> Scope;
  bool HasError;

  enum ErrorType { Twice, Not };

  void error(ErrorType ET, llvm::StringRef V) {
    llvm::errs() << "Variable " << V << " "
                 << (ET == Twice ? "already" : "not")
                 << " declared\n";
    HasError = true;
  }

public:
  DeclCheck() : HasError(false) {}

  bool hasError() { return HasError; }
```

As in the `Parser` class, a flag is used to indicate that an error occurred. The names are stored in a set called `Scope`. On a `Factor` node that holds a variable name, it is checked that the variable name is in the set:

```
virtual void visit(Factor &Node) override {
  if (Node.getKind() == Factor::Ident) {
    if (Scope.find(Node.getVal()) == Scope.end())
      error(Not, Node.getVal());
  }
};
```

For a `BinaryOp` node, there is nothing to check other than that both sides exist and are visited:

```
virtual void visit(BinaryOp &Node) override {
  if (Node.getLeft())
    Node.getLeft()->accept(*this);
  else
    HasError = true;
  if (Node.getRight())
    Node.getRight()->accept(*this);
  else
    HasError = true;
};
```

On a `WithDecl` node, the set is populated and the walk over the expression is started:

```
virtual void visit(WithDecl &Node) override {
  for (auto I = Node.begin(), E = Node.end(); I != E;
       ++I) {
    if (!Scope.insert(*I).second)
      error(Twice, *I);
  }
  if (Node.getExpr())
    Node.getExpr()->accept(*this);
  else
    HasError = true;
  };
};
}
```

The `semantic()` method only starts the tree walk and returns the error flag:

```
bool Sema::semantic(AST *Tree) {
  if (!Tree)
    return false;
  DeclCheck Check;
  Tree->accept(Check);
  return Check.hasError();
}
```

If required, much more could be done here. It would also be possible to print a warning if a declared variable is not used. We leave this for you to implement as an exercise. If the semantic analysis finishes without error, then we can generate the LLVM IR from the AST. This is done in the next section.

Generating code with the LLVM backend

The task of the backend is to create optimized machine code from the LLVM IR of a module. The IR is the interface to the backend and can be created using a C++ interface or in textual form. Again, the IR is generated from the AST.

Textual representation of LLVM IR

Before trying to generate the LLVM IR, it should be clear what we want to generate. For our example expression language, the high-level plan is as follows:

1. Ask the user for the value of each variable.

2. Calculate the value of the expression.

3. Print the result.

To ask the user to provide a value for a variable and to print the result, two library functions are used: `calc_read()` and `calc_write()`. For the `with a: 3*a` expression, the generated IR is as follows:

1. The library functions must be declared, like in C. The syntax also resembles C. The type before the function name is the return type. The type names surrounded by parenthesis are the argument types. The declaration can appear anywhere in the file:

   ```
   declare i32 @calc_read(ptr)
   declare void @calc_write(i32)
   ```

2. The `calc_read()` function takes the variable name as a parameter. The following construct defines a constant, holding a and the null byte used as a string terminator in C:

   ```
   @a.str = private constant [2 x i8] c"a\00"
   ```

3. It follows the `main()` function. The parameter names are omitted because they are not used. Just as in C, the body of the function is enclosed in braces:

   ```
   define i32 @main(i32, ptr) {
   ```

4. Each basic block must have a label. Because this is the first basic block of the function, we name it `entry`:

   ```
   entry:
   ```

5. The `calc_read()` function is called to read the value for the a variable. The nested `getelemenptr` instruction performs an index calculation to compute the pointer to the first element of the string constant. The function result is assigned to the unnamed `%2` variable.

   ```
   %2 = call i32 @calc_read(ptr @a.str)
   ```

6. Next, the variable is multiplied by 3:

   ```
   %3 = mul nsw i32 3, %2
   ```

7. The result is printed on the console via a call to the `calc_write()` function:

   ```
   call void @calc_write(i32 %3)
   ```

8. Last, the `main()` function returns 0 to indicate a successful execution:

   ```
   ret i32 0
   }
   ```

Each value in the LLVM IR is typed, with `i32` denoting the 32-bit bit integer type and `ptr` denoting a pointer.

> **Note**
>
> Previous versions of LLVM used typed pointers. For example, a pointer to a byte was expressed as i8* in LLVM. Since LLVM 16, **opaque pointers** are the default. An opaque pointer is just a pointer to memory, without carrying any type information about it. The notation in LLVM IR is `ptr`.

Since it is now clear what the IR looks like, let's generate it from the AST.

Generating the IR from the AST

The interface, provided in the `CodeGen.h` header file, is very small:

```
#ifndef CODEGEN_H
#define CODEGEN_H

#include "AST.h"

class CodeGen
{
public:
  void compile(AST *Tree);
};
#endif
```

Because the AST contains the information, the basic idea is to use a visitor to walk the AST. The `CodeGen.cpp` file is implemented as follows:

1. The required includes are at the top of the file:

    ```
    #include "CodeGen.h"
    #include "llvm/ADT/StringMap.h"
    #include "llvm/IR/IRBuilder.h"
    #include "llvm/IR/LLVMContext.h"
    #include "llvm/Support/raw_ostream.h"
    ```

2. The namespace of the LLVM libraries is used for name lookups:

    ```
    using namespace llvm;
    ```

3. First, some private members are declared in the visitor. Each compilation unit is represented in LLVM by the `Module` class and the visitor has a pointer to the module called M. For easy IR generation, the `Builder` (of type `IRBuilder<>`) is used. LLVM has a class hierarchy to represent types in IR. You can look up the instances for basic types such as i32 from the LLVM context.

These basic types are used very often. To avoid repeated lookups, we cache the needed type instances: `VoidTy`, `Int32Ty`, `PtrTy`, and `Int32Zero`. The V member is the current calculated value, which is updated through the tree traversal. And last, `nameMap` maps a variable name to the value returned from the `calc_read()` function:

```
namespace {
class ToIRVisitor : public ASTVisitor {
  Module *M;
  IRBuilder<> Builder;
  Type *VoidTy;
  Type *Int32Ty;
  PointerType *PtrTy;
  Constant *Int32Zero;
  Value *V;
  StringMap<Value *> nameMap;
```

4. The constructor initializes all members:

```
public:
  ToIRVisitor(Module *M) : M(M), Builder(M->getContext())
  {
    VoidTy = Type::getVoidTy(M->getContext());
    Int32Ty = Type::getInt32Ty(M->getContext());
    PtrTy = PointerType::getUnqual(M->getContext());
    Int32Zero = ConstantInt::get(Int32Ty, 0, true);
  }
```

5. For each function, a `FunctionType` instance must be created. In C++ terminology, this is a function prototype. A function itself is defined with a `Function` instance. The `run()` method defines the `main()` function in the LLVM IR first:

```
void run(AST *Tree) {
  FunctionType *MainFty = FunctionType::get(
      Int32Ty, {Int32Ty, PtrTy}, false);
  Function *MainFn = Function::Create(
      MainFty, GlobalValue::ExternalLinkage,
      "main", M);
```

6. Then we create the BB basic block with the `entry` label, and attach it to the IR builder:

```
  BasicBlock *BB = BasicBlock::Create(M->getContext(),
                                      "entry", MainFn);
  Builder.SetInsertPoint(BB);
```

7. With this preparation done, the tree traversal can begin:

    ```
    Tree->accept(*this);
    ```

8. After the tree traversal, the computed value is printed via a call to the `calc_write()` function. Again, a function prototype (an instance of `FunctionType`) has to be created. The only parameter is the current value, V:

    ```
    FunctionType *CalcWriteFnTy =
        FunctionType::get(VoidTy, {Int32Ty}, false);
    Function *CalcWriteFn = Function::Create(
        CalcWriteFnTy, GlobalValue::ExternalLinkage,
        "calc_write", M);
    Builder.CreateCall(CalcWriteFnTy, CalcWriteFn, {V});
    ```

9. The generation finishes by returning 0 from the `main()` function:

    ```
    Builder.CreateRet(Int32Zero);
    }
    ```

10. A `WithDecl` node holds the names of the declared variables. First, we create a function prototype for the `calc_read()` function:

    ```
    virtual void visit(WithDecl &Node) override {
    FunctionType *ReadFty =
        FunctionType::get(Int32Ty, {PtrTy}, false);
    Function *ReadFn = Function::Create(
        ReadFty, GlobalValue::ExternalLinkage,
        "calc_read", M);
    ```

11. The method loops through the variable names:

    ```
    for (auto I = Node.begin(), E = Node.end(); I != E;
        ++I) {
    ```

12. For each variable, a string with a variable name is created:

    ```
    StringRef Var = *I;
    Constant *StrText = ConstantDataArray::getString(
        M->getContext(), Var);
    GlobalVariable *Str = new GlobalVariable(
        *M, StrText->getType(),
        /*isConstant=*/true,
        GlobalValue::PrivateLinkage,
        StrText, Twine(Var).concat(".str"));
    ```

13. Then the IR code to call the `calc_read()` function is created. The string created in the previous step is passed as a parameter:

```
CallInst *Call =
    Builder.CreateCall(ReadFty, ReadFn, {Str});
```

14. The returned value is stored in the `mapNames` map for later use:

```
    nameMap[Var] = Call;
}
```

15. The tree traversal continues with the expression:

```
    Node.getExpr()->accept(*this);
};
```

16. A `Factor` node is either a variable name or a number. For a variable name, the value is looked up in the `mapNames` map. For a number, the value is converted to an integer and turned into a constant value:

```
virtual void visit(Factor &Node) override {
  if (Node.getKind() == Factor::Ident) {
    V = nameMap[Node.getVal()];
  } else {
    int intval;
    Node.getVal().getAsInteger(10, intval);
    V = ConstantInt::get(Int32Ty, intval, true);
  }
};
```

17. And last, for a `BinaryOp` node, the right calculation operation must be used:

```
virtual void visit(BinaryOp &Node) override {
  Node.getLeft()->accept(*this);
  Value *Left = V;
  Node.getRight()->accept(*this);
  Value *Right = V;
  switch (Node.getOperator()) {
  case BinaryOp::Plus:
    V = Builder.CreateNSWAdd(Left, Right); break;
  case BinaryOp::Minus:
    V = Builder.CreateNSWSub(Left, Right); break;
  case BinaryOp::Mul:
    V = Builder.CreateNSWMul(Left, Right); break;
  case BinaryOp::Div:
    V = Builder.CreateSDiv(Left, Right); break;
```

```
        }
      };
    };
  }
```

18. With this, the visitor class is complete. The `compile()` method creates the global context and the module, runs the tree traversal, and dumps the generated IR to the console:

```
void CodeGen::compile(AST *Tree) {
  LLVMContext Ctx;
  Module *M = new Module("calc.expr", Ctx);
  ToIRVisitor ToIR(M);
  ToIR.run(Tree);
  M->print(outs(), nullptr);
}
```

We now have implemented the frontend of the compiler, from reading the source up to generating the IR. Of course, all these components must work together on user input, which is the task of the compiler driver. We also need to implement the functions needed at runtime. Both are topics of the next section.

The missing pieces – the driver and the runtime library

All the phases from the previous sections are glued together by the `Calc.cpp` driver, which we implement as follows: a parameter for the input expression is declared, LLVM is initialized, and all the phases from the previous sections are called:

1. First, we include the required header files:

```
#include "CodeGen.h"
#include "Parser.h"
#include "Sema.h"
#include "llvm/Support/CommandLine.h"
#include "llvm/Support/InitLLVM.h"
#include "llvm/Support/raw_ostream.h"
```

2. LLVM comes with its own system for declaring command-line options. You only need to declare a static variable for each option you need. In doing so, the option is registered with a global command line parser. The advantage of this approach is that each component can add command-line options when needed. We declare an option for the input expression:

```
static llvm::cl::opt<std::string>
    Input(llvm::cl::Positional,
          llvm::cl::desc("<input expression>"),
          llvm::cl::init(""));
```

3. Inside the `main()` function, the LLVM libraries are initialized first. You need to call the `ParseCommandLineOptions()` function to handle the options given on the command line. This also handles the printing of help information. In the event of an error, this method exits the application:

```
int main(int argc, const char **argv) {
  llvm::InitLLVM X(argc, argv);
  llvm::cl::ParseCommandLineOptions(
      argc, argv, "calc - the expression compiler\n");
```

4. Next, we call the lexer and the parser. After the syntactical analysis, we check whether any errors occurred. If this is the case, then we exit the compiler with a return code indicating a failure:

```
Lexer Lex(Input);
Parser Parser(Lex);
AST *Tree = Parser.parse();
if (!Tree || Parser.hasError()) {
  llvm::errs() << "Syntax errors occured\n";
  return 1;
}
```

5. And we do the same if there was a semantic error:

```
Sema Semantic;
if (Semantic.semantic(Tree)) {
  llvm::errs() << "Semantic errors occured\n";
  return 1;
}
```

6. As the last step in the driver, the code generator is called:

```
CodeGen CodeGenerator;
CodeGenerator.compile(Tree);
return 0;
}
```

Now we have successfully created some IR code for the user input. We delegate the object code generation to the LLVM `llc` static compiler, so this finishes the implementation of our compiler. We link all the components together to create the `calc` application.

The runtime library consists of a single file, `rtcalc.c`. It has the implementation for the `calc_read()` and `calc_write()` functions, written in C:

```
#include <stdio.h>
#include <stdlib.h>
```

```
void calc_write(int v)
{
  printf("The result is: %d\n", v);
}
```

`calc_write()` only writes the result value to the terminal:

```
int calc_read(char *s)
{
  char buf[64];
  int val;
  printf("Enter a value for %s: ", s);
  fgets(buf, sizeof(buf), stdin);
  if (EOF == sscanf(buf, "%d", &val))
  {
    printf("Value %s is invalid\n", buf);
    exit(1);
  }
  return val;
}
```

`calc_read()` reads an integer number from the terminal. Nothing prevents the user from entering letters or other characters, so we must carefully check the input. If the input is not a number, we exit the application. A more complex approach would be to make the user aware of the problem and ask for a number again.

The next step is to build and try out our compiler, `calc`, which is an application that creates IR from an expression.

Building and testing the calc application

In order to build `calc`, we first need to create a new `CMakeLists.txt` file outside of the original `src` directory that contains all of the source file implementation:

1. First, we set the minimum required CMake version to the number required by LLVM, and give the project the name `calc`:

    ```
    cmake_minimum_required (VERSION 3.20.0)
    project ("calc")
    ```

2. Next, the LLVM package needs to be loaded, and we add the directory of the CMake modules provided by LLVM to the search path:

    ```
    find_package(LLVM REQUIRED CONFIG)
    message("Found LLVM ${LLVM_PACKAGE_VERSION}, build type ${LLVM_
    ```

```
BUILD_TYPE}")
list(APPEND CMAKE_MODULE_PATH ${LLVM_DIR})
```

3. We also need to add the definitions and the include path from LLVM. The used LLVM components are mapped to the library names with a function call:

```
separate_arguments(LLVM_DEFINITIONS_LIST NATIVE_COMMAND ${LLVM_
DEFINITIONS})
add_definitions(${LLVM_DEFINITIONS_LIST})
include_directories(SYSTEM ${LLVM_INCLUDE_DIRS})
llvm_map_components_to_libnames(llvm_libs Core)
```

4. Lastly, we indicate that we need to include the `src` subdirectory in our build, as this is where all of the C++ implementation that was done within this chapter resides:

```
add_subdirectory ("src")
```

There also needs to be a new `CMakeLists.txt` file inside of the `src` subdirectory. This CMake description inside the `src` directory appears as follows. We simply define the name of the executable, called `calc`, then list the source files to compile and the library to link against:

```
add_executable (calc
  Calc.cpp CodeGen.cpp Lexer.cpp Parser.cpp Sema.cpp)
target_link_libraries(calc PRIVATE ${llvm_libs})
```

Finally, we can begin building the `calc` application. Outside of the `src` directory, we create a new build directory and change into it. Afterwards, we can run the CMake and build invocation as follows:

```
$ cmake -GNinja -DCMAKE_C_COMPILER=clang -DCMAKE_CXX_COMPILER=clang++
 -DLLVM_DIR=<path to llvm installation configuration> ../
$ ninja
```

We now should have a newly built, functional `calc` application that can generate LLVM IR code. This can further be used with `llc`, which is the LLVM static backend compiler, to compile the IR code into an object file.

You can then use your favorite C compiler to link against the small runtime library. On Unix on X86, you can type the following:

```
$ calc "with a: a*3" | llc -filetype=obj \
  -relocation-model=pic  -o=expr.o
$ clang -o expr expr.o rtcalc.c
$ expr
Enter a value for a: 4
The result is: 12
```

On other Unix platforms such as AArch64 or PowerPC, you have to remove the `-relocation-model=pic` option.

On Windows, you need to use the `cl` compiler as follows:

```
$ calc "with a: a*3" | llc -filetype=obj -o=expr.obj
$ cl expr.obj rtcalc.c
$ expr
Enter a value for a: 4
The result is: 12
```

You have now created your first LLVM-based compiler! Please take some time to play around with various expressions. Especially check that multiplicative operators are evaluated before additive operators and that using parentheses changes the evaluation order, as we would expect from a basic calculator.

Summary

In this chapter, you learned about the typical components of a compiler. An arithmetic expression language was used to introduce you to the grammar of programming languages. You learned how to develop the typical components of a frontend for this language: a lexer, a parser, a semantic analyzer, and a code generator. The code generator only produced LLVM IR, and the LLVM `llc` static compiler was used to create object files from it. You have now developed your first LLVM-based compiler!

In the next chapter, you will deepen this knowledge, constructing the frontend for a programming language.

Part 2:
From Source to
Machine Code Generation

In this section, you will continue to learn how to develop your own compiler. You will begin by constructing the frontend, which reads the source file and creates an abstract syntax tree of it. Then, you will learn how to generate LLVM IR from the source file. Using the optimization capabilities of LLVM, you will then create optimized machine code. Additionally, you will explore several advanced topics, including generating LLVM IR for object-oriented language constructs and adding debug metadata.

This section comprises the following chapters:

- *Chapter 3, Turning the Source File into an Abstract Syntax Tree*
- *Chapter 4, Basics of IR Generation*
- *Chapter 5, IR Generation for High-Level Language Constructs*
- *Chapter 6, Advanced IR Generation*
- *Chapter 7, Optimizing IR*

3

Turning the Source File into an Abstract Syntax Tree

As we learned in the previous chapter, a compiler is typically divided into two parts – the frontend and the backend. In this chapter, we will implement the frontend of a programming language – that is, the part that mainly deals with the source language. We will learn about the techniques that real-world compilers use and apply them to our programming languages.

Our journey will begin with us defining our programming language's grammar and end with an **abstract syntax tree** (**AST**), which will become the base for code generation. You can use this approach for every programming language for which you would like to implement a compiler.

In this chapter, you will learn about the following:

- Defining a real programming language, where you will learn about the `tinylang` language, which is a subset of a real programming language, and for which you will implement a compiler frontend

- Organizing the directory structure of a compiler project

- Knowing how to handle multiple input files for the compiler

- The skill of handling user messages and informing them of issues in a pleasant manner

- Building the lexer using modular pieces

- Constructing a recursive descent parser from the rules derived from a grammar to perform syntax analysis

- Performing semantic analysis by creating an AST and analyzing its characteristics

With the skills you'll acquire in this chapter, you'll be able to build a compiler frontend for any programming language.

Defining a real programming language

Real programming brings up more challenges than the simple calc language from the previous chapter. To have a look at the details, we will be using a tiny subset of *Modula-2* in this and the following chapters. Modula-2 is well-designed and optionally supports **generics** and **object-orientated programming (OOP)**. However, we are not going to create a complete Modula-2 compiler in this book. Therefore, we will call the subset `tinylang`.

Let's begin with an example of what a program in `tinylang` looks like. The following function computes the greatest common divisor using the *Euclidean algorithm*:

```
MODULE Gcd;

PROCEDURE GCD(a, b: INTEGER) : INTEGER;
VAR t: INTEGER;
BEGIN
  IF b = 0 THEN
    RETURN a;
  END;
  WHILE b # 0 DO
    t := a MOD b;
    a := b;
    b := t;
  END;
  RETURN a;
END GCD;

END Gcd.
```

Now that we have a feeling for how a program in the language looks, let's take a quick tour of the `tinylang` subset's grammar as used in this chapter. In the next few sections, we'll use this grammar to derive the lexer and the parser from it:

```
compilationUnit
  : "MODULE" identifier ";" ( import )* block identifier "." ;
Import : ( "FROM" identifier )? "IMPORT" identList ";" ;
Block
  : ( declaration )* ( "BEGIN" statementSequence )? "END" ;
```

A compilation unit in Modula-2 begins with the MODULE keyword, followed by the name of the module. The content of a module can have a list of imported modules, declarations, and a block containing statements that run at initialization time:

```
declaration
    : "CONST" ( constantDeclaration ";" )*
    | "VAR" ( variableDeclaration ";" )*
    | procedureDeclaration ";" ;
```

A declaration introduces constants, variables, and procedures. The declaration of constants is prefixed with the CONST keyword. Similarly, variable declarations begin with the VAR keyword. The declaration of a constant is very simple:

```
constantDeclaration : identifier "=" expression ;
```

The identifier is the name of the constant. The value is derived from an expression, which must be computable at compile time. The declaration of variables is a bit more complex:

```
variableDeclaration : identList ":" qualident ;
qualident : identifier ( "." identifier )* ;
identList : identifier ( "," identifier)* ;
```

To be able to declare more than one variable in one go, a list of identifiers is used. The type name can potentially come from another module and is prefixed by the module name in this case. This is called a **qualified identifier**. A procedure requires the most details:

```
procedureDeclaration
    : "PROCEDURE" identifier ( formalParameters )? ";"
      block identifier ;
formalParameters
    : "(" ( formalParameterList )? ")" ( ":" qualident )? ;
formalParameterList
    : formalParameter (";" formalParameter )* ;
formalParameter : ( "VAR" )? identList ":" qualident ;
```

The preceding code shows how constants, variables, and procedures are declared. Procedures can have parameters and a return type. Normal parameters are passed as values, and VAR parameters are passed by reference. The other part missing from the block rule is statementSequence, which is a list of single statements:

```
statementSequence
    : statement ( ";" statement )* ;
```

A statement is delimited by a semicolon if it is followed by another statement. Again, only a subset of the *Modula-2* statements is supported:

```
statement
  : qualident ( ":=" expression | ( "(" ( expList )? ")" )? )
  | ifStatement | whileStatement | "RETURN" ( expression )? ;
```

The first part of this rule describes an assignment or a procedure call. A qualified identifier followed by : = is an assignment. If it is followed by (, then it is a procedure call. The other statements are the usual control statements:

```
ifStatement
  : "IF" expression "THEN" statementSequence
    ( "ELSE" statementSequence )? "END" ;
```

The IF statement also has a simplified syntax as it can only have a single ELSE block. With that statement, we can conditionally guard a statement:

```
whileStatement
  : "WHILE" expression "DO" statementSequence "END" ;
```

The WHILE statement describes a loop that's guarded by a condition. Together with the IF statement, this enables us to write simple algorithms in tinylang. Finally, the definition of an expression is missing:

```
expList
  : expression ( "," expression )* ;
expression
  : simpleExpression ( relation simpleExpression )? ;
relation
  : "=" | "#" | "<" | "<=" | ">" | ">=" ;
simpleExpression
  : ( "+" | "-" )? term ( addOperator term )* ;
addOperator
  : "+" | "-" | "OR" ;
term
  : factor ( mulOperator factor )* ;
mulOperator
  : "*" | "/" | "DIV" | "MOD" | "AND" ;
factor
  : integer_literal | "(" expression ")" | "NOT" factor
  | qualident ( "(" ( expList )? ")" )? ;
```

The expression syntax is very similar to that of calc in the previous chapter. Only the INTEGER and BOOLEAN data types are supported.

Additionally, the identifier and integer_literal tokens are used. An **identifier** is a name that begins with a letter or an underscore, followed by letters, digits, and underscores. An **integer literal** is either a sequence of decimal digits or a sequence of hexadecimal digits, followed by the letter H.

These are already a lot of rules, and we're only covering a part of Modula-2! Nevertheless, it is possible to write small applications in this subset. Let's implement a compiler for tinylang!

Creating the project layout

The project layout for tinylang follows the approach we laid out in *Chapter 1, Installing LLVM*. The source code for each component is in a subdirectory of the lib directory, and the header files are in a subdirectory of include/tinylang. The subdirectory is named after the component. In *Chapter 1, Installing LLVM*, we only created the Basic component.

From the previous chapter, we know that we need to implement a lexer, a parser, an AST, and a semantic analyzer. Each is a component of its own, called Lexer, Parser, AST, and Sema, respectively. The directory layout that will be used in this chapter looks like this:

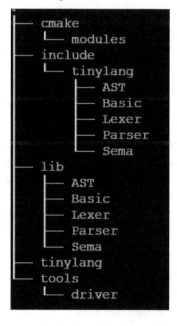

Figure 3.1 – The directory layout of the tinylang project

The components have clearly defined dependencies. Lexer depends only on Basic. Parser depends on Basic, Lexer, AST, and Sema. Sema only depends on Basic and AST. The well-defined dependencies help us reuse the components.

Let's have a closer look at the implementation!

Managing the input files for the compiler

A real compiler has to deal with many files. Usually, the developer calls the compiler with the name of the main compilation unit. This compilation unit can refer to other files – for example, via `#include` directives in C or `import` statements in Python or Modula-2. An imported module can import other modules, and so on. All these files must be loaded into memory and run through the analysis stages of the compiler. During development, a developer may make syntactical or semantical errors. When detected, an error message, including the source line and a marker, should be printed. This essential component is not trivial.

Luckily, LLVM comes with a solution: the `llvm::SourceMgr` class. A new source file is added to `SourceMgr` with a call to the `AddNewSourceBuffer()` method. Alternatively, a file can be loaded with a call to the `AddIncludeFile()` method. Both methods return an ID to identify the buffer. You can use this ID to retrieve a pointer to the memory buffer of the associated file. To define a location in the file, you can use the `llvm::SMLoc` class. This class encapsulates a pointer to the buffer. Various `PrintMessage()` methods allow you to emit errors and other informational messages to the user.

Handling messages for the user

Only a centralized definition of messages is missing. In a large piece of software (such as a compiler), you do not want to sprinkle message strings all over the place. If there is a request to change messages or translate them into another language, then you better have them in a central place!

A simple approach is that each message has an ID (an `enum` member), a severity level such as `Error` or `Warning`, and a string containing the messages. In your code, you only refer to the message ID. The severity level and message string are only used when the message is printed. These three items (the ID, the security level, and the message) must be managed consistently. The LLVM libraries use the preprocessor to solve this. The data is stored in a file with the `.def` suffix and is wrapped in a macro name. That file is usually included several times, with different definitions for the macro. The definition is in the `include/tinylang/Basic/Diagnostic.def` file path and looks as follows:

```
#ifndef DIAG
#define DIAG(ID, Level, Msg)
#endif

DIAG(err_sym_declared, Error, "symbol {0} already declared")

#undef DIAG
```

The first macro parameter, ID, is the enumeration label, the second parameter, Level, is the severity, and the third parameter, Msg, is the message text. With this definition at hand, we can define a DiagnosticsEngine class to emit error messages. The interface is in the include/tinylang/Basic/Diagnostic.h file:

```
#ifndef TINYLANG_BASIC_DIAGNOSTIC_H
#define TINYLANG_BASIC_DIAGNOSTIC_H

#include "tinylang/Basic/LLVM.h"
#include "llvm/ADT/StringRef.h"
#include "llvm/Support/FormatVariadic.h"
#include "llvm/Support/SMLoc.h"
#include "llvm/Support/SourceMgr.h"
#include "llvm/Support/raw_ostream.h"
#include <utility>

namespace tinylang {
```

After including the necessary header files, Diagnostic.def can be used to define the enumeration. To not pollute the global namespace, a nested namespace called diag is used:

```
namespace diag {
enum {
#define DIAG(ID, Level, Msg) ID,
#include "tinylang/Basic/Diagnostic.def"
};
} // namespace diag
```

The DiagnosticsEngine class uses a SourceMgr instance to emit the messages via the report() method. Messages can have parameters. To implement this facility, the variadic-format support provided by LLVM is used. The message text and the severity level are retrieved with the help of the static method. As a bonus, the number of emitted error messages is also counted:

```
class DiagnosticsEngine {
  static const char *getDiagnosticText(unsigned DiagID);
  static SourceMgr::DiagKind
  getDiagnosticKind(unsigned DiagID);
```

The message string is returned by getDiagnosticText(), while the level is returned by getDiagnosticKind(). Both methods are later implemented in the .cpp file:

```
  SourceMgr &SrcMgr;
  unsigned NumErrors;
```

```
public:
  DiagnosticsEngine(SourceMgr &SrcMgr)
      : SrcMgr(SrcMgr), NumErrors(0) {}

  unsigned nunErrors() { return NumErrors; }
```

As messages can have a variable number of parameters, the solution in C++ is to use a variadic template. Of course, this is also used by the `formatv()` function provided by LLVM. To get the formatted message, we just need to forward the template parameters:

```
template <typename... Args>
void report(SMLoc Loc, unsigned DiagID,
            Args &&... Arguments) {
  std::string Msg =
      llvm::formatv(getDiagnosticText(DiagID),
                    std::forward<Args>(Arguments)...)
          .str();
  SourceMgr::DiagKind Kind = getDiagnosticKind(DiagID);
  SrcMgr.PrintMessage(Loc, Kind, Msg);
  NumErrors += (Kind == SourceMgr::DK_Error);
  }
};


} // namespace tinylang


#endif
```

With that, we have implemented most of the class. Only `getDiagnosticText()` and `getDiagnosticKind()` are missing. They are defined in the `lib/Basic/Diagnostic.cpp` file and also make use of the `Diagnostic.def` file:

```
#include "tinylang/Basic/Diagnostic.h"

using namespace tinylang;

namespace {
const char *DiagnosticText[] = {
#define DIAG(ID, Level, Msg) Msg,
#include "tinylang/Basic/Diagnostic.def"
};
```

As in the header file, the DIAG macro is defined to retrieve the desired part. Here, we define an array that holds the text messages. Therefore, the DIAG macro only returns the Msg part. We use the same approach for the level:

```
SourceMgr::DiagKind DiagnosticKind[] = {
#define DIAG(ID, Level, Msg) SourceMgr::DK_##Level,
include "tinylang/Basic/Diagnostic.def"
};
} // namespace
```

Not surprisingly, both functions simply index the array to return the desired data:

```
const char *
DiagnosticsEngine::getDiagnosticText(unsigned DiagID) {
   return DiagnosticText[DiagID];
}


SourceMgr::DiagKind
DiagnosticsEngine::getDiagnosticKind(unsigned DiagID) {
   return DiagnosticKind[DiagID];
}
```

The combination of the SourceMgr and DiagnosticsEngine classes provides a good basis for the other components. We'll use them in the lexer first!

Structuring the lexer

As we know from the previous chapter, we need a Token class and a Lexer class. Additionally, a TokenKind enumeration is required to give each token class a unique number. Having an all-in-one header and implementation file does not scale, so let's move the items. TokenKind can be used universally and is placed in the Basic component. The Token and Lexer classes belong to the Lexer component but are placed in different headers and implementation files.

There are three different classes of tokens: **keywords**, **punctuators**, and **tokens**, which represent sets of many values. Examples are the CONST keyword, the ; delimiter, and the ident token, respectively, each of which represents identifiers in the source. Each token needs a member name for the enumeration. Keywords and punctuators have natural display names that can be used for messages.

Like in many programming languages, the keywords are a subset of the identifiers. To classify a token as a keyword, we need a keyword filter, which checks if the found identifier is indeed a keyword. This is the same behavior as in C or C++, where keywords are also a subset of identifiers. Programming languages evolve and new keywords may be introduced. As an example, the original K&R C language had no enumerations defined with the enum keyword. Due to this, a flag indicating the language level of a keyword should be present.

We collected several pieces of information, all of which belong to a member of the `TokenKind` enumeration: the label for the enumeration member, the spelling of punctuators, and a flag for keywords. For the diagnostic messages, we centrally store the information in a `.def` file called `include/tinylang/Basic/TokenKinds.def`, which looks like this. One thing to note is that keywords are prefixed with kw_:

```
#ifndef TOK
#define TOK(ID)
#endif
#ifndef PUNCTUATOR
#define PUNCTUATOR(ID, SP) TOK(ID)
#endif
#ifndef KEYWORD
#define KEYWORD(ID, FLAG) TOK(kw_ ## ID)
#endif

TOK(unknown)
TOK(eof)
TOK(identifier)
TOK(integer_literal)

PUNCTUATOR(plus,                  "+")
PUNCTUATOR(minus,                 "-")
// …

KEYWORD(BEGIN                             , KEYALL)
KEYWORD(CONST                             , KEYALL)
// …
#undef KEYWORD
#undef PUNCTUATOR
#undef TOK
```

With these centralized definitions, it's easy to create the `TokenKind` enumeration in the `include/tinylang/Basic/TokenKinds.h` file. Again, the enumeration is put into its own namespace, `tok`:

```
#ifndef TINYLANG_BASIC_TOKENKINDS_H
#define TINYLANG_BASIC_TOKENKINDS_H

namespace tinylang {

  namespace tok {
    enum TokenKind : unsigned short {
#define TOK(ID) ID,
```

```
#include "TokenKinds.def"
      NUM_TOKENS
    };
```

The pattern to fill the array should be familiar by now. The TOK macro is defined to only return ID. As a useful addition, we also define NUM_TOKENS as the last member of the enumeration, which denotes the number of defined tokens:

```
    const char *getTokenName(TokenKind Kind);
    const char *getPunctuatorSpelling(TokenKind Kind);
    const char *getKeywordSpelling(TokenKind Kind);
  }
}

#endif
```

The implementation file, lib/Basic/TokenKinds.cpp, also uses the .def file to retrieve the names:

```
#include "tinylang/Basic/TokenKinds.h"
#include "llvm/Support/ErrorHandling.h"

using namespace tinylang;

static const char * const TokNames[] = {
#define TOK(ID) #ID,
#define KEYWORD(ID, FLAG) #ID,
#include "tinylang/Basic/TokenKinds.def"
  nullptr
};
```

The textual name of a token is derived from its enumeration label, ID. There are two particularities:

- First, we need to define the TOK and KEYWORD macros since the default definition of KEYWORD does not use the TOK macro

- Second, a nullptr value is added at the end of the array, accounting for the added NUM_TOKENS enumeration member:

```
    const char *tok::getTokenName(TokenKind Kind) {
      return TokNames[Kind];
    }
```

We take a slightly different approach in the `getPunctuatorSpelling()` and `getKeywordSpelling()` functions. These functions only return a meaningful value for a subset of the enumeration. This can be realized with a `switch` statement, returning a `nullptr` value by default:

```
const char *tok::getPunctuatorSpelling(TokenKind Kind) {
  switch (Kind) {
#define PUNCTUATOR(ID, SP) case ID: return SP;
#include "tinylang/Basic/TokenKinds.def"
    default: break;
  }
  return nullptr;
}

const char *tok::getKeywordSpelling(TokenKind Kind) {
  switch (Kind) {
#define KEYWORD(ID, FLAG) case kw_ ## ID: return #ID;
#include "tinylang/Basic/TokenKinds.def"
    default: break;
  }
  return nullptr;
}
```

> **Tip**
>
> Note how the macros are defined to retrieve the necessary piece of information from the file.

In the previous chapter, the `Token` class was declared in the same header file as the `Lexer` class. To make this more versatile, we will put the `Token` class into its own header file in `include/Lexer/Token.h`. As before, `Token` stores a pointer to the start of the token, its length, and the token kind, as defined previously:

```
class Token {
  friend class Lexer;

  const char *Ptr;
  size_t Length;
  tok::TokenKind Kind;

public:
  tok::TokenKind getKind() const { return Kind; }
  size_t getLength() const { return Length; }
```

The SMLoc instance, which denotes the source position in messages, is created from the pointer to the token:

```
SMLoc getLocation() const {
  return SMLoc::getFromPointer(Ptr);
}
```

The getIdentifier() and getLiteralData() methods allow access to the text of the token for identifiers and literal data. It is not necessary to access the text for any other token type as this is implied by the token type:

```
StringRef getIdentifier() {
  assert(is(tok::identifier) &&
         "Cannot get identfier of non-identifier");
  return StringRef(Ptr, Length);
}

StringRef getLiteralData() {
  assert(isOneOf(tok::integer_literal,
                 tok::string_literal) &&
         "Cannot get literal data of non-literal");
  return StringRef(Ptr, Length);
}
};
```

We declare the Lexer class in the include/Lexer/Lexer.h header file and put the implementation in the lib/Lexer/lexer.cpp file. The structure is the same as for the calc language from the previous chapter. We need to take a closer look at two details here:

- First, some operators share the same prefix – for example, < and <=. When the current character we look at is <, then we must check the next character before deciding which token we found. Remember that the input needs to end with a null byte. Therefore, the next character can always be used if the current character is valid:

```
case '<':
  if (*(CurPtr + 1) == '=')
    formTokenWithChars(token, CurPtr + 2,
                       tok::lessequal);
  else
    formTokenWithChars(token, CurPtr + 1, tok::less);
  break;
```

- The other detail is that there are far more keywords now. How can we handle this? A simple and fast solution is to populate a hash table with the keywords, which are all stored in the TokenKinds.def file. This can be done during the instantiation of the Lexer class. With this approach, it is also possible to support different levels of the language as the keywords can be filtered with the attached flag. Here, this flexibility is not needed yet. In the header file, the keyword filter is defined as follows, using an instance of llvm::StringMap for the hash table:

```cpp
class KeywordFilter {
  llvm::StringMap<tok::TokenKind> HashTable;
  void addKeyword(StringRef Keyword,
                  tok::TokenKind TokenCode);

public:
  void addKeywords();
```

The getKeyword() method returns the token kind of the given string, or a default value if the string does not represent a keyword:

```cpp
tok::TokenKind getKeyword(
    StringRef Name,
    tok::TokenKind DefaultTokenCode = tok::unknown) {
  auto Result = HashTable.find(Name);
  if (Result != HashTable.end())
    return Result->second;
  return DefaultTokenCode;
  }
};
```

In the implementation file, the keyword table is filled:

```cpp
void KeywordFilter::addKeyword(StringRef Keyword,
                               tok::TokenKind TokenCode) {
  HashTable.insert(std::make_pair(Keyword, TokenCode));
}

void KeywordFilter::addKeywords() {
#define KEYWORD(NAME, FLAGS)                              \
  addKeyword(StringRef(#NAME), tok::kw_##NAME);
#include "tinylang/Basic/TokenKinds.def"
}
```

With the techniques you've just learned about, it's not difficult to write an efficient lexer class. As compilation speed matters, many compilers use a handwritten lexer, with one example being clang.

Constructing a recursive descent parser

As shown in the previous chapter, the parser is derived from the grammar. Let's recall all the *construction rules*. For each rule of the grammar, you create a method named after the non-terminal on the left-hand side of the rule to parse the right-hand side of the rule. Following the definition of the right-hand side, you do the following:

- For each non-terminal, the corresponding method is called

- Each token is consumed

- For alternatives and optional or repeating groups, the look-ahead token (the next unconsumed token) is examined to decide where to continue

Let's apply these construction rules to the following rule of grammar:

```
ifStatement
  : "IF" expression "THEN" statementSequence
    ( "ELSE" statementSequence )? "END" ;
```

We can easily translate this into the following C++ method:

```cpp
void Parser::parseIfStatement() {
  consume(tok::kw_IF);
  parseExpression();
  consume(tok::kw_THEN);
  parseStatementSequence();
  if (Tok.is(tok::kw_ELSE)) {
    advance();
    parseStatementSequence();
  }
  consume(tok::kw_END);
}
```

The whole grammar of `tinylang` can be turned into C++ in this way. In general, you have to be careful to avoid some pitfalls since most grammars that you will find on the internet are not suitable for this kind of construction.

Grammars and parsers

There are two different types of parsers: top-down parsers and bottom-up parsers. Their names are derived from the order in which a rule is handled during parsing. The input for a parser is the sequence of tokens generated by the lexer.

A top-down parser expands the leftmost symbol in a rule until a token is matched. Parsing is successful if all tokens are consumed and all symbols are expanded. This is exactly how the parser for tinylang works.

A bottom-up parser does the opposite: it looks at the sequence of tokens and tries to replace the tokens with a symbol of the grammar. For example, if the next tokens are IF, 3, +, and 4, then a bottom-up parser replaces the 3 + 4 token with the expression symbol, resulting in the IF expression sequence. When all tokens that belong to the IF statement are seen, then this sequence of tokens and symbols is replaced by the ifStatement symbol.

The parsing is successful if all tokens are consumed and the only symbol left is the start symbol. While top-down parsers can easily be constructed by hand, this is not the case for bottom-up parsers.

A different way to characterize both types of parsers is by which symbols are expanded first. Both read the input from left to right, but a top-down parser expands the leftmost symbol first while a bottom-up parser expands the rightmost symbol. Because of this, a top-down parser is also called an LL parser, while a bottom-up parser is called an LR parser.

A grammar must have certain properties so that either an LL or an LR parser can be derived from it. The grammars are named accordingly: you need an LL grammar to construct an LL parser.

You can find more details in university textbooks about compiler construction, such as Wilhelm, Seidl, and Hack: *Compiler Design. Syntactic and Semantic Analysis*, Springer 2013, and Grune and Jacobs: *Parsing Techniques, A practical guide*, Springer 2008.

One issue to look for is left-recursive rules. A rule is called **left-recursive** if the right-hand side begins with the same terminal that's on the left-hand side. A typical example can be found in grammars for expressions:

```
expression : expression "+" term ;
```

If it's not already clear from the grammar, then the translation to C++ makes it obvious that this results in infinite recursion:

```
Void Parser::parseExpression() {
  parseExpression();
  consume(tok::plus);
  parseTerm();
}
```

Left recursion can also occur indirectly and involve more rules, which is much more difficult to spot. That's why an algorithm exists that can detect and eliminate left recursion.

> **Note**
>
> Left-recursive rules are only a problem for LL parsers, such as the recursive-descent parser for `tinylang`. The reason is that these parsers expand the leftmost symbol first. In contrast, if you use a parser generator to generate an LR parser, which expands the rightmost symbol first, then you should avoid right-recursive rules.

At each step, the parser decides how to continue by just using the look-ahead token. The grammar is said to have conflicts if this decision cannot be made deterministically. To illustrate this, have a look at the `using` statement in C#. Like in C++, the `using` statement can be used to make a symbol visible in a namespace, such as in `using Math;`. It is also possible to define an alias name for the imported symbol with `using M = Math;`. In a grammar, this can be expressed as follows:

```
usingStmt : "using" (ident "=")? ident ";"
```

There's a problem here: after the parser consumes the `using` keyword, the look-ahead token is `ident`. However, this information is not enough for us to decide if the optional group must be skipped or parsed. This situation always arises if the set of tokens, with which the optional group can begin, overlaps with the set of tokens that follow the optional group.

Let's rewrite the rule with an alternative instead of an optional group:

```
usingStmt : "using" ( ident "=" ident | ident ) ";" ;
```

Now, there is a different conflict: both alternatives begin with the same token. Looking only at the look-ahead token, the parser can't decide which of the alternatives is the right one.

These conflicts are very common. Therefore, it's good to know how to handle them. One approach is to rewrite the grammar in such a way that the conflict disappears. In the previous example, both alternatives begin with the same token. This can be factored out, resulting in the following rule:

```
usingStmt : "using" ident ("=" ident)? ";" ;
```

This formulation has no conflict, but it should also be noted that it is less expressive. In the two other formulations, it is obvious which `ident` is the alias name and which `ident` is the namespace name. In the conflict-free rule, the leftmost `ident` changes its role. First, it is the namespace name, but if an equals sign follows, then it turns into the alias name.

The second approach is to add a predicate to distinguish between both cases. This predicate, often called a **resolver**, could use context information for the decision (such as a name lookup in a symbol table) or it could have a look at more than one token. Let's assume that the lexer has a method called `Token &peek(int n)` that returns the *n*th token after the current look-ahead token. Here, the existence of an equals sign can be used as an additional predicate in the decision:

```
if (Tok.is(tok::ident) && Lex.peek(0).is(tok::equal)) {
    advance();
    consume(tok::equal);
}
consume(tok::ident);
```

A third approach is to use backtracking. For this, you need to save the current state. Then, you must try to parse the conflicting group. If this does not succeed, then you need to go back to the saved state and try the other path. Here, you are searching for the correct rule to apply, which is not as efficient as the other methods. Therefore, you should only use this approach as a last resort.

Now, let's incorporate error recovery. In the previous chapter, I introduced the so-called *panic mode* as a technique for error recovery. The basic idea is to skip tokens until one is found that is suitable for continuing parsing. For example, in `tinylang`, a statement is followed by a semicolon (`:`).

If there is a syntax problem in an `IF` statement, then you skip all tokens until you find a semicolon. Then, you continue with the next statement. Instead of using an ad hoc definition for the token set, it's better to use a systematic approach.

For each non-terminal, you compute the set of tokens that can follow the non-terminal anywhere (called the **FOLLOW set**). For the non-terminal statement, the `;`, `ELSE`, and `END` tokens can follow. So, you must use this set in the error recovery part of `parseStatement()`. This method assumes that a syntax error can be handled locally. In general, this is not possible. Because the parser skips tokens, so many could be skipped that the end of input is reached. At this point, local recovery is not possible.

To prevent meaningless error messages, the calling method needs to be informed that an error recovery is still not finished. This can be done with `bool`. If it returns `true`, this means that error recovery hasn't finished yet, while `false` means that parsing (including a possible error recovery) was successful.

There are numerous ways to extend this error recovery scheme. Using the `FOLLOW` sets of active callers is a popular approach. As a simple example, assume that `parseStatement()` was called by `parseStatementSequence()`, which was itself called by `parseBlock()` and that from `parseModule()`.

Here, each of the corresponding non-terminals has a FOLLOW set. If the parser detects a syntax error in parseStatement(), then tokens are skipped until the token is in at least one of the FOLLOW sets of the active callers. If the token is in the FOLLOW set of the statement, then the error was recovered locally and a false value is returned to the caller. Otherwise, a true value is returned, meaning that error recovery must continue. A possible implementation strategy for this extension is passing std::bitset or std::tuple to represent the union of the current FOLLOW sets to the callee.

One last question is still open: how can we call the error recovery? In the previous chapter, goto was used to jump to the error recovery block. This works but is not a pleasing solution. Given what we discussed earlier, we can skip tokens in a separate method. Clang has a method has skipUntil() for this purpose; we also use this for tinylang.

Because the next step is to add semantic actions to the parser, it would also be nice to have a central place to put cleanup code if necessary. A nested function would be ideal for this. C++ does not have a nested function. Instead, a Lambda function can serve a similar purpose. The parseIfStatement() method, which we looked at initially, looks as follows when the complete error recovery code is added:

```cpp
bool Parser::parseIfStatement() {
  auto _errorhandler = [this] {
    return skipUntil(tok::semi, tok::kw_ELSE, tok::kw_END);
  };

  if (consume(tok::kw_IF))
    return _errorhandler();
  if (parseExpression(E))
    return _errorhandler();
  if (consume(tok::kw_THEN))
    return _errorhandler();
  if (parseStatementSequence(IfStmts))
    return _errorhandler();
  if (Tok.is(tok::kw_ELSE)) {
    advance();
    if (parseStatementSequence(ElseStmts))
      return _errorhandler();
  }
  if (expect(tok::kw_END))
    return _errorhandler();
  return false;
}
```

Parser and lexer generators

Manually constructing a parser and a lexer can be a tedious task, especially if you try to invent a new programming language and change the grammar very often. Luckily, some tools automate this task.

The classic Linux tools are **flex** (`https://github.com/westes/flex`) and **bison** (`https://www.gnu.org/software/bison/`). flex generates a lexer from a set of regular expressions, while bison generates an **LALR(1)** parser from a grammar description. Both tools generate C/C+ source code and can be used together.

Another popular tool is **AntLR** (`https://www.antlr.org/`). AntLR can generate a lexer, a parser, and an AST from a grammar description. The generated parser belongs to the **LL(*)** class, which means it is a top-down parser that uses a variable number of lookaheads to solve conflicts. The tool is written in Java but can generate source code for many popular languages, including C/C++.

All these tools require some library support. If you are looking for a tool that generates a self-contained lexer and parser, then **Coco/R** (`https://ssw.jku.at/Research/Projects/Coco/`) may be the tool for you. Coco/R generates a lexer and a recursive-descent parser from an **LL(1)** grammar description, similar to the one used in this book. The generated files are based on a template file that you can change if needed. The tool is written in C# but ports to C++, Java, and other languages.

There are many other tools available, and they vary a lot in terms of the features and output languages they support. Of course, when choosing a tool, there are also trade-offs to consider. An LALR(1) parser generator such as bison can consume a wide range of grammars, and free grammars you can find on the internet are often LALR(1) grammars.

As a downside, these generators generate a state machine that needs to be interpreted at runtime, which can be slower than a recursive descent parser. Error handling is also more complicated. bison has basic support for handling syntax errors, but the correct use requires a deep understanding of how the parser works. Compared to this, AntLR consumes a slightly smaller grammar class but automatically generates error handling, and can also generate an AST. So, rewriting grammar so that it can be used with AntLR may speed up development later.

Performing semantic analysis

The parser we constructed in the previous section only checks the syntax of the input. The next step is to add the ability to perform semantic analysis. In the calc example in the previous chapter, the parser constructed an AST. In a separate phase, the semantic analyzer worked on this tree. This approach can always be used. In this section, we will use a slightly different approach and intertwine the parser and the semantic analyzer more.

What does the semantic analyzer need to do? Let's take a look:

- For each declaration, the names of variables, objects, and more must be checked to ensure they have not been declared elsewhere.

- For each occurrence of a name in an expression or statement, it must be checked that the name is declared and that the desired use fits the declaration.

- For each expression, the resulting type must be computed. It is also necessary to compute if the expression is constant and if so, which value it has.

- For assignment and parameter passing, we must check that the types are compatible. Further, we must check that the conditions in IF and WHILE statements are of the BOOLEAN type.

That's already a lot to check for such a small subset of a programming language!

Handling the scope of names

Let's have a look at the scope of names first. The scope of a name is the range where the name is visible. Like C, tinylang uses a declare-before-use model. For example, the B and X variables are declared at the module level to be of the INTEGER type:

```
VAR B, X: INTEGER;
```

Before the declaration, the variables are not known and cannot be used. That's only possible after the declaration. Inside a procedure, more variables can be declared:

```
PROCEDURE Proc;
VAR B: BOOLEAN;
BEGIN
   (* Statements *)
END Proc;
```

Inside the procedure, at the point where the comment is, a use of B refers to the B local variable while a use of X refers to the X global variable. The scope of the local variable, B, is Proc. If a name cannot be found in the current scope, then the search continues in the enclosing scope. Therefore, the X variable can be used inside the procedure. In tinylang, only modules and procedures open a new scope. Other language constructs, such as structs and classes, usually also open a scope. Predefined entities such as the INTEGER type and the TRUE literal are declared in a global scope, enclosing the scope of the module.

In tinylang, only the name is crucial. Therefore, a scope can be implemented as a mapping from a name to its declaration. A new name can only be inserted if it is not already present. For the lookup, the enclosing or parent scope must also be known. The interface (in the include/tinylang/ Sema/Scope.h file) looks as follows:

```
#ifndef TINYLANG_SEMA_SCOPE_H
#define TINYLANG_SEMA_SCOPE_H

#include "tinylang/Basic/LLVM.h"
#include "llvm/ADT/StringMap.h"
#include "llvm/ADT/StringRef.h"

namespace tinylang {

class Decl;

class Scope {
  Scope *Parent;
  StringMap<Decl *> Symbols;

public:
  Scope(Scope *Parent = nullptr) : Parent(Parent) {}

  bool insert(Decl *Declaration);
  Decl *lookup(StringRef Name);

  Scope *getParent() { return Parent; }
};
} // namespace tinylang
#endif
```

The implementation in the lib/Sema/Scope.cpp file looks as follows:

```
#include "tinylang/Sema/Scope.h"
#include "tinylang/AST/AST.h"

using namespace tinylang;

bool Scope::insert(Decl *Declaration) {
  return Symbols
      .insert(std::pair<StringRef, Decl *>(
          Declaration->getName(), Declaration))
      .second;
}
```

Please note that the `StringMap::insert()` method does not override an existing entry. The `second` member of the resulting `std::pair` indicates if the table was updated. This information is returned to the caller.

To implement the search for the declaration of a symbol, the `lookup()` method searches in the current scope and, if nothing is found, searches the scopes linked by the `parent` member:

```
Decl *Scope::lookup(StringRef Name) {
  Scope *S = this;
  while (S) {
    StringMap<Decl *>::const_iterator I =
        S->Symbols.find(Name);
    if (I != S->Symbols.end())
      return I->second;
    S = S->getParent();
  }
  return nullptr;
}
```

The variable declaration is then processed as follows:

- The current scope is the module scope.

- The `INTEGER` type declaration is looked up. It's an error if no declaration is found or if it is not a type declaration.

- A new AST node called `VariableDeclaration` is instantiated, with the important attributes being the name, B, and the type.

- The name, B, is inserted into the current scope, mapping to the declaration instance. If the name is already present in the scope, then this is an error. The content of the current scope is not changed in this case.

- The same is done for the X variable.

Two tasks are performed here. As in the calc example, AST nodes are constructed. At the same time, attributes of the node, such as the type, are computed. Why is this possible?

The semantic analyzer can fall back on two different sets of attributes. The scope is inherited from the caller. The type declaration can be computed (or synthesized) by evaluating the name of the type declaration. The language is designed in such a way that these two sets of attributes are sufficient to compute all attributes of the AST node.

An important aspect is the *declare-before-use* model. If a language allows the use of names before declaration, such as members inside a class in C++, then it is not possible to compute all attributes of an AST node at once. In such a case, the AST node must be constructed with only partially computed attributes or just with plain information (such as in the calc example).

The AST must then be visited one or more times to determine the missing information. In the case of tinylang (and Modula-2), it would be possible to dispense with the AST construction – the AST is indirectly represented through the call hierarchy of the parseXXX() methods. Code generation from an AST is much more common, so we construct an AST here, too.

Before we put the pieces together, we need to understand the LLVM style of using **runtime type information** (**RTTI**).

Using an LLVM-style RTTI for the AST

Naturally, the AST nodes are a part of a class hierarchy. A declaration always has a name. Other attributes depend on what is being declared. If a variable is declared, then a type is required. A constant declaration needs a type, a value, and so on. Of course, at runtime, you need to find out which kind of declaration you are working with. The dynamic_cast<> C++ operator could be used for this. The problem is that the required RTTI is only available if the C++ class has a virtual table attached – that is, it uses virtual functions. Another disadvantage is that C++ RTTI is bloated. To avoid these disadvantages, the LLVM developers introduced a self-made RTTI style, which is used throughout the LLVM libraries.

The (abstract) base class of our hierarchy is Decl. To implement the LLVM-style RTTI, a public enumeration containing a label for each subclass must be added. Also, a private member of this type and a public getter are required. The private member is usually called Kind. In our case, this looks as follows:

```
class Decl {
public:
  enum DeclKind { DK_Module, DK_Const, DK_Type,
                  DK_Var, DK_Param, DK_Proc };
private:
  const DeclKind Kind;
public:
  DeclKind getKind() const { return Kind; }
};
```

Each subclass now needs a special function member called classof. The purpose of this function is to determine if a given instance is of the requested type. For VariableDeclaration, it is implemented as follows:

```
static bool classof(const Decl *D) {
  return D->getKind() == DK_Var;
}
```

Now, you can use the special templates, `llvm::isa<>`, to check if an object is of the requested type and `llvm::dyn_cast<>` to dynamically cast the object. More templates exist, but these two are the most commonly used ones. For the other templates, see `https://llvm.org/docs/ProgrammersManual.html#the-isa-cast-and-dyn-cast-templates` and for more information about the LLVM style, including more advanced uses, see `https://llvm.org/docs/HowToSetUpLLVMStyleRTTI.html`.

Creating the semantic analyzer

Equipped with this knowledge, we can now implement all the parts. First, we must create the definition of the AST node for a variable that's stored in the `include/llvm/tinylang/AST/AST.h` file. Besides support for the LLVM-style RTTI, the base class stores the name of the declaration, the location of the name, and a pointer to the enclosing declaration. The latter is required during code generation of nested procedures. The `Decl` base class is declared as follows:

```
class Decl {
public:
  enum DeclKind { DK_Module, DK_Const, DK_Type,
                  DK_Var, DK_Param, DK_Proc };

private:
  const DeclKind Kind;

protected:
  Decl *EnclosingDecL;
  SMLoc Loc;
  StringRef Name;

public:
  Decl(DeclKind Kind, Decl *EnclosingDecL, SMLoc Loc,
       StringRef Name)
    : Kind(Kind), EnclosingDecL(EnclosingDecL), Loc(Loc),
      Name(Name) {}

  DeclKind getKind() const { return Kind; }
  SMLoc getLocation() { return Loc; }
  StringRef getName() { return Name; }
  Decl *getEnclosingDecl() { return EnclosingDecL; }
};
```

The declaration for a variable only adds a pointer to the type declaration:

```
class TypeDeclaration;

class VariableDeclaration : public Decl {
  TypeDeclaration *Ty;

public:
  VariableDeclaration(Decl *EnclosingDecL, SMLoc Loc,
                      StringRef Name, TypeDeclaration *Ty)
      : Decl(DK_Var, EnclosingDecL, Loc, Name), Ty(Ty) {}

  TypeDeclaration *getType() { return Ty; }

  static bool classof(const Decl *D) {
    return D->getKind() == DK_Var;
  }
};
```

The method in the parser needs to be extended with a semantic action and variables for collected information:

```
bool Parser::parseVariableDeclaration(DeclList &Decls) {
  auto _errorhandler = [this] {
    while (!Tok.is(tok::semi)) {
      advance();
      if (Tok.is(tok::eof)) return true;
    }
    return false;
  };

  Decl *D = nullptr; IdentList Ids;
  if (parseIdentList(Ids)) return _errorhandler();
  if (consume(tok::colon)) return _errorhandler();
  if (parseQualident(D)) return _errorhandler();
  Actions.actOnVariableDeclaration(Decls, Ids, D);
  return false;
}
```

DeclList is a list of declarations, `std::vector<Decl*>`, and IdentList is a list of locations and identifiers, `std::vector<std::pair<SMLoc, StringRef>>`.

The `parseQualident()` method returns a declaration, which in this case is expected to be a type declaration.

The parser class knows an instance of the semantic analyzer class, `Sema`, that's stored in the `Actions` member. A call to `actOnVariableDeclaration()` runs the semantic analyzer and the AST construction. The implementation is in the `lib/Sema/Sema.cpp` file:

```
void Sema::actOnVariableDeclaration(DeclList &Decls,
                                    IdentList &Ids,
                                    Decl *D) {
  if (TypeDeclaration *Ty = dyn_cast<TypeDeclaration>(D)) {
    for (auto &[Loc, Name] : Ids) {
      auto *Decl = new VariableDeclaration(CurrentDecl, Loc,
                                           Name, Ty);
      if (CurrentScope->insert(Decl))
        Decls.push_back(Decl);
      else
        Diags.report(Loc, diag::err_symbold_declared, Name);
    }
  } else if (!Ids.empty()) {
    SMLoc Loc = Ids.front().first;
    Diags.report(Loc, diag::err_vardecl_requires_type);
  }
}
```

The type declaration is checked with `llvm::dyn_cast<TypeDeclaration>`. If it is not a type declaration, then an error message is printed. Otherwise, for each name in the `Ids` list, `VariableDeclaration` is instantiated and added to the list of declarations. If adding the variable to the current scope fails because the name is already declared, then an error message is printed as well.

Most of the other entities are constructed in the same way – the complexity of the semantic analysis is the only difference. More work is required for modules and procedures because they open a new scope. Opening a new scope is easy: only a new `Scope` object must be instantiated. As soon as the module or procedure has been parsed, the scope must be removed.

This must be done reliably because we do not want to add names to the wrong scope in case of a syntax error. This is a classic use of the **Resource Acquisition Is Initialization (RAII)** idiom in C++. Another complication comes from the fact that a procedure can recursively call itself. Therefore, the name of the procedure must be added to the current scope before it can be used. The semantic analyzer has two methods to enter and leave a scope. The scope is associated with a declaration:

```
void Sema::enterScope(Decl *D) {
  CurrentScope = new Scope(CurrentScope);
  CurrentDecl = D;
```

```
}

void Sema::leaveScope() {
  Scope *Parent = CurrentScope->getParent();
  delete CurrentScope;
  CurrentScope = Parent;
  CurrentDecl = CurrentDecl->getEnclosingDecl();
}
```

A simple helper class is used to implement the RAII idiom:

```
class EnterDeclScope {
  Sema &Semantics;

public:
  EnterDeclScope(Sema &Semantics, Decl *D)
      : Semantics(Semantics) {
    Semantics.enterScope(D);
  }
  ~EnterDeclScope() { Semantics.leaveScope(); }
};
```

When parsing a module or procedure, two interactions occur with the semantic analyzer. The first is after the name is parsed. Here, the (almost empty) AST node is constructed and a new scope is established:

```
bool Parser::parseProcedureDeclaration(/* … */) {
  /* … */
  if (consume(tok::kw_PROCEDURE)) return _errorhandler();
  if (expect(tok::identifier)) return _errorhandler();
  ProcedureDeclaration *D =
      Actions.actOnProcedureDeclaration(
          Tok.getLocation(), Tok.getIdentifier());
  EnterDeclScope S(Actions, D);
  /* … */
}
```

The semantic analyzer checks the name in the current scope and returns the AST node:

```
ProcedureDeclaration *
Sema::actOnProcedureDeclaration(SMLoc Loc, StringRef Name) {
  ProcedureDeclaration *P =
      new ProcedureDeclaration(CurrentDecl, Loc, Name);
  if (!CurrentScope->insert(P))
```

```
      Diags.report(Loc, diag::err_symbold_declared, Name);
   return P;
}
```

The real work is done after all the declarations and the procedure body have been parsed. You only need to check if the name at the end of the procedure declaration is equal to the name of the procedure and if the declaration used for the return type is a type declaration:

```
void Sema::actOnProcedureDeclaration(
    ProcedureDeclaration *ProcDecl, SMLoc Loc,
    StringRef Name, FormalParamList &Params, Decl *RetType,
    DeclList &Decls, StmtList &Stmts) {

  if (Name != ProcDecl->getName()) {
    Diags.report(Loc, diag::err_proc_identifier_not_equal);
    Diags.report(ProcDecl->getLocation(),
                 diag::note_proc_identifier_declaration);
  }
  ProcDecl->setDecls(Decls);
  ProcDecl->setStmts(Stmts);

  auto *RetTypeDecl =
      dyn_cast_or_null<TypeDeclaration>(RetType);
  if (!RetTypeDecl && RetType)
    Diags.report(Loc, diag::err_returntype_must_be_type,
                 Name);
  else
    ProcDecl->setRetType(RetTypeDecl);
}
```

Some declarations are inherently present and cannot be defined by the developer. This includes the BOOLEAN and INTEGER types and the TRUE and FALSE literals. These declarations exist in the global scope and must be added programmatically. Modula-2 also predefines some procedures, such as INC or DEC, that can be added to the global scope. Given our classes, initializing the global scope is simple:

```
void Sema::initialize() {
  CurrentScope = new Scope();
  CurrentDecl = nullptr;
  IntegerType =
      new TypeDeclaration(CurrentDecl, SMLoc(), "INTEGER");
  BooleanType =
      new TypeDeclaration(CurrentDecl, SMLoc(), "BOOLEAN");
```

```
    TrueLiteral = new BooleanLiteral(true, BooleanType);
    FalseLiteral = new BooleanLiteral(false, BooleanType);
    TrueConst = new ConstantDeclaration(CurrentDecl, SMLoc(),
                                    "TRUE", TrueLiteral);
    FalseConst = new ConstantDeclaration(
        CurrentDecl, SMLoc(), "FALSE", FalseLiteral);
    CurrentScope->insert(IntegerType);
    CurrentScope->insert(BooleanType);
    CurrentScope->insert(TrueConst);
    CurrentScope->insert(FalseConst);
}
```

With this scheme, all required calculations for `tinylang` can be done. For example, let's look at how to compute if an expression results in a constant value:

- We must ensure literal or a reference to a constant declaration is a constant

- If both sides of an expression are constant, then applying the operator also yields a constant

These rules are embedded into the semantic analyzer while creating the AST nodes for an expression. Likewise, the type and the constant value can be computed.

It should be noted that not all kinds are computation can be done in this way. For example, to detect the use of uninitialized variables, a method called *symbolic interpretation* can be used. In its general form, the method requires a special walk order through the AST, which is not possible during construction time. The good news is that the presented approach creates a fully decorated AST that is ready for code generation. This AST can be used for further analysis, given that costly analysis can be turned on or off on demand.

To play around with the frontend, you also need to update the driver. Since the code generation is missing, a correct `tinylang` program produces no output. Still, it can be used to explore error recovery and provoke semantic errors:

```
#include "tinylang/Basic/Diagnostic.h"
#include "tinylang/Basic/Version.h"
#include "tinylang/Parser/Parser.h"
#include "llvm/Support/InitLLVM.h"
#include "llvm/Support/raw_ostream.h"

using namespace tinylang;

int main(int argc_, const char **argv_) {
  llvm::InitLLVM X(argc_, argv_);
```

```
llvm::SmallVector<const char *, 256> argv(argv_ + 1,
                                          argv_ + argc_);

llvm::outs() << "Tinylang "
             << tinylang::getTinylangVersion() << "\n";

for (const char *F : argv) {
  llvm::ErrorOr<std::unique_ptr<llvm::MemoryBuffer>>
      FileOrErr = llvm::MemoryBuffer::getFile(F);
  if (std::error_code BufferError =
          FileOrErr.getError()) {
    llvm::errs() << "Error reading " << F << ": "
                 << BufferError.message() << "\n";
    continue;
  }

  llvm::SourceMgr SrcMgr;
  DiagnosticsEngine Diags(SrcMgr);
  SrcMgr.AddNewSourceBuffer(std::move(*FileOrErr),
                            llvm::SMLoc());
  auto TheLexer = Lexer(SrcMgr, Diags);
  auto TheSema = Sema(Diags);
  auto TheParser = Parser(TheLexer, TheSema);
  TheParser.parse();
  }
}
```

Congratulations! You've finished implementing the frontend for tinylang! You can use the example program, Gcd.mod, provided in the *Defining a real programming language* section to run the frontend:

```
$ tinylang Gcd.mod
```

Of course, this is a valid program, and it looks like nothing happens. Be sure to modify the file and provoke some error messages. We'll continue with the fun in the next chapter by adding code generation.

Summary

In this chapter, you learned about the techniques that a real-world compiler uses in the frontend. Starting with the project layout, you created separate libraries for the lexer, the parser, and the semantic analyzer. To output messages to the user, you extended an existing LLVM class, allowing the messages to be stored centrally. The lexer is now separated into several interfaces.

Then, you learned how to construct a recursive descent parser from a grammar description, looked at what pitfalls to avoid, and learned how to use generators to do the job. The semantic analyzer you constructed performs all the semantic checks required by the language while being intertwined with the parser and AST construction.

The result of your coding effort is a fully decorated AST. You'll use this in the next chapter to generate IR code and, finally, object code.s

4

Basics of IR Code Generation

Having created a decorated **abstract syntax tree** (**AST**) for your programming language, the next task is to generate the LLVM IR code from it. LLVM IR code resembles a three-address code with a human-readable representation. Therefore, we need a systematic approach to translate language concepts such as control structures into the lower level of LLVM IR.

In this chapter, you will learn about the basics of LLVM IR and how to generate IR for control flow structures from the AST. You will also learn how to generate LLVM IR for expressions in **static single assignment** (**SSA**) form using a modern algorithm. Finally, you will learn how to emit assembler text and object code.

This chapter will cover the following topics:

- Generating IR from the AST
- Using AST numbering to generate IR code in SSA form
- Setting up the module and the driver

By the end of this chapter, you will know how to create a code generator for your programming language and how to integrate it into your compiler.

Generating IR from the AST

The LLVM code generator takes a module in LLVM IR as input and turns it into object code or assembly text. We need to transform the AST representation into IR. To implement an IR code generator, we will look at a simple example first and then develop the classes needed for the code generator. The complete implementation will be divided into three classes:

- `CodeGenerator`
- `CGModule`
- `CGProcedure`

The `CodeGenerator` class is the general interface used by the compiler driver. The `CGModule` and `CGProcedure` classes hold the state required for generating the IR code for a compilation unit and a single function.

We'll begin by looking at the Clang-generated IR.

Understanding the IR code

Before generating the IR code, it's good to know the main elements of the IR language. In *Chapter 2, The Structure of a Compiler*, we had a brief look at IR. An easy way to get more knowledge of IR is to study the output from `clang`. For example, save this C source code, which implements the Euclidean algorithm for calculating the greatest common divisor of two numbers, as `gcd.c`:

```
unsigned gcd(unsigned a, unsigned b) {
  if (b == 0)
    return a;
  while (b != 0) {
    unsigned t = a % b;
    a = b;
    b = t;
  }
  return a;
}
```

Then, you can create the `gcd.ll` IR file by using `clang` and the following command:

```
$ clang --target=aarch64-linux-gnu -O1 -S -emit-llvm gcd.c
```

The IR code is not target-independent, even if it often looks like it is. The preceding command compiles the source file for an ARM 64-bit CPU on Linux. The `-S` option instructs `clang` to output an assembly file, and with the additional specification of `-emit-llvm`, an IR file is created. The optimization level, `-O1`, is used to get an easily readable IR code. Clang has many more options, all of which are documented in the command-line argument reference at https://clang.llvm.org/docs/ClangCommandLineReference.html. Let's have a look at the generated file and understand how the C source maps to the LLVM IR.

A C file is translated into a **module**, which holds the functions and the data objects. A function has at least one **basic block**, and a basic block contains instructions. This hierarchical structure is also reflected in the C++ API. All data elements are typed. Integer types are represented by the letter i, followed by the number of bits. For example, the 64-bit integer type is written as $i64$. The most basic float types are `float` and `double`, denoting the 32-bit and 64-bit IEEE floating-point types. It is also possible to create aggregate types such as vectors, arrays, and structures.

Here is what the LLVM IR looks like. At the top of the file, some basic properties are established:

```
; ModuleID = 'gcd.c'
source_filename = "gcd.c"
target datalayout = "e-m:e-i8:8:32-i16:16:32-i64:64-i128:128-
n32:64-S128"
target triple = "aarch64-unknown-linux-gnu"
```

The first line is a comment informing you about which module identifier was used. In the following line, the filename of the source file is named. With `clang`, both are the same.

The `target datalayout` string establishes some basic properties. The different parts are separated by `-`. The following information is included:

- A small `e` means that bytes in memory are stored using the little-endian schema. To specify a big endian, you must use a big `E`.

- `M:` specifies the name mangling that's applied to symbols. Here, `m:e` means that ELF name mangling is used.

- The entries in `iN:A:P` form, such as `i8:8:32`, specify the alignment of data, given in bits. The first number is the alignment required by the ABI, and the second number is the preferred alignment. For bytes (`i8`), the ABI alignment is 1 byte (`8`) and the preferred alignment is 4 bytes (`32`).

- `n` specifies which native register sizes are available. `n32:64` means that 32-bit and 64-bit wide integers are natively supported.

- `S` specifies the alignment of the stack, again in bits. `S128` means that the stack maintains a 16-byte alignment.

> **Note**
>
> The provided target data layout must match what the backend expects. Its purpose is to communicate the captured information to the target-independent optimization passes. For example, an optimization pass can query the data layout to get the size and alignment of a pointer. However, changing the size of a pointer in the data layout does not change the code generation in the backend.
>
> A lot more information is provided with the target data layout. You can find more information in the reference manual at `https://llvm.org/docs/LangRef.html#data-layout`.

Last, the `target triple` string specifies the architecture we are compiling for. This reflects the information we gave on the command line. The triple is a configuration string that usually consists of the CPU architecture, the vendor, and the operating system. More information about the environment is often added. For example, the `x86_64-pc-win32` triple is used for a Windows system running on a 64-bit X86 CPU. `x86_64` is the CPU architecture, `pc` is a generic vendor, and `win32` is the operating system. The parts are connected by a hyphen. A Linux system running on an ARMv8 CPU uses `aarch64-unknown-linux-gnu` as its triple. `aarch64` is the CPU architecture, while the operating system is `linux` running a `gnu` environment. There is no real vendor for a Linux-based system, so this part is `unknown`. Parts that are not known or unimportant for a specific purpose are often omitted: the `aarch64-linux-gnu` triple describes the same Linux system.

Next, the `gcd` function is defined in the IR file:

```
define i32 @gcd(i32 %a, i32 %b) {
```

This resembles the function signature in the C file. The `unsigned` data type is translated into the 32-bit integer type, `i32`. The function name is prefixed with `@`, and the parameter names are prefixed with `%`. The body of the function is enclosed in curly braces. The code of the body follows:

```
entry:
  %cmp = icmp eq i32 %b, 0
  br i1 %cmp, label %return, label %while.body
```

The IR code is organized into so-called **basic blocks**. A well-formed basic block is a linear sequence of instructions, which begins with an optional label and ends with a terminator instruction. So, each basic block has one entry point and one exit point. LLVM allows malformed basic blocks at construction time. The label of the first basic block is `entry`. The code in the block is simple: the first instruction compares the `%b` parameter against `0`. The second instruction branches to the `return` label if the condition is `true` and to the `while.body` label if the condition is `false`.

Another characteristic of the IR code is that it is in a **static single assignment (SSA)** form. The code uses an unlimited number of virtual registers, but each register is only written once. The result of the comparison is assigned to the named virtual register, `%cmp`. This register is then used, but it is never written again. Optimizations such as constant propagation and common-sub-expression elimination work very well with the SSA form and all modern compilers are using it.

> **SSA**
>
> The SSA form was developed in the late 1980s. Since then, it has been widely used in compilers because it simplifies data flow analysis and optimizations. For example, the identification of common sub-expressions inside a loop becomes much easier if the IR is in SSA form. A basic property of SSA is that it establishes def-use and use-def chains: for a single definition, you know of all uses (def-use), and for each use, you know the unique definition (use-def). This knowledge is used a lot, such as in constant propagation: if a definition is determined to be a constant, then all uses of this value can be easily replaced with that constant value.
>
> To construct the SSA form, the algorithm from Cytron et al. (1989) is very popular, and it is also used in the LLVM implementation. Other algorithms have been developed too. An early observation is that these algorithms become simpler if the source language does not have a goto statement.
>
> An in-depth treatment of SSA can be found in the book *SSA-based Compiler Design*, by F. Rastello and F. B. Tichadou, Springer 2022.

The next basic block is the body of the while loop:

```
while.body:
    %b.loop = phi i32 [ %rem, %while.body ],
                      [ %b, %entry ]
    %a.loop = phi i32 [ %b.loop, %while.body ],
                      [ %a, %entry ]
    %rem = urem i32 %a.loop, %b.loop
    %cmp1 = icmp eq i32 %rem, 0
    br i1 %cmp1, label %return, label %while.body
```

Inside the loop of gcd, the a and b parameters are assigned new values. If a register can be only written once, then this is not possible. The solution is to use the special phi instruction. The phi instruction has a list of basic blocks and values as parameters. A basic block presents the incoming edge from that basic block, and the value is the value from that basic block. At runtime, the phi instruction compares the label of the previously executed basic block with the labels in the parameter list.

The value of the instruction is the value that's associated with the label. For the first phi instruction, the value is the %rem register if the previously executed basic block was while.body. The value is %b if entry was the previously executed basic block. The values are the ones at the start of the basic block. The %b.loop register gets a value from the first phi instruction. The same register is used in the parameter list of the second phi instruction, but the value is assumed to be the one before it is changed through the first phi instruction.

After the loop's body, the return value must be chosen:

```
return:
  %retval = phi i32 [ %a, %entry ],
                    [ %b.loop, %while.body ]
  ret i32 %retval
}
```

Again, a phi instruction is used to select the desired value. The ret instruction not only ends this basic block but also denotes the end of this function at runtime. It has the return value as a parameter.

There are some restrictions on the use of phi instructions. They must be the first instructions of a basic block. The first basic block is special: it has no previously executed block. Therefore, it cannot begin with a phi instruction.

> **LLVM IR reference**
>
> We've only touched the very basics of the LLVM IR. Please visit the LLVM Language Reference Manual at https://llvm.org/docs/LangRef.html to look up all the details.

The IR code itself looks a lot like a mix of C and assembly language. Despite this familiar style, it is not clear how we can easily generate the IR code from an AST. The phi instruction in particular looks difficult to generate. But don't be scared – in the next section, we'll implement a simple algorithm to just do that!

Learning about the load-and-store approach

All local optimizations in LLVM are based on the SSA form shown here. For global variables, memory references are used. The IR language knows load and store instructions, which are used to fetch and store those values. You can use this for local variables too. These instructions do not belong to the SSA form, and LLVM knows how to convert them into the required SSA form. Therefore, you can allocate memory slots for each local variable and use load and store instructions to change their value. All you need to remember is the pointer to the memory slot where a variable is stored. The clang compiler uses this approach.

Let's look at the IR code for load and store. Compile gcd.c again, but this time without enabling optimization:

```
$ clang --target=aarch64-linux-gnu -S -emit-llvm gcd.c
```

The gcd function now looks different. This is the first basic block:

```
define i32 @gcd(i32, i32) {
  %3 = alloca i32, align 4
  %4 = alloca i32, align 4
```

```
%5 = alloca i32, align 4
%6 = alloca i32, align 4
store i32 %0, ptr %4, align 4
store i32 %1, ptr %5, align 4
%7 = load i32, ptr %5, align 4
%8 = icmp eq i32 %7, 0
br i1 %8, label %9, label %11
```

The IR code now relies on the automatic numbering of registers and labels. The names of the parameters are not specified. Implicitly, they are %0 and %1. The basic block has no label, so 2 is assigned. The first few instructions allocate memory for the four 32-bit values. After that, the %0 and %1 parameters are stored in the memory slots pointed to by registers %4 and %5. To compare %1 to 0, the value is explicitly loaded from the memory slot. With this approach, you do not need to use the phi instruction! Instead, you load a value from a memory slot, perform a calculation on it, and store the new value back in the memory slot. The next time you read the memory slot, you get the last computed value. All the other basic blocks for the gcd function follow this pattern.

The advantage of using load and store instructions in this way is that it is fairly easy to generate the IR code. The disadvantage is that you generate a lot of IR instructions that LLVM will remove with the mem2reg pass in the very first optimization step, after converting the basic block into SSA form. Therefore, we generate the IR code in SSA form directly.

We'll start developing IR code generation by mapping the control flow to basic blocks.

Mapping the control flow to basic blocks

The conceptual idea of a basic block is that it is a *linear sequence of instructions* that are executed in that order. A basic block has exactly one entry at the beginning, and it ends with a terminator instruction, which is an instruction that transfers the control flow to another basic block, such as a branch instruction, a switch instruction, or a return instruction. See https://llvm.org/docs/LangRef.html#terminator-instructions for a complete list of terminator instructions. A basic block can begin with phi instructions, but inside a basic block, neither phi nor branch instructions are allowed. In other words, you can only enter a basic block at the first instruction, and you can only leave a basic block at the last instruction, which is the terminator instruction. It is not possible to branch to an instruction inside a basic block or to branch to another basic block from the middle of a basic block. Please note that a simple function call with the call instruction can occur inside a basic block. Each basic block has exactly one label, marking the first instruction of the basic block. Labels are the targets of branch instructions. You can view branches as directed edges between two basic blocks, resulting in the **control flow graph** (**CFG**). A basic block can have **predecessors** and **successors**. The first basic block of a function is special in the sense that no predecessors are allowed.

As a consequence of these restrictions, control statements of the source language, such as WHILE and IF, produce several basic blocks. Let's look at the WHILE statement. The condition of the WHILE statement controls if the loop body or the next statement is executed. The condition must be generated in a basic block of its own because there are two predecessors:

- The basic block resulting from the statement before WHILE

- The branch from the end of the loop body back to the condition

There are also two successors:

- The beginning of the loop body

- The basic block resulting from the statement following WHILE

The loop body itself has at least one basic block:

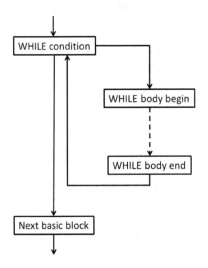

Figure 4.1 – The basic blocks of a WHILE statement

The IR code generation follows this structure. We store a pointer to the current basic block in the CGProcedure class and use an instance of llvm::IRBuilder<> to insert instructions into the basic block. First, we create the basic blocks:

```
void emitStmt(WhileStatement *Stmt) {
    llvm::BasicBlock *WhileCondBB = llvm::BasicBlock::Create(
        CGM.getLLVMCtx(), "while.cond", Fn);
    llvm::BasicBlock *WhileBodyBB = llvm::BasicBlock::Create(
        CGM.getLLVMCtx(), "while.body", Fn);
    llvm::BasicBlock *AfterWhileBB = llvm::BasicBlock::Create(
        CGM.getLLVMCtx(), "after.while", Fn);
```

The Fn variable denotes the current function and `getLLVMCtx()` returns the LLVM context. Both are set later. We end the current basic block with a branch to the basic block, which will hold the condition:

```
Builder.CreateBr(WhileCondBB);
```

The basic block for the condition becomes the new current basic block. We generate the condition and end the block with a conditional branch:

```
setCurr(WhileCondBB);
llvm::Value *Cond = emitExpr(Stmt->getCond());
Builder.CreateCondBr(Cond, WhileBodyBB, AfterWhileBB);
```

Next, we generate the loop body. Finally, we add a branch back to the basic block of the condition:

```
setCurr(WhileBodyBB);
emit(Stmt->getWhileStmts());
Builder.CreateBr(WhileCondBB);
```

With that, we have generated the WHILE statement. Now that we've generated the WhileCondBB and Curr blocks, we can seal them:

```
sealBlock(WhileCondBB);
sealBlock(Curr);
```

The empty basic block for statements following WHILE becomes the new current basic block:

```
  setCurr(AfterWhileBB);
}
```

Following this schema, you can create an `emit()` method for each statement of the source language.

Using AST numbering to generate IR code in SSA form

To generate IR code in SSA form from the AST, we can use an approach called **AST numbering**. The basic idea is that for each basic block, we store the current value of local variables written in this basic block.

> **Note**
>
> The implementation is based on the paper *Simple and Efficient Construction of Static Single Assignment Form*, by Braun et al., International Conference on CompilerConstruction 2013 (CC 2013), Springer (see http://individual.utoronto.ca/dfr/ece467/braun13.pdf). In its presented form, it only works for IR code that has a structured controlled flow. The paper also describes the necessary extensions if you need to support arbitrary control flow – for example, a goto statement.

Although it is simple, we will still need several steps. We will introduce the required data structure first, and after that, we will learn how to read and write values local to a basic block. Then, we will handle values that are used in several basic blocks and conclude by optimizing the created phi instructions.

Defining the data structure to hold values

We use the `BasicBlockDef` struct to hold the information for a single block:

```
struct BasicBlockDef {
  llvm::DenseMap<Decl *, llvm::TrackingVH<llvm::Value>> Defs;
  // ...
};
```

The `llvm::Value` LLVN class represents a value in SSA form. The `Value` class acts like a label on the result of a computation. It is created once, usually through an IR instruction, and then used. Various changes can occur during optimizations. For example, if the optimizer detects that the `%1` and `%2` values are always the same, then it can replace the use of `%2` with `%1`. This changes the label but not the computation.

To be aware of such changes, we cannot use the `Value` class directly. Instead, we need a value handle. There are value handles with different functionality. To track replacement, we can use the `llvm::TrackingVH<>` class. As a result, the `Defs` member maps a declaration of the AST (a variable or a formal parameter) to its current value. Now, we need to store this information for each basic block:

```
llvm::DenseMap<llvm::BasicBlock *, BasicBlockDef> CurrentDef;
```

With this data structure, we are now able to handle local values.

Reading and writing values local to a basic block

To store the current value of a local variable in a basic block, we will create an entry in the maps:

```
void writeLocalVariable(llvm::BasicBlock *BB, Decl *Decl,
                        llvm::Value *Val) {
  CurrentDef[BB].Defs[Decl] = Val;
}
```

The lookup of a variable's value is a bit more complicated because the value might not be in the basic block. In this case, we need to extend the search to the predecessors using a possible recursive search:

```
llvm::Value *
readLocalVariable(llvm::BasicBlock *BB, Decl *Decl) {
  auto Val = CurrentDef[BB].Defs.find(Decl);
  if (Val != CurrentDef[BB].Defs.end())
```

```
        return Val->second;
    return readLocalVariableRecursive(BB, Decl);
}
```

The real work is searching the predecessors, which we'll implement in the next section.

Searching the predecessor blocks for a value

If the current basic block we are looking at has only one predecessor, then we search there for the value of the variable. If the basic block has several predecessors, then we need to search for the value in all these blocks and combine the results. To illustrate this situation, you can look at the basic block with the condition of a WHILE statement from the previous section.

This basic block has two predecessors – the one resulting from the statement before the WHILE statement and the one resulting from the branch for the end of the body of the WHILE loop. A variable that's used in the condition should have some initial value and will most likely be changed in the body of the loop. So, we need to collect these definitions and create a phi instruction from it. The basic blocks that are created from the WHILE statement contain a cycle.

Because we recursively search the predecessor blocks, we must break this cycle. To do so, we can use a simple trick: we can insert an empty phi instruction and record this as the current value of the variable. If we see this basic block again in our search, then we'll see that the variable has a value that we can use. The search stops at this point. Once we've collected all the values, we must update the phi instruction.

However, we will still face a problem. At the time of the lookup, not all predecessors of a basic block may be known. How can this happen? Look at the creation of the basic blocks for the WHILE statement. The IR for the condition of the loop is generated first. However, the branch from the end of the body that goes back to the basic block, which contains the condition, can only be added after the IR for the body is generated. This is because this basic block is not known earlier. If we need to read the value of a variable in the condition, then we are stuck, because not all predecessors are known.

To solve this situation, we must do a little more:

1. First, we must attach a Sealed flag to the basic block.
2. Then, we must define a basic block as sealed if we know all the predecessors of the basic block. If the basic block is not sealed and we need to look up the value of the variable not yet defined in this basic block, then we must insert an empty phi instruction and use it as the value.

3. We also need to remember this instruction. If the block is later sealed, then we need to update the instruction with the real values. To implement this, we must add two more members to struct `BasicBlockDef`: the `IncompletePhis` map, which records the `phi` instructions we need later to update, and the `Sealed` flag, which indicates if the basic block is sealed:

```
llvm::DenseMap<llvm::PHINode *, Decl *> IncompletePhis;
unsigned Sealed : 1;
```

4. Then, the method can be implemented, as discussed at the beginning of this section:

```
llvm::Value *CGProcedure::readLocalVariableRecursive(
    llvm::BasicBlock *BB, Decl *Decl) {
  llvm::Value *Val = nullptr;
  if (!CurrentDef[BB].Sealed) {
    llvm::PHINode *Phi = addEmptyPhi(BB, Decl);
    CurrentDef[BB].IncompletePhis[Phi] = Decl;
    Val = Phi;
  } else if (auto *PredBB = BB->getSinglePredecessor()) {
    Val = readLocalVariable(PredBB, Decl);
  } else {
    llvm::PHINode *Phi = addEmptyPhi(BB, Decl);
    writeLocalVariable(BB, Decl, Phi);
    Val = addPhiOperands(BB, Decl, Phi);
  }
  writeLocalVariable(BB, Decl, Val);
  return Val;
}
```

5. The `addEmptyPhi()` method inserts an empty `phi` instruction at the beginning of the basic block:

```
llvm::PHINode *
CGProcedure::addEmptyPhi(llvm::BasicBlock *BB,
                         Decl *Decl) {
  return BB->empty()
            ? llvm::PHINode::Create(mapType(Decl), 0,
                                    "", BB)
            : llvm::PHINode::Create(mapType(Decl), 0,
                                    "", &BB->front());
}
```

6. To add the missing operands to the `phi` instruction, first, we must search all the predecessors of the basic block and add the operand pair value and basic block to the `phi` instruction. Then, we must try to optimize the instruction:

```
llvm::Value *
CGProcedure::addPhiOperands(llvm::BasicBlock *BB,
                            Decl *Decl,
                            llvm::PHINode *Phi) {
  for (auto *PredBB : llvm::predecessors(BB))
    Phi->addIncoming(readLocalVariable(PredBB, Decl),
                     PredBB);
  return optimizePhi(Phi);
}
```

This algorithm can generate unneeded `phi` instructions. One approach to optimize these will be implemented in the next section.

Optimizing the generated phi instructions

How can we optimize a `phi` instruction and why should we do it? Although the SSA form is advantageous for many optimizations, the `phi` instruction is often not interpreted by the algorithms and thus hinders the optimization in general. Therefore, the fewer `phi` instructions we generate, the better. Let's take a closer look:

1. If the instruction has only one operand or all operands have the same value, then we replace the instruction with this value. If the instruction has no operand, then we replace the instruction with the special `Undef` value. Only if the instruction has two or more distinct operands do we have to keep the instruction:

```
llvm::Value *
CGProcedure::optimizePhi(llvm::PHINode *Phi) {
  llvm::Value *Same = nullptr;
  for (llvm::Value *V : Phi->incoming_values()) {
    if (V == Same || V == Phi)
      continue;
    if (Same && V != Same)
      return Phi;
    Same = V;
  }
  if (Same == nullptr)
    Same = llvm::UndefValue::get(Phi->getType());
```

2. Removing a `phi` instruction may lead to optimization opportunities in other `phi` instructions. Fortunately, LLVM keeps track of the users and the use of values (which is the use-def chain mentioned in the definition of SSA). We must search for all uses of the value in other `phi` instructions and try to optimize these instructions too:

```
llvm::SmallVector<llvm::PHINode *, 8> CandidatePhis;
for (llvm::Use &U : Phi->uses()) {
  if (auto *P =
          llvm::dyn_cast<llvm::PHINode>(U.getUser()))
    if (P != Phi)
      CandidatePhis.push_back(P);
}
Phi->replaceAllUsesWith(Same);
Phi->eraseFromParent();
for (auto *P : CandidatePhis)
  optimizePhi(P);
return Same;
}
```

If we want, we can improve this algorithm even further. Instead of always iterating the list of values for each `phi` instruction, we can pick and remember two distinct values. Then, in the `optimizePhi` function, we can check if these two values are still in the list of the `phi` instruction. If that is the case, then we know that there is nothing to optimize. But even without this optimization, this algorithm runs very fast, so we are not going to implement this now.

We are almost done. The only thing we haven't done is implement the operation to seal a basic block. We will do this in the next section.

Sealing a block

As soon as we know that all the predecessors of a block are known, we can seal the block. If the source language contains only structured statements such as `tinylang`, then it is easy to determine where a block can be sealed. Take another look at the basic blocks that are generated for the WHILE statement.

The basic block that contains the condition can be sealed after the branch from the end of the body is added because this was the last missing predecessor. To seal a block, we can simply add the missing operands to the incomplete `phi` instructions and set the flag:

```
void CGProcedure::sealBlock(llvm::BasicBlock *BB) {
  for (auto PhiDecl : CurrentDef[BB].IncompletePhis) {
    addPhiOperands(BB, PhiDecl.second, PhiDecl.first);
  }
  CurrentDef[BB].IncompletePhis.clear();
  CurrentDef[BB].Sealed = true;
}
```

With these methods, we are now ready to generate the IR code for expressions.

Creating the IR code for expressions

In general, you translate expressions, as shown in *Chapter 2, The Structure of a Compiler*. The only interesting part is how to access variables. The previous section treated local variables, but there are other kinds of variables we can consider. Let's discuss what we need to do:

- For a local variable of the procedure, we use the `readLocalVariable()` and `writeLocalVariable()` methods from the previous section.

- For a local variable in an enclosing procedure, we need a pointer to the frame of the enclosing procedure. This will be handled later in this chapter.

- For a global variable, we generate load and store instructions.

- For a formal parameter, we have to differentiate between passing by value and passing by reference (the VAR parameter in tinylang). A parameter that's passed by value is treated as a local variable, and a parameter passed by reference is treated as a global variable.

Putting it all together, we get the following code for reading a variable or formal parameter:

```
llvm::Value *CGProcedure::readVariable(llvm::BasicBlock *BB,
                                       Decl *D) {
  if (auto *V = llvm::dyn_cast<VariableDeclaration>(D)) {
    if (V->getEnclosingDecl() == Proc)
      return readLocalVariable(BB, D);
    else if (V->getEnclosingDecl() ==
             CGM.getModuleDeclaration()) {
      return Builder.CreateLoad(mapType(D),
                                CGM.getGlobal(D));
    } else
      llvm::report_fatal_error(
          "Nested procedures not yet supported");
  } else if (auto *FP =
                 llvm::dyn_cast<FormalParameterDeclaration>(
                 D)) {
    if (FP->isVar()) {
      return Builder.CreateLoad(mapType(FP, false),
                                FormalParams[FP]);
    } else
      return readLocalVariable(BB, D);
  } else
    llvm::report_fatal_error("Unsupported declaration");
}
```

Writing to a variable or formal parameter is symmetrical – we just need to exchange the method to read with the one to write and use a `store` instruction instead of a `load` instruction.

Next, these functions are applied while generating the IR code for the functions.

Emitting the IR code for a function

Most of the IR code will live in a function. A function in IR code resembles a function in C. It specifies in the name, the types of parameters, the return value, and other attributes. To call a function in a different compilation unit, you need to declare the function. This is similar to a prototype in C. If you add basic blocks to the function, then you define the function. We will do all this in the next few sections, but first, we will discuss the visibility of symbol names.

Controlling visibility with linkage and name mangling

Functions (and also global variables) have a linkage style attached. With the linkage style, we define the visibility of a symbol name and what should happen if more than one symbol has the same name. The most basic linkage styles are `private` and `external`. A symbol with `private` linkage is only visible in the current compilation unit, while a symbol with `external` linkage is globally available.

For a language without a proper module concept, such as C, this is adequate. With modules, we need to do more. Let's assume that we have a module called `Square` that provides a `Root()` function and a `Cube` module, which also provides a `Root()` function. If the functions are private, then there is no problem. The function gets the name `Root` and private linkage. The situation is different if the function is exported so that it can be called from other modules. Using the function name alone is not enough, because this name is not unique.

The solution is to tweak the name to make it globally unique. This is called name **mangling**. How this is done depends on the requirements and characteristics of the language. In our case, the base idea is to use a combination of the module and the function name to create a globally unique name. Using `Square.Root` as the name looks like an obvious solution, but it may lead to problems with assemblers as the dot may have a special meaning. Instead of using a delimiter between the name components, we can get a similar effect by prefixing the name components with their length: `6Square4Root`. This is no legal identifier for LLVM, but we can fix this by prefixing the whole name with `_t` (with t for `tinylang`): `_t6Square4Root`. In this way, we can create unique names for exported symbols:

```
std::string CGModule::mangleName(Decl *D) {
  std::string Mangled("_t");
  llvm::SmallVector<llvm::StringRef, 4> List;
  for (; D; D = D->getEnclosingDecl())
    List.push_back(D->getName());
  while (!List.empty()) {
    llvm::StringRef Name = List.pop_back_val();
    Mangled.append(
```

```
                llvm::Twine(Name.size()).concat(Name).str());
    }
    return Mangled;
}
```

If your source language supports type overloading, then you need to extend this scheme with type names. For example, to distinguish between the int root(int) and double root(double) C++ functions, the type of the parameter and the return value must be added to the function name.

You also need to think about the length of the generated name since some linkers place restrictions on the length. With nested namespaces and classes in C++, the mangled names can be rather long. There, C++ defines a compression scheme to avoid repeating name components over and over again.

Next, we'll look at how to treat parameters.

Converting a type from an AST description into LLVM types

The parameters of a function also need some consideration. First, we need to map the types of the source language to an LLVM type. As tinylang currently has only two types, this is easy:

```
llvm::Type *CGModule::convertType(TypeDeclaration *Ty) {
    if (Ty->getName() == "INTEGER")
      return Int64Ty;
    if (Ty->getName() == "BOOLEAN")
      return Int1Ty;
    llvm::report_fatal_error("Unsupported type");
}
```

Int64Ty, Int1Ty, and VoidTy are class members that hold the type representation of the i64, i1, and void LLVM types.

For a formal parameter passed by reference, this is not enough. The LLVM type of this parameter is a pointer. However, when we want to use the value of the formal parameter, we need to know the underlying type. This is controlled by the HonorReference flag, which has a default value of true. We generalize the function and take the formal parameter into account:

```
llvm::Type *CGProcedure::mapType(Decl *Decl,
                                 bool HonorReference) {
    if (auto *FP = llvm::dyn_cast<FormalParameterDeclaration>(
            Decl)) {
      if (FP->isVar() && HonorReference)
        return llvm::PointerType::get(CGM.getLLVMCtx(),
                                      /*AddressSpace=*/0);
      return CGM.convertType(FP->getType());
    }
```

```
    if (auto *V = llvm::dyn_cast<VariableDeclaration>(Decl))
      return CGM.convertType(V->getType());
    return CGM.convertType(llvm::cast<TypeDeclaration>(Decl));
  }
```

With these helpers at hand, we can create the LLVM IR function.

Creating the LLVM IR function

To emit a function in LLVM IR, a function type is needed, which is similar to a prototype in C. Creating the function type involves mapping the types and then calling the factory method to create the function type:

```
llvm::FunctionType *CGProcedure::createFunctionType(
    ProcedureDeclaration *Proc) {
  llvm::Type *ResultTy = CGM.VoidTy;
  if (Proc->getRetType()) {
    ResultTy = mapType(Proc->getRetType());
  }
  auto FormalParams = Proc->getFormalParams();
  llvm::SmallVector<llvm::Type *, 8> ParamTypes;
  for (auto FP : FormalParams) {
    llvm::Type *Ty = mapType(FP);
    ParamTypes.push_back(Ty);
  }
  return llvm::FunctionType::get(ResultTy, ParamTypes,
                                 /*IsVarArgs=*/false);
}
```

Based on the function type, we also create the LLVM function. This associates the function type with the linkage and the mangled name:

```
llvm::Function *
CGProcedure::createFunction(ProcedureDeclaration *Proc,
                            llvm::FunctionType *FTy) {
  llvm::Function *Fn = llvm::Function::Create(
      Fty, llvm::GlobalValue::ExternalLinkage,
      CGM.mangleName(Proc), CGM.getModule());
```

The getModule() method returns the current LLVM module, which we'll set up a bit later.

With the function created, we can add some more information about it:

- First, we can give the parameter's names. This makes the IR more readable.

- Second, we can add attributes to the function and to the parameters to specify some characteristics. As an example, we will do this for parameters passed by reference.

At the LLVM level, these parameters are pointers. But from the source language design, these are very restricted pointers. Analogous to references in C++, we always need to specify a variable for a VAR parameter. So, by design, we know that this pointer will never be null and that it is always dereferenceable, meaning that we can read the value that's being pointed to without risking a general protection fault. Also, by design, this pointer cannot be passed around – in particular, there are no copies of the pointer that outlive the call to the function. Therefore, the pointer is said to not be captured.

The `llvm::AttributeBuilder` class is used to build the set of attributes for a formal parameter. To get the storage size of a parameter type, we can simply query the data layout object:

```
for (auto [Idx, Arg] : llvm::enumerate(Fn->args())) {
  FormalParameterDeclaration *FP =
      Proc->getFormalParams()[Idx];
  if (FP->isVar()) {
    llvm::AttrBuilder Attr(CGM.getLLVMCtx());
    llvm::TypeSize Sz =
        CGM.getModule()->getDataLayout().getTypeStoreSize(
            CGM.convertType(FP->getType()));
    Attr.addDereferenceableAttr(Sz);
    Attr.addAttribute(llvm::Attribute::NoCapture);
    Arg.addAttrs(Attr);
  }
  Arg.setName(FP->getName());
}
return Fn;
}
```

With that, we have created the IR function. In the next section, we'll add the basic blocks of the function body to the function.

Emitting the function body

We are almost done with emitting the IR code for a function! We only need to put the pieces together to emit a function, including its body:

1. Given a procedure declaration from `tinylang`, first, we will create the function type and the function:

    ```
    void CGProcedure::run(ProcedureDeclaration *Proc) {
      this->Proc = Proc;
      Fty = createFunctionType(Proc);
      Fn = createFunction(Proc, Fty);
    ```

2. Next, we will create the first basic block of the function and make it the current one:

    ```
    llvm::BasicBlock *BB = llvm::BasicBlock::Create(
        CGM.getLLVMCtx(), "entry", Fn);
    setCurr(BB);
    ```

3. Then, we must step through all formal parameters. To handle VAR parameters correctly, we need to initialize the `FormalParams` member (used in `readVariable()`). In contrast to local variables, formal parameters have a value in the first basic block, so we must make these values known:

    ```
    for (auto [Idx, Arg] : llvm::enumerate(Fn->args())) {
      FormalParameterDeclaration *FP =
          Proc->getFormalParams()[Idx];
      FormalParams[FP] = &Arg;
      writeLocalVariable(Curr, FP, &Arg);
    }
    ```

4. After this setup, we can call the `emit()` method to start generating the IR code for statements:

    ```
    auto Block = Proc->getStmts();
    emit(Proc->getStmts());
    ```

5. The last block after generating the IR code may not be sealed yet, so we must call `sealBlock()` now. A procedure in `tinylang` may have an implicit return, so we must also check if the last basic block has a proper terminator, and add one if not:

    ```
    if (!Curr->getTerminator()) {
      Builder.CreateRetVoid();
    }
    sealBlock(Curr);
    }
    ```

With that, we've finished generating IR code for functions. However, we still need to create the LLVM module, which holds all the IR code together. We'll do this in the next section.

Setting up the module and the driver

We collect all the functions and global variables of a compilation unit in an LLVM module. To ease the IR generation process, we can wrap all the functions from the previous sections into a code generator class. To get a working compiler, we also need to define the target architecture for which we want to generate code, and also add the passes that emit the code. We will implement this in this and the next few chapters, starting with the code generator.

Wrapping all in the code generator

The IR module is the brace around all elements we generate for a compilation unit. At the global level, we iterate through the declarations at the module level, create global variables, and call the code generation for procedures. A global variable in tinylang is mapped to an instance of the llvm::GobalValue class. This mapping is saved in Globals and made available to the code generation for procedures:

```
void CGModule::run(ModuleDeclaration *Mod) {
  for (auto *Decl : Mod->getDecls()) {
    if (auto *Var =
            llvm::dyn_cast<VariableDeclaration>(Decl)) {
      // Create global variables
      auto *V = new llvm::GlobalVariable(
          *M, convertType(Var->getType()),
          /*isConstant=*/false,
          llvm::GlobalValue::PrivateLinkage, nullptr,
          mangleName(Var));
      Globals[Var] = V;
    } else if (auto *Proc =
                   llvm::dyn_cast<ProcedureDeclaration>(
                       Decl)) {
      CGProcedure CGP(*this);
      CGP.run(Proc);
    }
  }
}
```

The module also holds the LLVMContext class and caches the most commonly used LLVM types. The latter ones need to be initialized, for example, for the 64-bit integer type:

```
Int64Ty = llvm::Type::getInt64Ty(getLLVMCtx());
```

The `CodeGenerator` class initializes the LLVM IR module and calls the code generation for the module. Most importantly, this class must know for which target architecture we'd like to generate code. This information is passed in the `llvm::TargetMachine` class, which is set up in the driver:

```
std::unique_ptr<llvm::Module>
CodeGenerator::run(ModuleDeclaration *Mod,
                   std::string FileName) {
  std::unique_ptr<llvm::Module> M =
      std::make_unique<llvm::Module>(FileName, Ctx);
  M->setTargetTriple(TM->getTargetTriple().getTriple());
  M->setDataLayout(TM->createDataLayout());
  CGModule CGM(M.get());
  CGM.run(Mod);
  return M;
}
```

For ease of use, we must also introduce a factory method for the code generator:

```
CodeGenerator *
CodeGenerator::create(llvm::LLVMContext &Ctx,
                      llvm::TargetMachine *TM) {
  return new CodeGenerator(Ctx, TM);
}
```

The `CodeGenerator` class provides a small interface to create IR code, which is ideal for use in the compiler driver. Before we integrate it, we need to implement the support for machine code generation.

Initializing the target machine class

Now, only the target machine is missing. With the target machine, we define the CPU architecture we'd like to generate code for. For each CPU, there are features available that can be used to influence the code generation process. For example, a newer CPU of a CPU architecture family can support vector instructions. With features, we can toggle the use of vector instructions on or off. To support setting all these options from the command line, LLVM provides some supporting code. In the `Driver` class, we can add the following `include` variable:

```
#include "llvm/CodeGen/CommandFlags.h"
```

This `include` variable adds common command-line options to our compiler driver. Many LLVM tools also use these command-line options, which have the benefit of providing a common interface to the user. Only the option to specify a target triple is missing. As this is very useful, we'll add this ourselves:

```
static llvm::cl::opt<std::string> MTriple(
    "mtriple",
    llvm::cl::desc("Override target triple for module"));
```

Let's create the target machine:

1. To display error messages, the name of the application must be passed to the function:

    ```
    llvm::TargetMachine *
    createTargetMachine(const char *Argv0) {
    ```

2. First, we must collect all the information provided by the command line. These are options for the code generator – that is, the name of the CPU and possible features that should be activated or deactivated, and the triple of the target:

    ```
    llvm::Triple Triple = llvm::Triple(
        !MTriple.empty()
            ? llvm::Triple::normalize(MTriple)
            : llvm::sys::getDefaultTargetTriple());

    llvm::TargetOptions TargetOptions =
        codegen::InitTargetOptionsFromCodeGenFlags(Triple);
    std::string CPUStr = codegen::getCPUStr();
    std::string FeatureStr = codegen::getFeaturesStr();
    ```

3. Then, we must look up the target in the target registry. If an error occurs, then we will display the error message and bail out. A possible error would be an unsupported triple specified by the user:

    ```
    std::string Error;
    const llvm::Target *Target =
        llvm::TargetRegistry::lookupTarget(
            codegen::getMArch(), Triple, Error);

    if (!Target) {
      llvm::WithColor::error(llvm::errs(), Argv0) << Error;
      return nullptr;
    }
    ```

4. With the help of the `Target` class, we can configure the target machine using all the known options requested by the user:

    ```
    llvm::TargetMachine *TM = Target->createTargetMachine(
        Triple.getTriple(), CPUStr, FeatureStr,
        TargetOptions, std::optional<llvm::Reloc::Model>(
            codegen::getRelocModel()));
    return TM;
    }
    ```

With the target machine instance, we can generate IR code that targets a CPU architecture of our choice. What is missing is the translation to assembly text or the generation of object code files. We'll add this support in the next section.

Emitting assembler text and object code

In LLVM, the IR code is run through a pipeline of passes. Each pass performs a single task, such as removing dead code. We'll learn more about passes in *Chapter 7, Optimizing IR*. Outputting assembler code or an object file is implemented as a pass too. Let's add basic support for it!

We need to include even more LLVM header files. First, we need the `llvm::legacy::PassManager` class to hold the passes to emit code to a file. We also want to be able to output LLVM IR code, so we also need a pass to emit this. Finally, we'll use the `llvm::ToolOutputFile` class for the file operations:

```
#include "llvm/IR/IRPrintingPasses.h"
#include "llvm/IR/LegacyPassManager.h"
#include "llvm/MC/TargetRegistry.h"
#include "llvm/Pass.h"
#include "llvm/Support/ToolOutputFile.h"
```

Another command-line option for outputting LLVM IR is also needed:

```
static llvm::cl::opt<bool> EmitLLVM(
    "emit-llvm",
    llvm::cl::desc("Emit IR code instead of assembler"),
    llvm::cl::init(false));
```

Finally, we want to be able to give the output file a name:

```
static llvm::cl::opt<std::string>
    OutputFilename("o",
                    llvm::cl::desc("Output filename"),
                    llvm::cl::value_desc("filename"));
```

The first task in the new `emit()` method is to deal with the name of the output file if it's not given by the user on the command line. If the input is read from `stdin`, indicated by the use of the minus symbol, -, then we output the result to `stdout`. The `ToolOutputFile` class knows how to handle the special filename, -:

```
bool emit(StringRef Argv0, llvm::Module *M,
          llvm::TargetMachine *TM,
          StringRef InputFilename) {
  CodeGenFileType FileType = codegen::getFileType();
  if (OutputFilename.empty()) {
```

```
if (InputFilename == "-") {
  OutputFilename = "-";
}
```

Otherwise, we drop a possible extension of the input filename and append .ll, .s, or .o as an extension, depending on the command-line options given by the user. The FileType option is defined in the llvm/CodeGen/CommandFlags.inc header file, which we included earlier. This option doesn't support emitting IR code, so we've added the new-emit-llvm option, which only takes effect if it's used together with the assembly file type:

```
else {
  if (InputFilename.endswith(".mod"))
    OutputFilename =
        InputFilename.drop_back(4).str();
  else
    OutputFilename = InputFilename.str();
  switch (FileType) {
  case CGFT_AssemblyFile:
    OutputFilename.append(EmitLLVM ? ".ll" : ".s");
    break;
  case CGFT_ObjectFile:
    OutputFilename.append(".o");
    break;
  case CGFT_Null:
    OutputFilename.append(".null");
    break;
  }
}
}
```

Some platforms distinguish between text and binary files, so we have to provide the right open flags when opening the output file:

```
std::error_code EC;
sys::fs::OpenFlags OpenFlags = sys::fs::OF_None;
if (FileType == CGFT_AssemblyFile)
  OpenFlags |= sys::fs::OF_TextWithCRLF;
auto Out = std::make_unique<llvm::ToolOutputFile>(
    OutputFilename, EC, OpenFlags);
if (EC) {
  WithColor::error(llvm::errs(), Argv0)
      << EC.message() << '\n';
  return false;
}
```

Now, we can add the required passes to `PassManager`. The `TargetMachine` class has a utility method that adds the requested classes. Therefore, we only need to check if the user requests to output the LLVM IR code:

```
legacy::PassManager PM;
if (FileType == CGFT_AssemblyFile && EmitLLVM) {
  PM.add(createPrintModulePass(Out->os()));
} else {
  if (TM->addPassesToEmitFile(PM, Out->os(), nullptr,
                              FileType)) {
    WithColor::error(llvm::errs(), Argv0)
        << "No support for file type\n";
    return false;
  }
}
```

With all this preparation done, emitting the file boils down to a single function call:

```
PM.run(*M);
```

The `ToolOutputFile` class automatically deletes the file if we do not explicitly request that we want to keep it. This makes error handling easier as there are potentially many places where we need to handle errors and only one place is reached if everything goes well. We successfully emitted the code, so we want to keep the file:

```
Out->keep();
```

Finally, we must report success to the caller:

```
    return true;
}
```

Calling the `emit()` method with `llvm::Module`, which we created with a call to the `CodeGenerator` class, emits the code as requested.

Suppose you have the greatest common divisor algorithm in `tinylang` stored in the `Gcd.mod` file:

```
MODULE Gcd;

PROCEDURE GCD(a, b: INTEGER) : INTEGER;
VAR t: INTEGER;
BEGIN
  IF b = 0 THEN
    RETURN a;
  END;
  WHILE b # 0 DO
```

```
      t  :=  a  MOD  b;
      a  :=  b;
      b  :=  t;
    END;
    RETURN a;
END  GCD;


END Gcd.
```

To translate this to the `Gcd.o` object file, type the following:

```
$ tinylang --filetype=obj Gcd.mod
```

If you'd like to inspect the generated IR code directly on the screen, type the following:

```
$ tinylang --filetype=asm --emit-llvm -o - Gcd.mod
```

With the current state of the implementation, it is not possible to create a complete program in `tinylang`. However, you can use a small C program called `callgcd.c` to test the generated object file. Note the use of the mangled name to call the GCD function:

```c
#include <stdio.h>

extern long _t3Gcd3GCD(long, long);

int main(int argc, char *argv[]) {
    printf(„gcd(25, 20) = %ld\n", _t3Gcd3GCD(25, 20));
    printf(„gcd(3, 5) = %ld\n", _t3Gcd3GCD(3, 5));
    printf(„gcd(21, 28) = %ld\n", _t3Gcd3GCD(21, 28));
    return 0;
}
```

To compile and run the whole application with `clang`, type the following:

```
$ tinylang --filetype=obj Gcd.mod
$ clang callgcd.c Gcd.o -o gcd
$ gcd
```

Let's celebrate! At this point, we have created a complete compiler by reading the source language up and emitting assembler code or an object file.

Summary

In this chapter, you learned how to implement a code generator for LLVM IR code. Basic blocks are important data structures that hold all the instructions and express branches. You learned how to create basic blocks for the control statements of the source language and how to add instructions to a basic block. You applied a modern algorithm to handle local variables in functions, leading to less IR code. The goal of a compiler is to generate assembler text or an object file for the input, so you also added a simple compilation pipeline. With this knowledge, you will be able to generate LLVM IR code and assembler text or object code for your language compiler.

In the next chapter, you'll learn how to deal with aggregate data structures and how to ensure that function calls comply with the rules of your platform.

5

IR Generation for High-Level Language Constructs

High-level languages today usually make use of aggregate data types and **object-oriented programming (OOP)** constructs. LLVM IR has some support for aggregate data types, and OOP constructs such as classes must be implemented on their own. Adding aggregate types raises the question of how the parameters of an aggregate type are passed. Different platforms have different rules, and this is also reflected in the IR. Complying with the calling convention also ensures that system functions can be called.

In this chapter, you will learn how to translate aggregate data types and pointers to LLVM IR and how to pass parameters to a function in a system-compliant way. You will also learn how to implement classes and virtual functions in LLVM IR.

This chapter will cover the following topics:

- Working with arrays, structs, and pointers

- Getting the **application binary interface (ABI)** right

- Creating IR code for classes and virtual functions

By the end of the chapter, you will have acquired the knowledge to create LLVM IR for aggregate data types and OOP constructs. You will also know how to pass aggregate data types according to the rules of the platform.

Technical requirements

The code used in this chapter can be found at https://github.com/PacktPublishing/Learn-LLVM-17/tree/main/Chapter05.

Working with arrays, structs, and pointers

For almost all applications, basic types such as INTEGER are not sufficient. For example, to represent mathematical objects such as a matrix or a complex number, you must construct new data types based on existing ones. These new data types are generally known as **aggregate** or **composite**.

Arrays are a sequence of elements of the same type. In LLVM, arrays are always static, which means that the number of elements is constant. The tinylang type ARRAY [10] OF INTEGER or the C type long[10] is expressed in IR as follows:

```
[10 x i64]
```

Structures are composites of different types. In programming languages, they are often expressed with named members. For example, in tinylang, a structure is written as RECORD x: REAL; color: INTEGER; y: REAL; END; and the same structure in C is struct { float x; long color; float y; };. In LLVM IR, only the type names are listed:

```
{ float, i64, float }
```

To access a member, a numerical index is used. Like arrays, the first element has an index number of 0.

The members of this structure are arranged in memory according to the specification in the data layout string. For more information regarding the data layout string within LLVM, *Chapter 4, Basics of IR Code Generation*, describes these details.

Furthermore, if necessary, unused padding bytes are inserted. If you need to take control of the memory layout, then you can use a packed structure in which all elements have a 1-byte alignment. Within C, we utilize the __packed__ attribute in the struct in the following way:

```
struct __attribute__((__packed__)) { float x; long long color; float y; }
```

Likewise, the syntax within LLVM IR is slightly different and looks like the following:

```
<{ float, i64, float }>
```

Loaded into a register, arrays, and structs are treated as a unit. It is not possible to refer to a single element of array-valued register %x as %x[3], for example. This is due to the SSA form because it is not possible to tell if %x[i] and %x[j] refer to the same element or not. Instead, we need special instructions to extract and insert single-element values into an array. To read the second element, we use the following:

```
%el2 = extractvalue [10 x i64] %x, 1
```

We can also update an element such as the first one:

```
%xnew = insertvalue [10 x i64] %x, i64 %el2, 0
```

Both instructions work on structures, too. For example, to access the `color` member from register `%pt`, you write the following:

```
%color = extractvalue { float, float, i64 } %pt, 2
```

There exists an important limitation on both instructions: the index must be a constant. For structures, this is easily explainable. The index number is only a substitute for the name, and languages such as C have no notion of dynamically computing the name of a struct member. For arrays, it is simply that it can't be implemented efficiently. Both instructions have value in specific cases when the number of elements is small and known. For example, a complex number could be modeled as an array of two floating-point numbers. It's reasonable to pass this array around, and it is always clear which part of the array must be accessed during a computation.

For general use in the front end, we have to resort to pointers to memory. All global values in LLVM are expressed as pointers. Let's declare a `@arr` global variable as an array of eight `i64` elements. This is the equivalent of the `long arr[8]` C declaration:

```
@arr = common global [8 x i64] zeroinitializer
```

To access the second element of the array, an address calculation must be performed to determine the address of the indexed element. Then the value can then be loaded from that address and put into a function `@second`, this looks like this:

```
define i64 @second() {
    %1 = load i64, ptr getelementptr inbounds ([8 x i64], ptr @arr, i64
0, i64 1)
    ret i64 %1
}
```

The `getelementptr` instruction is the workhorse for address calculations. As such, it needs some more explanation. The first operand, `[8 x i64]`, is the base type the instruction is operating on. The second operand, `ptr @arr`, specifies the base pointer. Please note the subtle difference here: we declared an array of eight elements, but because all global values are treated as pointers, we have a pointer to the array. In C syntax, we really work with `long (*arr)[8]`! The consequence is that we first have to dereference the pointer before we can index the element, such as `arr[0][1]` in C. The third operand, `i64 0`, dereferences the pointer, and the fourth operand, `i64 1`, is the element index. The result of this computation is the address of the indexed element. Please note that no memory is touched by this instruction.

Except for structs, the index parameters do not need to be constant. Therefore, the `getelementptr` instruction can be used in a loop to retrieve the elements of an array. Structs are treated differently here: only constants can be used, and the type must be `i32`.

With this knowledge, arrays are easily integrated into the code generator from *Chapter 4, Basics of IR Code Generation*. The `convertType()` method must be extended to create the type. If the `Arr` variable holds the type denoter of an array, and assuming the number of elements within an array is an integer literal, we then can add the following to the `convertType()` method to handle arrays:

```
if (auto *ArrayTy =
                llvm::dyn_cast<ArrayTypeDeclaration>(Ty)) {
    llvm::Type *Component =
        convertType(ArrayTy->getType());
    Expr *Nums = ArrayTy->getNums();
    uint64_t NumElements =
        llvm::cast<IntegerLiteral>(Nums)
            ->getValue()
            .getZExtValue();
    llvm::Type *T =
        llvm::ArrayType::get(Component, NumElements);
    // TypeCache is a mapping between the original
    // TypeDeclaration (Ty) and the current Type (T).
    return TypeCache[Ty] = T;
}
```

This type can be used to declare global variables. For local variables, we need to allocate memory for the array. We do this in the first basic block of the procedure:

```
for (auto *D : Proc->getDecls()) {
    if (auto *Var =
            llvm::dyn_cast<VariableDeclaration>(D)) {
        llvm::Type *Ty = mapType(Var);
        if (Ty->isAggregateType()) {
            llvm::Value *Val = Builder.CreateAlloca(Ty);
            // The following method requires a BasicBlock (Curr),
            // a VariableDeclation (Var), and an llvm::Value (Val)
            writeLocalVariable(Curr, Var, Val);
        }
    }
}
```

To read and write an element, we have to generate the `getelementptr` instruction. This is added to the `emitExpr()` (reading a value) and `emitStmt()` (writing a value) methods. To read an element of an array, the value of the variable is read first. Then, the selectors of the variable are processed. For each index, the expression is evaluated and the value is stored. Based on this list, the address of the referenced element is calculated and the value is loaded:

```
auto &Selectors = Var->getSelectors();
for (auto I = Selectors.begin(), E = Selectors.end();
     I != E; ) {
  if (auto *IdxSel =
        llvm::dyn_cast<IndexSelector>(*I)) {
    llvm::SmallVector<llvm::Value *, 4> IdxList;
    while (I != E) {
      if (auto *Sel =
            llvm::dyn_cast<IndexSelector>(*I)) {
        IdxList.push_back(emitExpr(Sel->getIndex()));
        ++I;
      } else
        break;
    }
    Val = Builder.CreateInBoundsGEP(Val->getType(), Val, IdxList);
    Val = Builder.CreateLoad(
        Val->getType(), Val);
  }
  // . . . Check for additional selectors and handle
  // appropriately by generating getelementptr and load.
  else {
    llvm::report_fatal_error("Unsupported selector");
  }
}
```

Writing to an array element uses the same code, with the exception that you do not generate a `load` instruction. Instead, you use the pointer as the target in a `store` instruction. For records, you use a similar approach. The selector for a record member contains the constant field index, named `Idx`. You convert this constant into a constant LLVM value:

```
llvm::Value *FieldIdx = llvm::ConstantInt::get(Int32Ty, Idx);
```

Then you can use value in the `Builder.CreateGEP()` methods as in for arrays.

Now, you should know how to translate aggregate data types to LLVM IR. Passing values of those types in a system-compliant way requires some care, and you will learn to implement it correctly in the next section.

Getting the application binary interface right

With the addition of arrays and records to the code generator, you can note that sometimes, the generated code does not execute as expected. The reason is that we have ignored the calling conventions of the platform so far. Each platform defines its own rules on how one function can call another function in the same program or library. These rules are summarized in the ABI documentation. Typical information includes the following:

- Are machine registers used for parameter passing? If yes, which ones?

- How are aggregates such as arrays and structs passed to a function?

- How are return values handled?

There is a wide variety in use. On some platforms, aggregates are always passed indirectly, meaning that a copy of the aggregate is placed on the stack and only a pointer to the copy is passed as a parameter. On other platforms, a small aggregate (say 128 or 256 bit wide) is passed in registers, and only above that threshold is indirect parameter passing used. Some platforms also use floating-point and vector registers for parameter passing, while others demand that floating-point values be passed in integer registers.

Of course, this is all interesting low-level stuff. Unfortunately, it leaks into LLVM IR. At first, this is surprising. After all, we define the types of all parameters of a function in LLVM IR! It turns out that this is not enough. To understand this, let's consider complex numbers. Some languages have built-in data types for complex numbers. For example, C99 has `float _Complex` (among others). Older versions of C do not have complex number types, but you can easily define `struct Complex { float re, im; }` and create arithmetic operations on this type. Both types can be mapped to the `{ float, float }` LLVM IR type.

If the ABI now states that values of a built-in, complex-number type are passed in two floating-point registers, but user-defined aggregates are always passed indirectly, then the information given with the function is not enough for LLVM to decide how to pass this particular parameter. The unfortunate consequence is that we need to provide more information to LLVM, and this information is highly ABI-specific.

There are two ways to specify this information to LLVM: parameter attributes and type rewriting. What you need to use depends on the target platform and the code generator. The most commonly used parameter attributes are the following:

- `inreg` specifies that the parameter is passed in a register

- `byval` specifies that the parameter is passed by value. The parameter must be a pointer type. A hidden copy is made of the pointed-to data, and this pointer is passed to the called function.

- `zeroext` and `signext` specify that the passed integer value should be zero or sign extended.

- `sret` specifies that this parameter holds a pointer to memory which is used to return an aggregate type from the function.

While all code generators support `zeroext`, `signext`, and `sret` attributes, only some support `inreg` and `byval`. An attribute can be added to the argument of a function with the `addAttr()` method. For example, to set the `inreg` attribute on argument `Arg`, you call the following:

```
Arg->addAttr(llvm::Attribute::InReg);
```

To set multiple attributes, you can use the `llvm::AttrBuilder` class.

The other way to provide additional information is to use type rewriting. With this approach, you disguise the original types. You can do the following:

1. Split the parameter. For example, instead of passing one complex argument, you can pass two floating-point arguments.

2. Cast the parameter into a different representation, such as passing a floating-point value through an integer register.

To cast between types without changing the bits of the value, you use the `bitcast` instruction. The `bitcast` instruction can operate on simple data types such as integers and floating-point values. When floating-point values are passed via an integer register, the floating-point value must be cast to an integer. In LLVM, a 32-bit floating-point value is expressed as `float`, and a 32-bit bit integer is expressed as `i32`. The floating point value can be bitcasted to an integer in the following way:

```
%intconv = bitcast float %fp to i32
```

Additionally, the `bitcast` instruction requires that both types have the same size.

Adding attributes to an argument or changing the type is not complicated. But how do you know what you need to implement? First of all, you should get an overview of the calling convention used on your target platform. For example, the ELF ABI on Linux is documented for each supported CPU platform, so you can look up the document and make yourself comfortable with it.

There is also documentation about the requirements of the LLVM code generators. The source of information is the clang implementation, which you can find at `https://github.com/llvm/llvm-project/blob/main/clang/lib/CodeGen/TargetInfo.cpp`. This single file contains the ABI-specific actions for all supported platforms, and it is also where all information is collected.

In this section, you learned to generate the IR for function calls to be compliant with the ABI of your platform. The next section covers the different ways to create IR for classes and virtual functions.

Creating IR code for classes and virtual functions

Many modern programming languages support object orientation using classes. A **class** is a high-level language construct, and in this section, we will explore how we can map a class construct into LLVM IR.

Implementing single inheritance

A class is a collection of data and methods. A class can inherit from another class, potentially adding more data fields and methods, or overriding existing virtual methods. Let's illustrate this with classes in Oberon-2, which is also a good model for tinylang. A Shape class defines an abstract shape with a color and an area:

```
TYPE Shape = RECORD
            color: INTEGER;
            PROCEDURE (VAR s: Shape) GetColor(): INTEGER;
            PROCEDURE (VAR s: Shape) Area(): REAL;
          END;
```

The GetColor method only returns the color number:

```
PROCEDURE (VAR s: Shape) GetColor(): INTEGER;
BEGIN RETURN s.color; END GetColor;
```

The area of an abstract shape cannot be calculated, so this is an abstract method:

```
PROCEDURE (VAR s: Shape) Area(): REAL;
BEGIN HALT; END;
```

The Shape type can be extended to represent a Circle class:

```
TYPE Circle = RECORD (Shape)
            radius: REAL;
            PROCEDURE (VAR s: Circle) Area(): REAL;
          END;
```

For a circle, the area can be calculated:

```
PROCEDURE (VAR s: Circle) Area(): REAL;
BEGIN RETURN 2 * radius * radius; END;
```

The type can also be queried at runtime. If the shape is a variable of type Shape, then we can formulate a type test in this way:

```
IF shape IS Circle THEN (* … *) END;
```

The different syntax aside, this works much like it does in C++. One notable difference to C++ is that the Oberon-2 syntax makes the implicit this pointer explicit, calling it the receiver of a method.

The basic problems to solve are how to lay out a class in memory and how to implement the dynamic call of methods and run time-type checking. For the memory layout, this is quite easy. The Shape class has only one data member, and we can map it to a corresponding LLVM structure type:

```
@Shape = type { i64 }
```

The Circle class adds another data member. The solution is to append the new data member at the end:

```
@Circle = type { i64, float }
```

The reason is that a class can have many sub-classes. With this strategy, the data member of the common base class always has the same memory offset and also uses the same index to access the field via the getelementptr instruction.

To implement the dynamic call of a method, we must further extend the LLVM structure. If the Area() function is called on a Shape object, then the abstract method is called, causing the application to halt. If it is called on a Circle object, then the corresponding method to calculate the area of a circle is called. On the other hand, the GetColor() function can be called for objects of both classes.

The basic idea to implement this is to associate a table with function pointers with each object. Here, a table would have two entries: one for the GetColor() method and one for the Area() function. The Shape class and the Circle class each have such a table. The tables differ in the entry for the Area() function, which calls different code depending on the type of the object. This table is called the **virtual method table**, often abbreviated as **vtable**.

The vtable alone is not useful. We must connect it with an object. To do so, we always add a pointer to the vtable as the first data member to the structure. At the LLVM level, this is what becomes of the @Shape type:

```
@Shape = type { ptr, i64 }
```

The @Circle type is similarly extended.

The resulting memory structure is shown in *Figure 5.1*:

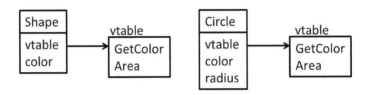

Figure 5.1 – Memory layout of the classes and the virtual method tables

In terms of LLVM IR, the vtable for the `Shape` class can be visualized as the following, where the two pointers correspond to the `GetColor()` and `GetArea()` methods, as represented in *Figure 5.1*:

```
@ShapeVTable = constant { ptr, ptr } { GetColor(), Area() }
```

Furthermore, LLVM does not have void pointers. Pointers to bytes are used instead. With the introduction of the hidden `vtable` field, there is now also the need to have a way to initialize it. In C++, this is part of calling the constructor. In Oberon-2, the field is initialized automatically when the memory is allocated.

A dynamic call to a method is then executed with the following steps:

1. Calculate the offset of the vtable pointer via the `getelementptr` instruction.
2. Load the pointer to the vtable.
3. Calculate the offset of the function in the vtable.
4. Load the function pointer.
5. Indirectly call the function via the pointer with the `call` instruction.

We can visualize the dynamic call to a virtual method, such as `Area()`, within LLVM IR, as well. First, we load a pointer from the corresponding designated location of the `Shape` class. The following load represents loading the pointer to the actual vtable for `Shape`:

```
// Load a pointer from the corresponding location.
%ptrToShapeObj = load ptr, ...
// Load the first element of the Shape class.
%vtable = load ptr, ptr %ptrToShapeObj, align 8
```

Following this, a `getelementptr` gets to the offset to call the `Area()` method:

```
%offsetToArea = getelementptr inbounds ptr, ptr %vtable, i64 1
```

Then, we load the function pointer to `Area()`:

```
%ptrToAreaFunction = load ptr, ptr %offsetToArea, align 8
```

Finally, the `Area()` function is called through the pointer with the call, similar to the general steps that are highlighted previously:

```
%funcCall = call noundef float %ptrToAreaFunction(ptr noundef
nonnull align 8 dereferenceable(12) %ptrToShapeObj)
```

As we can see, even in the case of a single inheritance, the LLVM IR that is generated can appear to be very verbose. Although the general procedure of generating a dynamic call to a method does not sound very efficient, most CPU architectures can perform this dynamic call with just two instructions.

Moreover, to turn a function into a method, a reference to the object's data is required. This is implemented by passing the pointer to the data as the first parameter of the method. In Oberon-2, this is the explicit receiver. In languages similar to C++, it is the implicit `this` pointer.

With the vtable, we have a unique address in memory for each class. Does this help with the *runtime-type test*, too? The answer is that it helps only in a limited way. To illustrate the problem, let's extend the class hierarchy with an `Ellipse` class, which inherits from the `Circle` class. This is not the classical *is-a* relationship in the mathematical sense.

If we have a `shape` variable of the `Shape` type, then we could implement the `shape IS Circle` type test as a comparison of the vtable pointer stored in the `shape` variable with the vtable pointer of the `Circle` class. This comparison only results in true if `shape` has the exact `Circle` type. However, if `shape` is indeed of the `Ellipse` type, then the comparison returns false, even if an object of the `Ellipse` type can be used in all places where only an object of the `Circle` type is required.

Clearly, we need to do more. The solution is to extend the virtual method table with runtime-type information. How much information you need to store depends on the source language. To support the runtime-type check, it is enough to store a pointer to the vtable of the base class, which then looks like in *Figure 5.2*:

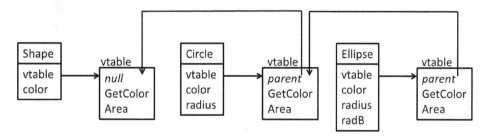

Figure 5.2 – Class and vtable layout supporting simple type tests

If the test fails as described earlier, then the test is repeated with the pointer to the vtable of the base class. This is repeated until the test yields true or, if there is no base class, false. In contrast to calling a dynamic function, the type test is a costly operation because, in the worst-case scenario, the inheritance hierarchy is walked up to the root class.

If you know the whole class hierarchy, then an efficient approach is possible: you number each member of the class hierarchy in a depth-first order. Then, the type test becomes compare-against-a-number or an interval, which can be done in constant time. In fact, that is the approach of LLVM's own runtime-type test, which we learned about in the previous chapter.

To couple runtime-type information with the vtable is a design decision, either mandated by the source language or just as an implementation detail. For example, if you need detailed runtime-type information because the source language supports reflection at runtime, and you have data types without a vtable, then coupling both is not a good idea. In C++, the coupling results in the fact that a class with virtual functions (and therefore no vtable) has no runtime-type data attached to it.

Often, programming languages support interfaces which are a collection of virtual methods. Interfaces are important because they add a useful abstraction. We will look at possible implementations of interfaces in the next section.

Extending single inheritance with interfaces

Languages such as **Java** support interfaces. An interface is a collection of abstract methods, comparable to a base class with no data members and only abstract methods defined. Interfaces pose an interesting problem because each class implementing an interface can have the corresponding method at a different position in the vtable. The reason is simply that the order of function pointers in the vtable is derived from the order of the functions in the class definition in the source language. The definition of the interface is independent of this, and different orders are the norm.

Because the methods defined in an interface can have a different order, we attach a table for each implemented interface to the class. For each method of the interface, this table can specify either the index of the method in the vtable or a copy of the function pointer stored in the vtable. If a method is called on the interface, then the corresponding vtable of the interface is searched, the pointer to the function is fetched, and the method is called. Adding two I1 and I2 interfaces to the Shape class results in the following layout:

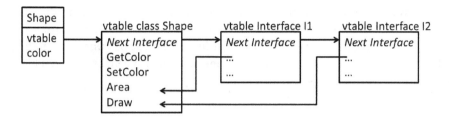

Figure 5.3 – Layout of vtables for interfaces

The caveat lies in the fact that we have to find the right vtable. We can use an approach similar to the runtime-type test: we can perform a linear search through the list of interface vtables. We can assign a unique number to each interface (for example, a memory address) and identify this vtable using this number. The disadvantage of this scheme is obvious: calling a method through an interface takes much more time than calling the same method on the class. There is no easy mitigation for this problem.

A good approach is to replace the linear search with a hash table. At compilation time, the interface that a class implements is known. Therefore, we can construct a perfect hash function, which maps the interface number to the vtable for the interface. A known unique number identifying an interface may be needed for the construction, so memory does not help, but there are other ways to compute a unique number. If the symbol names in the source are unique, then it is always possible to compute a cryptographic hash such as MD5 of the symbol, and use the hash as the number. The calculation occurs at compile time and therefore has no runtime cost.

The result is much faster than the linear search and only takes constant time. Still, it involves several arithmetic operations on a number and is slower than the method call of a class type.

Usually, interfaces also take part in runtime-type tests, making the list search even longer. Of course, if the hash-table approach is implemented, then it can also be used for the runtime-type test.

Some languages allow for more than one parent class. This has some interesting challenges for the implementation, and we will master this in the next section.

Adding support for multiple inheritance

Multiple inheritance adds another challenge. If a class inherits from two or more base classes, then we need to combine the data members in such a way that they are still accessible from the methods. Like in the single inheritance case, the solution is to append all data members, including the hidden vtable pointers.

The Circle class is not only a geometric shape but also a graphic object. To model this, we let the Circle class inherit from the Shape class and the GraphicObj class. In the class layout, the fields from the Shape class come first. Then, we append all fields of the GraphicObj class, including the hidden vtable pointer. After that, we add the new data members of the Circle class, resulting in the overall structure shown in *Figure 5.4*:

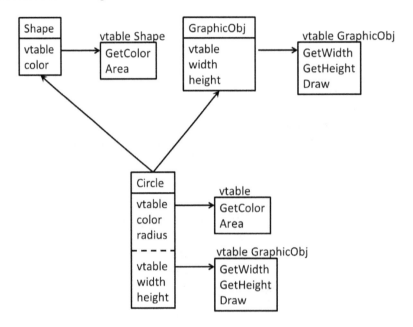

Figure 5.4 - Layout of classes and vtables with multiple inheritance

This approach has several implications. There can now be several pointers to the object. A pointer to the Shape or Circle class points to the top of the object, while a pointer to a GraphicObj class points to inside this object, the beginning of the embedded GraphicObj object. This has to be taken into account when comparing pointers.

Calling a virtual method is also affected. If a method is defined in the GraphicObj class, then this method expects the class layout of the GraphicObj class. If this method is not overridden in the Circle class, then there a two possibilities. The easy case is if the method call is done with a pointer to a GraphicObj instance: in this case, you look up the address of the method in the vtable of the GraphicObj class and call the function. The more complicated case is if you call the method with a pointer to the Circle class. Again, you can look up the address of the method in the vtable of the Circle class. The called method expects a this pointer to be an instance of the GraphicObj class, so we have to adjust that pointer, too. We can do this because we know the offset of the GraphicObj class inside the Circle class.

If a `GrapicObj` method is overridden in the `Circle` class, then nothing special needs to be done if the method is called through a pointer to the `Circle` class. However, if the method is called through a pointer to a `GraphicObj` instance, then we need to make another adjustment because the method needs a `this` pointer pointing to a `Circle` instance. At compilation time, we cannot compute this adjustment because we do not know whether or not this `GraphicObj` instance is part of a multiple inheritance hierarchy. To solve this, we store the adjustment we need to make to the `this` pointer before calling the method together with each function pointer in the vtable, as in *Figure 5.5*:

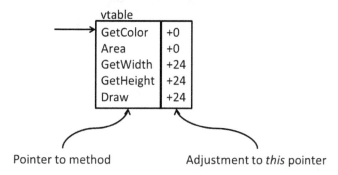

Figure 5.5 – vtable with adjustments to the this pointer

A method call now becomes the following:

1. Look up the function pointer in the vtable.

2. Adjust the `this` pointer.

3. Call the method.

This approach can also be used for implementing interfaces. As an interface only has methods, each implemented interface adds a new vtable pointer to the object. This is easier to implement and most likely faster, but it adds overhead to each object instance.

In the worst case, if your class has a single 64-bit data field but implements 10 interfaces, then your object requires 96 bytes in memory: eight bytes for the vtable pointer of the class itself, eight bytes for the data member, and 10 * 8 bytes for the vtable pointers of each interface.

To support meaningful comparisons to objects and to perform runtime-type tests, we need to normalize a pointer to an object first. If we add an additional field to the vtable, containing an offset to the top of the object, then we can always adjust the pointer to point to the real object. In the vtable of the `Circle` class, this offset is 0, but not in the vtable of the embedded `GraphicObj` class. Of course, whether this needs to be implemented depends on the semantics of the source language.

LLVM itself does not favor a special implementation of object-oriented features. As seen in this section, we can implement all approaches with the available LLVM data types. Additionally, as we have seen an example of LLVM IR with single inheritance, it is also worth noting that the IR can become more verbose when multiple inheritance is involved. If you want to try a new approach, then a good way is to do a prototype in C first. The required pointer manipulations are quickly translated to LLVM IR, but reasoning about the functionality is easier in a higher-level language.

With the knowledge acquired in this section, you can implement the lowering of all OOP constructs commonly found in programming languages into LLVM IR in your own code generator. You have recipes on how to represent single inheritance, single inheritance with interface, or multiple inheritance in memory, and also how to implement type tests and how to look up virtual functions, which are the core concepts of OOP languages.

Summary

In this chapter, you learned how to translate aggregate data types and pointers to LLVM IR code. You also learned about the intricacies of the application binary interface. Finally, you learned about the different approaches to translating classes and virtual functions to LLVM IR. With the knowledge of this chapter, you will be able to create an LLVM IR code generator for most real programming languages.

In the next chapter, you will learn some advanced techniques regarding IR generation. Exception handling is fairly common in modern programming languages, and LLVM has some support for it. Attaching type information to pointers can help with certain optimizations, so we will add this, too. Last but not least, the ability to debug an application is essential for many developers, so we will also add the generation of debugging metadata to our code generator.

Advanced IR Generation

With IR generation introduced in the previous chapters, you can already implement most of the functionality required in a compiler. In this chapter, we will look at some advanced topics that often arise in real-world compilers. For example, many modern languages make use of exception handling, so we'll look at how to translate this into LLVM IR.

To support the LLVM optimizer so that it can produce better code in certain situations, we must add additional type metadata to the IR code. Moreover, attaching debug metadata enables the compiler's user to take advantage of source-level debug tools.

In this chapter, we will cover the following topics:

- *Throwing and catching exceptions*: Here, you will learn how to implement exception handling in your compiler

- *Generating metadata for type-based alias analysis*: Here, you will attach additional metadata to LLVM IR, which helps LLVM to better optimize the code

- *Adding debug metadata*: Here, you will implement the support classes needed to add debug information to the generated IR code

By the end of this chapter, you will have learned about exception handling, as well as metadata for type-based alias analysis and debug information.

Throwing and catching exceptions

Exception handling in LLVM IR is closely tied to platform support. Here, we will look at the most common type of exception handling using `libunwind`. Its full potential is used by C++, so we will look at an example in C++ first, where the `bar()` function can throw an `int` or `double` value:

```
int bar(int x) {
   if (x == 1) throw 1;
   if (x == 2) throw 42.0;
```

```
    return x;
}
```

The `foo()` function calls `bar()`, but only handles a thrown `int`. It also declares that it only throws `int` values:

```
int foo(int x) {
    int y = 0;
    try {
        y = bar(x);
    }
    catch (int e) {
        y = e;
    }
    return y;
}
```

Throwing an exception requires two calls into the runtime library; this can be seen in the `bar()` function. First, memory for the exception is allocated with a call to `__cxa_allocate_exception()`. This function takes the number of bytes to allocate as a parameter. The exception payload (the `int` or `double` value in this example) is copied to the allocated memory. The exception is then raised with a call to `__cxa_throw()`. This function takes three arguments: the pointer to the allocated exception, type information about the payload, and a pointer to a destructor, in case the exception payload has one. The `__cxa_throw()` function initiates the stack unwinding process and never returns. In LLVM IR, this is done for the `int` value, as follows:

```
%eh = call ptr @__cxa_allocate_exception(i64 4)
store i32 1, ptr %eh
call void @__cxa_throw(ptr %eh, ptr @_ZTIi, ptr null)
unreachable
```

`_ZTIi` is the type information describing an `int` type. For a double type, it would be `_ZTId`.

So far, nothing LLVM-specific is done. This changes in the `foo()` function because the call to `bar()` can raise an exception. If it is an `int` type exception, then the control flow must be transferred to the IR code of the catch clause. To accomplish this, the `invoke` instruction must be used instead of the `call` instruction:

```
%y = invoke i32 @_Z3bari(i32 %x) to label %next
                        unwind label %lpad
```

The difference between both instructions is that `invoke` has two labels associated. The first label is where execution continues in case the called function ends normally, usually with a `ret` instruction. In the example code, this label is called `%next`. If an exception occurs, then execution continues at a so-called *landing pad*, with a label of `%lpad`.

The landing pad is a basic block that must begin with a `landingpad` instruction. The `landingpad` instruction gives LLVM information about the handled exception types. For example, a possible landing pad could look like this:

```
lpad:
%exc = landingpad { ptr, i32 }
            cleanup
            catch ptr @_ZTIi
            filter [1 x ptr] [ptr @_ZTIi]
```

There are three possible types of action here:

- `cleanup`: This denotes that code to clean up the current state is present. Usually, this is used to call destructors of local objects. If this marker is present, then the landing pad is always called during stack unwinding.

- `catch`: This is a list of type-value pairs and denotes the exception types that can be handled. The landing pad is called if the thrown exception type is found in this list. In the case of the `foo()` function, the value is the pointer to the C++ runtime type information for the `int` type, similar to the parameter of the `__cxa_throw()` function.

- `filter`: This specifies an array of exception types. The landing pad is called if the exception type of the current exception is not found in the array. This is used to implement the `throw()` specification. For the `foo()` function, the array has only one member – the type information for the `int` type.

The result type of the `landingpad` instruction is the `{ ptr, i32 }` structure. The first element is a pointer to the thrown exception, while the second is a type selector. Let's extract both from the structure:

```
%exc.ptr = extractvalue { ptr, i32 } %exc, 0
%exc.sel = extractvalue { ptr, i32 } %exc, 1
```

The *type selector* is a number that helps us identify the cause of *why the landing pad is called*. It is a positive value if the current exception type matches one of the exception types given in the `catch` part of the `landingpad` instruction. If the current exception type does not match any of the values given in the `filter` part, then the value is negative. It is 0 if the cleanup code should be called.

The type selector is an offset into a type information table, constructed from the values given in the `catch` and `filter` parts of the `landingpad` instruction. During optimization, multiple landing pads can be combined into one, which means that the structure of this table is not known at the IR level. To retrieve the type selector for a given type, we need to call the intrinsic `@llvm.eh.typeid.for` function. We need this to check if the type selector value corresponds to the type information for `int` so that we can execute the code in the `catch (int e) {}` block:

```
%tid.int = call i32 @llvm.eh.typeid.for(ptr @_ZTIi)
%tst.int = icmp eq i32 %exc.sel, %tid.int
    br i1 %tst.int, label %catchint, label %filterorcleanup
```

The handling of an exception is framed by calls to __cxa_begin_catch() and __cxa_end_ catch(). The __cxa_begin_catch() function needs one argument – the current exception – which is one of the values returned by the landingpad instruction. It returns a pointer to the exception payload – an int value in our case.

The __cxa_end_catch() function marks the end of exception handling and deallocates the memory allocated with __cxa_allocate_exception(). Please note that the runtime behavior is much more complicated if another exception is thrown inside the catch block. The exception is handled as follows:

```
catchint:
%payload = call ptr @__cxa_begin_catch(ptr %exc.ptr)
%retval = load i32, ptr %payload
call void @__cxa_end_catch()
br label %return
```

If the type of the current exception does not match the list in the throws() declaration, the unexpected exception handler is called. First, we need to check the type selector again:

```
filterorcleanup:
%tst.blzero = icmp slt i32 %exc.sel, 0
br i1 %tst.blzero, label %filter, label %cleanup
```

If the value of the type selector is lower than 0, then we call the handler:

```
filter:
call void @__cxa_call_unexpected(ptr %exc.ptr) #4
unreachable
```

Again, the handler is not expected to come back.

No cleanup work is needed in this case, so all the cleanup code does is resume the execution of the stack unwinder:

```
cleanup:
resume { ptr, i32 } %exc
```

One piece is still missing: libunwind drives the stack unwinding process, but it is not tied to a single language. Language-dependent handling is done in the personality function. For C++ on Linux, the personality function is called __gxx_personality_v0(). Depending on the platform or compiler, this name can vary. Each function that needs to take part in stack unwinding has a personality function attached. This personality function analyzes if the function catches an exception, has a non-matching filter list, or needs a cleanup call. It gives this information back to the unwinder, which acts accordingly. In LLVM IR, the pointer to the personality function is given as a part of the function definition:

```
define i32 @_Z3fooi(i32) personality ptr @__gxx_personality_v0
```

With this, the exception handling facility is complete.

To use exception handling in the compiler for your programming language, the simplest strategy is to piggyback on the existing C++ runtime functions. This also has the advantage that your exceptions are interoperable with C++. The disadvantage is that you tie some of the C++ runtime into the runtime of your language, most notably memory management. If you want to avoid this, then you need to create your own equivalents of the _cxa_ functions. Still, you will want to use `libunwind`, which provides the stack unwinding mechanism:

1. Let's look at how to create this IR. We created the `calc` expression compiler in *Chapter 2, The Structure of a Compiler*. Now, we will extend the code generator of the expression compiler to raise and handle an exception in case a division by zero is performed. The generated IR will check if the divisor of a division is `0`. If true, then an exception will be raised. We will also add a landing pad to the function, which catches the exception and prints `Divide by zero!` to the console and ends the calculation. Using exception handling is not necessary in this simple case, but it allows us to concentrate on the code generation process. We must add all the code to the `CodeGen.cpp` file. We begin by adding the required new fields and some helper methods. First of all, we need to store the LLVM declaration of the `__cxa_allocate_exception()` and `__cxa_throw()` functions, which consist of the function type and the function itself. A `GlobalVariable` instance is needed to hold the type information. We also need references to the basic blocks holding the landing pad and a basic block containing just an `unreachable` instruction:

    ```
    GlobalVariable *TypeInfo = nullptr;
    FunctionType *AllocEHFty = nullptr;
    Function *AllocEHFn = nullptr;
    FunctionType *ThrowEHFty = nullptr;
    Function *ThrowEHFn = nullptr;
    BasicBlock *LPadBB = nullptr;
    BasicBlock *UnreachableBB = nullptr;
    ```

2. We will also add a new helper function to create the IR for comparing two values. The `createICmpEq()` function takes the `Left` and `Right` values to compare as parameters. It creates a compare instruction testing for equality of the values, and a branch instruction to two basic blocks, for the equal and inequal cases. The two basic blocks are returned via references in the `TrueDest` and `FalseDest` parameters. Furthermore, a label for the new basic blocks can be given in the `TrueLabel` and `FalseLabel` parameters. The code is as follows:

    ```
    void createICmpEq(Value *Left, Value *Right,
                      BasicBlock *&TrueDest,
                      BasicBlock *&FalseDest,
                      const Twine &TrueLabel = "",
                      const Twine &FalseLabel = "") {
      Function *Fn =
    ```

```
        Builder.GetInsertBlock()->getParent();
    TrueDest = BasicBlock::Create(M->getContext(),
                                    TrueLabel, Fn);
    FalseDest = BasicBlock::Create(M->getContext(),
                                    FalseLabel, Fn);
    Value *Cmp = Builder.CreateCmp(CmpInst::ICMP_EQ,
                                    Left, Right);
    Builder.CreateCondBr(Cmp, TrueDest, FalseDest);
}
```

3. To use the functions from the runtime, we need to create several function declarations. In LLVM, a function type gives the signature, and the function itself must be constructed. We use the `createFunc()` method to create both objects. The functions need references to the `FunctionType` and `Function` pointers, the name of the newly declared function, and the result type. The parameter type list is optional, and the flag to indicate a variable parameter list is set to `false`, indicating that there is no variable part in the parameter list:

```
void createFunc(FunctionType *&Fty, Function *&Fn,
                const Twine &N, Type *Result,
                ArrayRef<Type *> Params = None,
                bool IsVarArgs = false) {
    Fty = FunctionType::get(Result, Params, IsVarArgs);
    Fn = Function::Create(
        Fty, GlobalValue::ExternalLinkage, N, M);
}
```

With these preparations done, we can generate the IR to raise an exception.

Raising an exception

To generate the IR code to raise an exception, we will add the `addThrow()` method. This new method needs to initialize the new fields and then generate the IR to raise an exception via the `___ cxa_throw()` function. The payload of the raised exception is of the `int` type and can be set to an arbitrary value. Here is what we need to code:

1. The new `addThrow()` method begins by checking if the `TypeInfo` field has been initialized. If it has not been initialized, then a global external constant of an `i8` pointer type called `_ZTIi` is created. This represents the C++ metadata describing the C++ `int` type:

```
void addThrow(int PayloadVal) {
    if (!TypeInfo) {
        TypeInfo = new GlobalVariable(
```

```
*M, Int8PtrTy,
/*isConstant=*/true,
GlobalValue::ExternalLinkage,
/*Initializer=*/nullptr, "_ZTIi");
```

2. The initialization continues with creating the IR declaration for the __cxa_allocate_
    exception() and __cxa_throw() functions using our helper createFunc() method:

```
createFunc(AllocEHFty, AllocEHFn,
          "__cxa_allocate_exception", Int8PtrTy,
          {Int64Ty});
createFunc(ThrowEHFty, ThrowEHFn, "__cxa_throw",
          VoidTy,
          {Int8PtrTy, Int8PtrTy, Int8PtrTy});
```

3. A function that uses exception handling needs a personality function, which helps with stack
    unwinding. We add the IR code to declare the __gxx_personality_v0() personality
    function from the C++ library and set it as the personality routine of the current function. The
    current function is not stored as a field, but we can use the Builder instance to query the
    current basic block, which has the function stored as a Parent field:

```
FunctionType *PersFty;
Function *PersFn;
createFunc(PersFty, PersFn,
          "__gxx_personality_v0", Int32Ty, std::nulopt,
          true);
Function *Fn =
    Builder.GetInsertBlock()->getParent();
Fn->setPersonalityFn(PersFn);
```

4. Next, we must create and populate the basic block for the landing pad. First, we need to save the
    pointer to the current basic block. Then, we must create a new basic block, set it in the builder so
    that it can be used as the basic block to insert instructions, and call the addLandingPad()
    method. This method generates the IR code for handling an exception and is described in the
    next section, *Catching an exception*. This code populates the basic block for the landing pad:

```
BasicBlock *SaveBB = Builder.GetInsertBlock();
LPadBB = BasicBlock::Create(M->getContext(),
                            "lpad", Fn);
Builder.SetInsertPoint(LPadBB);
addLandingPad();
```

5. The initialization part is completed by creating the basic block holding an unreachable instruction. Again, we create the basic block and set it as an insertion point at the builder. Then, we can add the unreachable instruction to it. Lastly, we can set the insertion point of the builder back to the saved SaveBB instance so that the following IR is added to the right basic block:

```
    UnreachableBB = BasicBlock::Create(
        M->getContext(), "unreachable", Fn);
    Builder.SetInsertPoint(UnreachableBB);
    Builder.CreateUnreachable();
    Builder.SetInsertPoint(SaveBB);
}
```

6. To raise an exception, we need to allocate memory for the exception and the payload via a call to the __cxa_allocate_exception() function. Our payload is of the C++ int type, which usually has a size of 4 bytes. We create a constant unsigned value for the size and call the function with it as a parameter. The function type and the function declaration are already initialized, so we only need to create the call instruction:

```
    Constant *PayloadSz =
        ConstantInt::get(Int64Ty, 4, false);
    CallInst *EH = Builder.CreateCall(
        AllocEHFty, AllocEHFn, {PayloadSz});
```

7. Next, we store the PayloadVal value in the allocated memory. To do so, we need to create an LLVM IR constant with a call to the ConstantInt::get() function. The pointer to the allocated memory is of an i8 pointer type; to store a value of the i32 type, we need to create a bitcast instruction to cast the type:

```
    Value *PayloadPtr =
        Builder.CreateBitCast(EH, Int32PtrTy);
    Builder.CreateStore(
        ConstantInt::get(Int32Ty, PayloadVal, true),
        PayloadPtr);
```

8. Finally, we must raise the exception with a call to the __cxa_throw() function. As this function raises an exception, which is also handled in the same function, we need to use the invoke instruction instead of the call instruction. Unlike the call instruction, the invoke instruction ends a basic block because it has two successor basic blocks. Here, these are the UnreachableBB and LPadBB basic blocks. If the function raises no exception, the control flow is transferred to the UnreachableBB basic blocks. Due to the design of the __cxa_throw() function, this will never happen because the control flow is transferred

to the LPadBB basic block to handle the exception. This finishes the implementation of the addThrow() method:

```
Builder.CreateInvoke(
    ThrowEHFty, ThrowEHFn, UnreachableBB, LPadBB,
    {EH,
     ConstantExpr::getBitCast(TypeInfo, Int8PtrTy),
     ConstantPointerNull::get(Int8PtrTy)});
}
```

Next, we'll add the code to generate the IR to handle the exception.

Catching an exception

To generate the IR code to catch an exception, we must add the addLandingPad() method. The generated IR extracts the type information from the exception. If it matches the C++ int type, then the exception is handled by printing Divide by zero! to the console and returning from the function. If the type does not match, we simply execute the resume instruction, which transfers control back to the runtime. As there are no other functions in the call hierarchy to handle this exception, the runtime will terminate the application. The following steps describe the code that is needed to generate the IR for catching an exception:

1. In the generated IR, we need to call the __cxa_begin_catch() and __cxa_end_catch() functions from the C++ runtime library. To print the error message, we will generate a call to the puts() function from the C runtime library. Furthermore, to get the type information from the exception, we must generate a call to the llvm.eh.typeid.for intrinsic. We also need the FunctionType and Function instances for all of them; we will take advantage of our createFunc() method to create them:

    ```
    void addLandingPad() {
      FunctionType *TypeIdFty; Function *TypeIdFn;
      createFunc(TypeIdFty, TypeIdFn,
                 "llvm.eh.typeid.for", Int32Ty,
                 {Int8PtrTy});
      FunctionType *BeginCatchFty; Function *BeginCatchFn;
      createFunc(BeginCatchFty, BeginCatchFn,
                 "__cxa_begin_catch", Int8PtrTy,
                 {Int8PtrTy});
      FunctionType *EndCatchFty; Function *EndCatchFn;
      createFunc(EndCatchFty, EndCatchFn,
                 "__cxa_end_catch", VoidTy);
      FunctionType *PutsFty; Function *PutsFn;
      createFunc(PutsFty, PutsFn, "puts", Int32Ty,
                 {Int8PtrTy});
    ```

2. The `landingpad` instruction is the first instruction we generate. The result type is a structure containing fields of an `i8` pointer and an `i32` type. This structure is generated with a call to the `StructType::get()` function. Moreover, since we need to handle an exception of a C++ `int` type, we need to also add this as a clause to the `landingpad` instruction, which must be a constant of an `i8` pointer type. This means that generating a `bitcast` instruction is required to convert the `TypeInfo` value into this type. After, we must store the value that's returned from the instruction for later use in the `Exc` variable:

    ```
    LandingPadInst *Exc = Builder.CreateLandingPad(
        StructType::get(Int8PtrTy, Int32Ty), 1, "exc");
    Exc->addClause(
        ConstantExpr::getBitCast(TypeInfo, Int8PtrTy));
    ```

3. Next, we extract the type selector from the returned value. With a call to the `llvm.eh.typeid.for` intrinsic, we retrieve the type ID for the `TypeInfo` field, representing the C++ `int` type. With this IR, we have generated the two values we need to compare to decide if we can handle the exception:

    ```
    Value *Sel =
        Builder.CreateExtractValue(Exc, {1}, "exc.sel");
    CallInst *Id =
        Builder.CreateCall(TypeIdFty, TypeIdFn,
                           {ConstantExpr::getBitCast(
                               TypeInfo, Int8PtrTy)});
    ```

4. To generate the IR for the comparison, we must call our `createICmpEq()` function. This function also generates two basic blocks, which we store in the `TrueDest` and `FalseDest` variables:

    ```
    BasicBlock *TrueDest, *FalseDest;
    createICmpEq(Sel, Id, TrueDest, FalseDest, "match",
                 "resume");
    ```

5. If the two values do not match, the control flow continues at the `FalseDest` basic block. This basic block only contains a `resume` instruction, to give control back to the C++ runtime:

    ```
    Builder.SetInsertPoint(FalseDest);
    Builder.CreateResume(Exc);
    ```

6. If the two values are equal, the control flow continues at the `TrueDest` basic block. First, we generate the IR code to extract the pointer to the exception from the return value of the `landingpad` instruction, stored in the `Exc` variable. Then, we generate a call to the `__cxa_begin_catch()` function, passing the pointer to the exception as a parameter. This indicates the beginning of handling the exception for the runtime:

    ```
    Builder.SetInsertPoint(TrueDest);
    Value *Ptr =
    ```

```
Builder.CreateExtractValue(Exc, {0}, "exc.ptr");
Builder.CreateCall(BeginCatchFty, BeginCatchFn,
                   {Ptr});
```

7. The exception is then handled by calling the `puts()` function to print a message to the console. For this, we generate a pointer to the string with a call to the `CreateGlobalStringPtr()` function, and then pass this pointer as a parameter in the generated call to the `puts()` function:

```
Value *MsgPtr = Builder.CreateGlobalStringPtr(
    "Divide by zero!", "msg", 0, M);
Builder.CreateCall(PutsFty, PutsFn, {MsgPtr});
```

8. Now that we've handled the exception, we must generate a call to the `__cxa_end_catch()` function to inform the runtime about it. Finally, we return from the function with a `ret` instruction:

```
Builder.CreateCall(EndCatchFty, EndCatchFn);
Builder.CreateRet(Int32Zero);
}
```

With the `addThrow()` and `addLandingPad()` functions, we can generate the IR to raise an exception and handle an exception. However, we still need to add the IR to check if the divisor is 0. We'll cover this in the next section.

Integrating the exception handling code into the application

The IR for the division is generated inside the `visit(BinaryOp &)` method. Instead of just generating a `sdiv` instruction, we must generate an IR to compare the divisor with 0. If the divisor is 0, then the control flow continues in a basic block, raising the exception. Otherwise, the control flow continues in a basic block with the `sdiv` instruction. With the help of the `createICmpEq()` and `addThrow()` functions, we can code this very easily:

```
case BinaryOp::Div:
    BasicBlock *TrueDest, *FalseDest;
    createICmpEq(Right, Int32Zero, TrueDest,
                 FalseDest, "divbyzero", "notzero");
    Builder.SetInsertPoint(TrueDest);
    addThrow(42); // Arbitrary payload value.
    Builder.SetInsertPoint(FalseDest);
    V = Builder.CreateSDiv(Left, Right);
    break;
```

The code generation part is now complete. To build the application, we must change into the build directory and run the `ninja` tool:

```
$ ninja
```

Once the build has finished, you can check the generated IR with the `with a: 3/a` expression:

```
$ src/calc "with a: 3/a"
```

You will see the additional IR needed to raise and catch the exception.

The generated IR now depends on the C++ runtime. The easiest way to link against the required libraries is to use the `clang++` compiler. Rename the `rtcalc.c` file with the runtime functions for the expression calculator to `rtcalc.cpp`, and add `extern "C"` in front of each function inside the file. Then, use the `llc` tool to turn the generated IR into an object file, and the `clang++` compiler to create an executable:

```
$ src/calc "with a: 3/a" | llc -filetype obj -o exp.o
$ clang++ -o exp exp.o ../rtcalc.cpp
```

Now, we can run the generated application with different values:

```
$ ./exp
Enter a value for a: 1
The result is: 3
$ ./exp
Enter a value for a: 0
Divide by zero!
```

In the second run, the input is 0, and this raises the exception. It works as expected!

In this section, we learned how to raise and catch exceptions. The code to generate the IR can be used as a blueprint for other compilers. Of course, the type information that's used and the number of catch clauses depends on the input to the compiler, but the IR we need to generate still follows the pattern presented in this section.

Adding metadata is another way to provide further information to LLVM. In the next section, we'll add type metadata to support the LLVM optimizer in certain situations.

Generating metadata for type-based alias analysis

Two pointers may point to the same memory cell, at which point they alias each other. Memory is not typed in the LLVM model, which makes it difficult for the optimizer to decide if two pointers alias each other or not. If the compiler can prove that two pointers do not alias each other, then more optimizations are possible. In the next section, we will have a closer look at the problem and investigate how adding additional metadata will help before we implement this approach.

Understanding the need for additional metadata

To demonstrate the problem, let's look at the following function:

```
void doSomething(int *p, float *q) {
  *p = 42;
  *q = 3.1425;
}
```

The optimizer cannot decide if the pointers, p and q, point to the same memory cell or not. During optimization, an important analysis can be performed called **alias analysis**. If p and q point to the same memory cell, then they are aliases. Moreover, if the optimizer can prove that both pointers never alias each other, this enables additional optimization opportunities. For example, in the doSomething() function, the stores can be reordered without altering the result in this case.

In addition, it depends on the definition of the source language if a variable of one type can be an alias of another variable of a different type. Please note that languages may also contain expressions that break the type-based alias assumption – for example, type casts between unrelated types.

The solution chosen by the LLVM developers is to add metadata to the load and store instructions. The added metadata serves two purposes:

- First, it defines the type hierarchy based on which type may alias another type
- Second, it describes the memory access in a load or store instruction

Let's have a look at the type hierarchy in C. Each type of hierarchy starts with a root node, either **named** or **anonymous**. LLVM assumes that root nodes with the same name describe the same type of hierarchy. You can use different type hierarchies in the same LLVM modules, and LLVM makes the safe assumption that these types may alias. Beneath the root node, there are the nodes for scalar types. Nodes for aggregate types are not attached to the root node, but they refer to scalar types and other aggregate types. Clang defines the hierarchy for C as follows:

- The root node is called Simple C/C++ TBAA.
- Beneath the root node is the node for the char types. This is a special type in C because all pointers can be converted into a pointer to char.
- Beneath the char node are the nodes for the other scalar types and a type for all pointers, called any pointer.

In addition to this, aggregate types are defined as a sequence of member types and offsets.

These metadata definitions are used in access tags attached to the `load` and `store` instructions. An access tag is made up of three parts: a base type, an access type, and an offset. Depending on the base type, there are two possible ways the access tag describes memory access:

1. If the base type is an aggregate type, then the access tag describes the memory access of a `struct` member with the necessary access type and is located at the given offset.

2. If the base type is a scalar type, then the access type must be the same as the base type and the offset must be 0.

With these definitions, we can now define a relation on the access tags, which is used to evaluate if two pointers may alias each other or not. Let's take a closer look at the options for the immediate parent of a (`base type`, `offset`) tuple:

1. If the base type is a scalar type and the offset is 0, then the immediate parent is (`parent type`, 0), with the parent type being the type of the parent node, as defined in the type hierarchy. If the offset is not 0, then the immediate parent is undefined.

2. If the base type is an aggregate type, then the immediate parent of the (`base type`, `offset`) tuple is the (`new type`, `new offset`) tuple, with the new type being the type of the member at offset. The new offset is the offset of the new type, adjusted to its new start.

The transitive closure of this relation is the parent relation. Two memory accesses, (base type 1, access type 1, offset 1) and (base type 2, access type 2, offset 2), may alias one another if (base type 1, offset 1) and (base type 2, offset 2) or vice versa are related in the parent relation.

Let's illustrate this with an example:

```
struct Point { float x, y; }
void func(struct Point *p, float *x, int *i, char *c) {
  p->x = 0; p->y = 0; *x = 0.0; *i = 0; *c = 0;
}
```

When using the memory access tag definition for scalar types, the access tag for the i parameter is (`int`, `int`, 0), while for the c parameter, it is (`char`, `char`, 0). In the type hierarchy, the parent of the node for the `int` type is the `char` node. Therefore, the immediate parent of (`int`, 0) is (`char`, 0) and both pointers can alias. The same is true for the x and c parameters. However, the x and i parameters are not related, so they do not alias each other. The access for the y member of `struct Point` is (`Point`, `float`, 4), with 4 being the offset of the y member in the struct. The immediate parent of (`Point`, 4) is (`float`, 0), so the access to p->y and x may alias, and with the same reasoning also with the c parameter.

Creating TBAA metadata in LLVM

To create the metadata, we must use the `llvm::MDBuilder` class, which is declared in the `llvm/IR/MDBuilder.h` header file. The data itself is stored in instances of the `llvm::MDNode` and `llvm::MDString` classes. Using the builder class shields us from the internal details of the construction.

A root node is created with a call to the `createTBAARoot()` method, which expects the name of the type hierarchy as a parameter and returns the root node. An anonymous, unique root node can be created with the `createAnonymousTBAARoot()` method.

A scalar type is added to the hierarchy with the `createTBAAScalarTypeNode()` method, which takes the name of the type and the parent node as a parameter.

On the other hand, adding a type node for an aggregate type is slightly more complex. The `createTBAAStructTypeNode()` method takes the name of the type and a list of the fields as parameters. Specifically, the fields are given as a `std::pair<llvm::MDNode*, uint64_t>` instance, where the first element indicates the type of the member and the second element represents the offset in `struct`.

An access tag is created with the `createTBAAStructTagNode()` method, which takes the base type, the access type, and the offset as parameters.

Lastly, the metadata must be attached to a `load` or `store` instruction. The `llvm::Instruction` class contains a method called `setMetadata()`, which is used to add various type-based alias analysis metadata. The first parameter must be of the `llvm::LLVMContext::MD_tbaa` type and the second must be the access tag.

Equipped with this knowledge, we must add metadata for **type-based alias analysis (TBAA)** to `tinylang`.

Adding TBAA metadata to tinylang

To support TBAA, we must add a new `CGTBAA` class. This class is responsible for generating the metadata nodes. Furthermore, we make the `CGTBAA` class a member of the `CGModule` class, calling it `TBAA`.

Every load and store instruction must be annotated. A new function is created for this purpose in the `CGModule` class called `decorateInst()`. This function tries to create the tag access information. If this is successful, the metadata is attached to the respective load or store instruction. Moreover, this design also allows us to turn off the metadata generation process in case we do not need it, such as in builds with optimizations turned off:

```
void CGModule::decorateInst(llvm::Instruction *Inst,
                            TypeDeclaration *Type) {
  if (auto *N = TBAA.getAccessTagInfo(Type))
    Inst->setMetadata(llvm::LLVMContext::MD_tbaa, N);
}
```

We put the declaration of the new CGTBAA class in the include/tinylang/CodeGen/
CGTBAA.h header file and the definition in the lib/CodeGen/CGTBAA.cpp file. Aside from the
AST definitions, the header file needs to include the files for defining the metadata nodes and builder:

```
#include "tinylang/AST/AST.h"
#include "llvm/IR/MDBuilder.h"
#include "llvm/IR/Metadata.h"
```

The CGTBAA class needs to store some data members. So, let's see how to do this step by step:

1. First of all, we need to cache the root of the type hierarchy:

    ```
    class CGTBAA {
      llvm::MDNode *Root;
    ```

2. To construct the metadata nodes, we need an instance of the MDBuilder class:

    ```
    llvm::MDBuilder MDHelper;
    ```

3. Lastly, we must store the metadata that's been generated for a type for reuse:

    ```
    llvm::DenseMap<TypeDenoter *, llvm::MDNode *> MetadataCache;
    // …
    };
    ```

Now that we've defined the variables that are required for the construction, we must add the methods
that are required to create the metadata:

1. The constructor initializes the data members:

    ```
    CGTBAA::CGTBAA(CGModule &CGM)
        : CGM(CGM),
          MDHelper(llvm::MDBuilder(CGM.getLLVMCtx())),
          Root(nullptr) {}
    ```

2. We must lazily instantiate the root of the type hierarchy, which we name Simple tinylang
    TBAA:

    ```
    llvm::MDNode *CGTBAA::getRoot() {
      if (!Root)
        Root = MDHelper.createTBAARoot("Simple tinylang TBAA");
      return Root;
    }
    ```

3. For a scalar type, we must create a metadata node with the help of the `MDBuilder` class based on the name of the type. The new metadata node is stored in the cache:

```
llvm::MDNode *
CGTBAA::createScalarTypeNode(TypeDeclaration *Ty,
                             StringRef Name,
                             llvm::MDNode *Parent) {
  llvm::MDNode *N =
      MDHelper.createTBAAScalarTypeNode(Name, Parent);
  return MetadataCache[Ty] = N;
}
```

4. The method to create the metadata for a record is more complicated as we have to enumerate all the fields of the record. Similar to scalar types, the new metadata node is stored in the cache:

```
llvm::MDNode *CGTBAA::createStructTypeNode(
    TypeDeclaration *Ty, StringRef Name,
    llvm::ArrayRef<std::pair<llvm::MDNode *, uint64_t>>
        Fields) {
  llvm::MDNode *N =
      MDHelper.createTBAAStructTypeNode(Name, Fields);
  return MetadataCache[Ty] = N;
}
```

5. To return the metadata for a `tinylang` type, we need to create the type hierarchy. Due to the type system of `tinylang` being very restricted, we can use a simple approach. Each scalar type is mapped to a unique type attached to the root node, and we map all pointers to a single type. Structured types then refer to these nodes. If we cannot map a type, then we return `nullptr`:

```
llvm::MDNode *CGTBAA::getTypeInfo(TypeDeclaration *Ty) {
  if (llvm::MDNode *N = MetadataCache[Ty])
    return N;

  if (auto *Pervasive =
          llvm::dyn_cast<PervasiveTypeDeclaration>(Ty)) {
    StringRef Name = Pervasive->getName();
    return createScalarTypeNode(Pervasive, Name, getRoot());
  }
  if (auto *Pointer =
          llvm::dyn_cast<PointerTypeDeclaration>(Ty)) {
    StringRef Name = "any pointer";
    return createScalarTypeNode(Pointer, Name, getRoot());
  }
  if (auto *Array =
          llvm::dyn_cast<ArrayTypeDeclaration>(Ty)) {
```

```
        StringRef Name = Array->getType()->getName();
        return createScalarTypeNode(Array, Name, getRoot());
    }
    if (auto *Record =
            llvm::dyn_cast<RecordTypeDeclaration>(Ty)) {
        llvm::SmallVector<std::pair<llvm::MDNode *, uint64_t>,
        4> Fields;
        auto *Rec =
            llvm::cast<llvm::StructType>(CGM.convertType(Record));
        const llvm::StructLayout *Layout =
            CGM.getModule()->getDataLayout().getStructLayout(Rec);

        unsigned Idx = 0;
        for (const auto &F : Record->getFields()) {
          uint64_t Offset = Layout->getElementOffset(Idx);
          Fields.emplace_back(getTypeInfo(F.getType()), Offset);
          ++Idx;
        }
        StringRef Name = CGM.mangleName(Record);
        return createStructTypeNode(Record, Name, Fields);
    }
    return nullptr;
}
```

6.   The general method to get the metadata is `getAccessTagInfo()`. To get the TBAA access tag information, a call to the `getTypeInfo()` function must be added. This function expects `TypeDeclaration` as its parameter, which is retrieved from the instructions we want to produce metadata for:

```
llvm::MDNode *CGTBAA::getAccessTagInfo(TypeDeclaration *Ty) {
    return getTypeInfo(Ty);
}
```

Finally, to enable the generation of TBAA metadata, we simply need to attach the metadata to all of the load and store instructions that we generate within `tinylang`.

For example, in `CGProcedure::writeVariable()`, a store to a global variable uses a store instruction:

```
Builder.CreateStore(Val, CGM.getGlobal(D));
```

To decorate this particular instruction, we need to replace this line with the following lines, where `decorateInst()` adds the TBAA metadata to this store instruction:

```
auto *Inst = Builder.CreateStore(Val, CGM.getGlobal(D));
// NOTE: V is of the VariableDeclaration class, and
// the getType() method in this class retrieves the
// TypeDeclaration that is needed for decorateInst().
CGM.decorateInst(Inst, V->getType());
```

With these changes in place, we have finished generating the TBAA metadata.

We can now compile a sample `tinylang` file into an LLVM intermediate representation to see our newly implemented TBAA metadata. For instance, consider the following file, `Person.mod`:

```
MODULE Person;

TYPE
  Person = RECORD
              Height: INTEGER;
              Age: INTEGER
           END;

PROCEDURE Set(VAR p: Person);
BEGIN
  p.Age := 18;
END Set;

END Person.
```

The `tinylang` compiler that is built in the build directory of this chapter can be used to generate the intermediate representation for this file:

```
$ tools/driver/tinylang -emit-llvm ../examples/Person.mod
```

In the newly generated `Person.ll` file, we can see that the store instruction is decorated with the TBAA metadata that we have generated within this chapter, where the metadata reflects the fields of the record type that was originally declared:

```
; ModuleID = '../examples/Person.mod'
source_filename = "../examples/Person.mod"
target datalayout = "e-m:o-i64:64-i128:128-n32:64-S128"
target triple = "arm64-apple-darwin22.6.0"

define void @_t6Person3Set(ptr nocapture dereferenceable(16) %p) {
entry:
```

```
    %0 = getelementptr inbounds ptr, ptr %p, i32 0, i32 1
    store i64 18, ptr %0, align 8, !tbaa !0
    ret void
}

!0 = !{!"_t6Person6Person", !1, i64 0, !1, i64 8}
!1 = !{!"INTEGER", !2, i64 0}
!2 = !{!"Simple tinylang TBAA"}
```

Now that we have learned how to generate TBAA metadata, we will explore a very similar topic in the next section: generating debug metadata.

Adding debug metadata

To allow source-level debugging, we have to add debug information. Support for debug information in LLVM uses debug metadata to describe the types of the source language and other static information, and intrinsics to track variable values. The LLVM core libraries generate debug information in the *DWARF format on Unix systems* and in *PDB format for Windows*. We'll look at the general structure in the next section.

Understanding the general structure of debug metadata

To describe the general structure, LLVM uses metadata similar to the metadata for type-based analysis. The static structure describes the file, the compilation unit, functions and lexical blocks, and the used data types.

The main class we use is `llvm::DIBuilder`, and we need to use the `llvm/IR/DIBuilder` header file to get the class declaration. This builder class provides an easy-to-use interface to create the debug metadata. Later, the metadata is either added to LLVM objects such as global variables, or is used in calls to debug intrinsics. Here's some important metadata that the builder class can create:

- `llvm::DIFile`: This describes a file using the filename and the absolute path of the directory containing the file. You use the `createFile()` method to create it. A file can contain the main compilation unit or it could contain imported declarations.

- `llvm::DICompileUnit`: This is used to describe the current compilation unit. Among other things, you specify the source language, a compiler-specific producer string, whether optimizations are enabled or not, and, of course, `DIFile`, in which the compilation unit resides. You create it with a call to `createCompileUnit()`.

- `llvm::DISubprogram`: This describes a function. The most important information here is the scope (usually `DICompileUnit` or `DISubprogram` for a nested function), the name of the function, the mangled name of the function, and the function type. It is created with a call to `createFunction()`.

- `llvm::DILexicalBlock`: This describes a lexical block and models the block scoping found in many high-level languages. You can create this with a call to `createLexicalBlock()`.

LLVM makes no assumptions about the language your compiler translates. As a consequence, it has no information about the data types of the language. To support source-level debugging, especially displaying variable values in a debugger, type information must be added too. Here are some important constructs:

- The `createBasicType()` function, which returns a pointer to the `llvm::DIBasicType` class, creates the metadata to describe a basic type such as `INTEGER` in `tinylang` or `int` in C++. Besides the name of the type, the required parameters are the size in bits and the encoding – for example, if it is a signed or unsigned type.

- There are several ways to construct the metadata for composite data types, as represented by the `llvm::DIComposite` class. You can use the `createArrayType()`, `createStructType()`, `createUnionType()`, and `createVectorType()` functions to instantiate the metadata for array, struct, union, and vector data types, respectively. These functions require the parameter you expect, such as the base type and the number of subscriptions for an array type or a list of the field members of a struct type.

- There are also methods to support enumerations, templates, classes, and so on.

The list of functions shows you that you have to add every detail of the source language to the debug information. Let's assume your instance of the `llvm::DIBuilder` class is called `DBuilder`. Let's also assume that you have some `tinylang` source in a file called `File.mod` in the `/home/llvmuser` folder. Inside this file is the `Func() : INTEGER` function at *line 5*, which contains a local `VAR i : INTEGER` declaration at *line 7*. Let's create the metadata for this, beginning with the information for the file. You need to specify the filename and the absolute path of the folder in which the file resides:

```
llvm::DIFile *DbgFile = DBuilder.createFile("File.mod",
                                            "/home/llvmuser");
```

The file is a module in `tinylang`, which makes it the compilation unit for LLVM. This carries a lot of information:

```
bool IsOptimized = false;
llvm::StringRef CUFlags;
unsigned ObjCRunTimeVersion = 0;
llvm::StringRef SplitName;
llvm::DICompileUnit::DebugEmissionKind EmissionKind =
      llvm::DICompileUnit::DebugEmissionKind::FullDebug;
llvm::DICompileUnit *DbgCU = DBuilder.createCompileUnit(
      llvm::dwarf::DW_LANG_Modula2, DbgFile, „tinylang",
```

```
        IsOptimized, CUFlags, ObjCRunTimeVersion, SplitName,
    EmissionKind);
```

Furthermore, the debugger needs to know the source language. The DWARF standard defines an enumeration with all the common values. One disadvantage of this is that you cannot simply add a new source language. To do that, you have to create a request at the DWARF committee. Be aware that the debugger and other debug tools also need support for a new language – just adding a new member to the enumeration is not enough.

In many cases, it is sufficient to choose a language that is close to your source language. In the case of `tinylang`, this is Modula-2, and we use `DW_LANG_Modula2` as the language identifier. A compilation unit resides in a file, which is identified by the `DbgFile` variable we created previously. Additionally, the debug information can carry information about the producer, which can be the name of the compiler and version information. Here, we just pass the `tinylang` string. If you do not want to add this information, then you can simply use an empty string as a parameter.

The next set of information includes the `IsOptimized` flag, which should indicate if the compiler has turned optimization on or not. Usually, this flag is derived from the `-O` command-line switch. You can pass additional parameter settings to the debugger with the `CUFlags` parameter. This is not used here, and we pass an empty string. We also do not use Objective-C, so we pass `0` as the Objective-C runtime version.

Normally, debug information is embedded in the object file we are creating. If we want to write the debug information into a separate file, then the `SplitName` parameter must contain the name of this file. Otherwise, simply passing an empty string is sufficient. Finally, you can define the level of debug information that should be emitted. The default is full debug information, as indicated by the use of the `FullDebug` enum value, but you can also choose the `LineTablesOnly` value if you want to emit only line numbers, or the `NoDebug` value for no debug information at all. For the latter, it is better to not create debug information in the first place.

Our minimalistic source only uses the `INTEGER` data type, which is a signed 32-bit value. Creating the metadata for this type is straightforward:

```
llvm::DIBasicType *DbgIntTy =
                    DBuilder.createBasicType("INTEGER", 32,
                            llvm::dwarf::DW_ATE_signed);
```

To create the debug metadata for the function, we have to create a type for the signature first, and then the metadata for the function itself. This is similar to the creation of IR for a function. The signature of the function is an array with all the types of parameters in source order and the return type of the function as the first element at index 0. Usually, this array is constructed dynamically. In our case, we can also construct the metadata statically. This is useful for internal functions, such as for module initializing. Typically, the parameters of these functions are always known, and the compiler writer can hard-code them:

```
llvm::Metadata *DbgSigTy = {DbgIntTy};
llvm::DITypeRefArray DbgParamsTy =
                    DBuilder.getOrCreateTypeArray(DbgSigTy);
llvm::DISubroutineType *DbgFuncTy =
                    DBuilder.createSubroutineType(DbgParamsTy);
```

Our function has the INTEGER return type and no further parameters, so the DbgSigTy array only contains the pointer to the metadata for this type. This static array is turned into a type array, which is then used to create the type for the function.

The function itself requires more data:

```
unsigned LineNo = 5;
unsigned ScopeLine = 5;
llvm::DISubprogram *DbgFunc = DBuilder.createFunction(
      DbgCU, "Func", "_t4File4Func", DbgFile, LineNo,
      DbgFuncTy, ScopeLine, llvm::DISubprogram::FlagPrivate,
      llvm::DISubprogram::SPFlagLocalToUnit);
```

A function belongs to a compilation unit, which in our case is stored in the DbgCU variable. We need to specify the name of the function in the source file, which is Func, and the mangled name is stored in the object file. This information helps the debugger locate the machine code of the function. The mangled name, based on the rules of tinylang, is _t4File4Func. We also have to specify the file that contains the function.

This may sound surprising at first, but think of the include mechanism in C and C++: a function can be stored in a different file, which is then included with #include in the main compilation unit. Here, this is not the case and we use the same file as the one the compilation unit uses. Next, the line number of the function and the function type are passed. The line number of the function may not be the line number where the lexical scope of the function begins. In this case, you can specify a different ScopeLine. A function also has protection, which we specify here with the FlagPrivate value to indicate a private function. Other possible values for function protection are FlagPublic and FlagProtected, for public and protected functions, respectively.

Besides the protection level, other flags can be specified here. For example, FlagVirtual indicates a virtual function and FlagNoReturn indicates that the function does not return to the caller. You can find the complete list of possible values in the LLVM include file – that is, llvm/include/llvm/IR/DebugInfoFlags.def.

Lastly, flags specific to a function can be specified. The most commonly used flag is the `SPFlagLocalToUnit` value, which indicates that the function is local to this compilation unit. The `MainSubprogram` value is also used often, indicating that this function is the main function of the application. The LLVM include file mentioned previously also lists all possible values related to flags specific to functions.

So far, we have only created the metadata referring to static data. Variables are dynamic, so we'll explore how to attach the static metadata to the IR code for accessing variables in the next section.

Tracking variables and their values

To be useful, the type metadata described in the previous section needs to be associated with variables of the source program. For a global variable, this is pretty easy. The `createGlobalVariableExpression()` function of the `llvm::DIBuilder` class creates the metadata to describe a global variable. This includes the name of the variable in the source, the mangled name, the source file, and so on. A global variable in LLVM IR is represented by an instance of the `GlobalVariable` class. This class has a method called `addDebugInfo()`, which associates the metadata node returned from `createGlobalVariableExpression()` with the global variable.

For local variables, we need to take another approach. LLVM IR does not know of a class representing a local variable as it only knows about values. The solution the LLVM community has developed is to insert calls to intrinsic functions into the IR code of a function. An **intrinsic function** is a function that LLVM knows about and, therefore, can do some magic with it. In most cases, intrinsic functions do not result in a subroutine call at the machine level. Here, the function call is a convenient vehicle to associate the metadata with a value. The most important intrinsic functions for debug metadata are `llvm.dbg.declare` and `llvm.dbg.value`.

The `llvm.dbg.declare` intrinsic provides information and is generated once by the frontend to declare a local variable. Essentially, this intrinsic describes the address of a local variable. During optimization, passes can replace this intrinsic with (possibly multiple) calls to `llvm.dbg.value` to preserve the debug information and to track the local source variables. After optimization, multiple calls to `llvm.dbg.declare` may be present as it is used to describe the program points where the local variables live within memory.

On the other hand, the `llvm.dbg.value` intrinsic is called whenever a local variable is set to a new value. This intrinsic describes the value of a local variable, not its address.

How does all of this work? The LLVM IR representation and the programmatic creation via the `llvm::DIBuilder` class differ a bit, so we will look at both.

Continuing with our example from the previous section, we'll allocate local storage for the `I` variable inside the `Func` function with the `alloca` instruction:

```
@i = alloca i32
```

After that, we must add a call to the `llvm.dbg.declare` intrinsic:

```
call void @llvm.dbg.declare(metadata ptr %i,
                        metadata !1, metadata !DIExpression())
```

The first parameter is the address to the local variable. The second parameter is the metadata describing the local variable, which is created by a call to either `createAutoVariable()` for a local variable or `createParameterVariable()` for a parameter of the `llvm::DIBuilder` class. Finally, the third parameter describes an address expression, which will be explained later.

Let's implement the IR creation. You can allocate the storage for the local `@i` variable with a call to the `CreateAlloca()` method of the `llvm::IRBuilder<>` class:

```
llvm::Type *IntTy = llvm::Type::getInt32Ty(LLVMCtx);
llvm::Value *Val = Builder.CreateAlloca(IntTy, nullptr, "i");
```

The `LLVMCtx` variable is the used context class, and `Builder` is the used instance of the `llvm::IRBuilder<>` class.

A local variable also needs to be described by metadata:

```
llvm::DILocalVariable *DbgLocalVar =
  Dbuilder.createAutoVariable(DbgFunc, "i", DbgFile,
                        7, DbgIntTy);
```

Using the values from the previous section, we can specify that the variable is part of the `DbgFunc` function, is called `i`, is defined in the `DbgFile` file at *line 7*, and is of the `DbgIntTy` type.

Finally, we associate the debug metadata with the address of the variable using the `llvm.dbg.declare` intrinsic. Using `llvm::DIBuilder` shields you from all of the details of adding a call:

```
llvm::DILocation *DbgLoc =
              llvm::DILocation::get(LLVMCtx, 7, 5, DbgFunc);
DBuilder.insertDeclare(Val, DbgLocalVar,
                    DBuilder.createExpression(), DbgLoc,
                    Val.getParent());
```

Again, we have to specify a source location for the variable. An instance of `llvm::DILocation` is a container that holds the line and column of a location associated with a scope. Furthermore, the `insertDeclare()` method adds the call to the intrinsic function of the LLVM IR. In terms of this function's parameters, it requires the address of the variable, stored in `Val`, and the debug metadata for the variable, stored in `DbgValVar`. We also pass an empty address expression and the debug location we created previously. As with normal instructions, we need to specify into which basic block the call is inserted. If we specify a basic block, then the call is inserted at the end. Alternatively, we can specify an instruction, and the call is inserted before that instruction. We also have the pointer to the `alloca` instruction, which is the last instruction that we inserted into the underlying basic block. Therefore, we can use this basic block, and the call gets appended after the `alloca` instruction.

If the value of a local variable changes, then a call to `llvm.dbg.value` must be added to the IR to set the new value of a local variable. The `insertValue()` method of the `llvm::DIBuilder` class can be used to achieve this.

When we implemented the IR generation for functions, we used an advanced algorithm that mainly used values and avoided allocating storage for local variables. In terms of adding debug information, this only means that we use `llvm.dbg.value` much more often than you see it in clang-generated IR.

What can we do if the variable does not have dedicated storage space but is part of a larger, aggregate type? One of the situations where this can arise is with the use of nested functions. To implement access to the stack frame of the caller, you must collect all used variables in a structure and pass a pointer to this record to the called function. Inside the called function, you can refer to the variables of the caller as if they are local to the function. What is different is that these variables are now part of an aggregate.

In the call to `llvm.dbg.declare`, you use an empty expression if the debug metadata describes the whole memory the first parameter is pointing to. However, if it only describes a part of the memory, then you need to add an expression indicating which part of the memory the metadata applies to.

In the case of the nested frame, you need to calculate the offset in the frame. You need access to a `DataLayout` instance, which you can get from the LLVM module into which you are creating the IR code. If the `llvm::Module` instance is named Mod, and the variable holding the nested frame structure is named Frame and is of the `llvm::StructType` type, you can access the third member of the frame in the following manner. This access gives you the offset of the member:

```
const llvm::DataLayout &DL = Mod->getDataLayout();
uint64_t Ofs = DL.getStructLayout(Frame)->getElementOffset(3);
```

Moreover, the expression is created from a sequence of operations. To access the third member of the frame, the debugger needs to add the offset to the base pointer. As an example, you need to create an array and this information like so:

```
llvm::SmallVector<int64_t, 2> AddrOps;
AddrOps.push_back(llvm::dwarf::DW_OP_plus_uconst);
AddrOps.push_back(Offset);
```

From this array, you can create the expression that you must then pass to `llvm.dbg.declare` instead of the empty expression:

```
llvm::DIExpression *Expr = DBuilder.createExpression(AddrOps);
```

It is important to note that you are not limited to this offset operation. DWARF knows many different operators, and you can create fairly complex expressions. You can find the complete list of operators in the LLVM include file, called `llvm/include/llvm/BinaryFormat/Dwarf.def`.

At this point, you can create debug information for variables. To enable the debugger to follow the control flow in the source, you also need to provide line number information. This is the topic of the next section.

Adding line numbers

A debugger allows a programmer to step through an application line by line. For this, the debugger needs to know which machine instructions belong to which line in the source. LLVM allows adding a source location to each instruction. In the previous section, we created location information of the `llvm::DILocation` type. A debug location provides more information than just the line, column, and scope. If needed, the scope into which this line is inlined can be specified. It is also possible to indicate that this debug location belongs to implicit code – that is, code that the frontend has generated but is not in the source.

Before this information can be attached to an instruction, we must wrap the debug location in a `llvm::DebugLoc` object. To do so, you must simply pass the location information obtained from the `llvm::DILocation` class to the `llvm::DebugLoc` constructor. With this wrapping, LLVM can track the location information. While the location in the source does not change, the generated machine code for a source-level statement or expression can be dropped during optimization. This encapsulation helps deal with these possible changes.

Adding line number information mostly boils down to retrieving the line number information from the AST and adding it to the generated instructions. The `llvm::Instruction` class has the `setDebugLoc()` method, which attaches the location information to the instruction.

In the next section, we'll learn how to generate debug information and add it to our `tinylang` compiler.

Adding debug support to tinylang

We encapsulate the generation of debug metadata in the new `CGDebugInfo` class. Additionally, we place the declaration in the `tinylang/CodeGen/CGDebugInfo.h` header file and the definition in the `tinylang/CodeGen/CGDebugInfo.cpp` file.

The `CGDebugInfo` class has five important members. We need a reference to the code generator for the module, CGM, because we need to convert types from AST representation into LLVM types. Of course, we also need an instance of the `llvm::DIBuilder` class called `Dbuilder`, as we did in the previous sections. A pointer to the instance of the compile unit is also needed; we store it in the CU member.

To avoid having to create the debug metadata for types again, we must also add a map to cache this information. The member is called `TypeCache`. Finally, we need a way to manage the scope information, for which we must create a stack based on the `llvm::SmallVector<>` class called `ScopeStack`. Thus, we have the following:

```
CGModule &CGM;
llvm::DIBuilder DBuilder;
llvm::DICompileUnit *CU;
llvm::DenseMap<TypeDeclaration *, llvm::DIType *>
    TypeCache;
llvm::SmallVector<llvm::DIScope *, 4> ScopeStack;
```

The following methods of the `CGDebugInfo` class make use of these members:

1. First, we need to create the compile unit, which we do in the constructor. We also create the file containing the compile unit here. Later, we can refer to the file through the CU member. The code for the constructor is as follows:

```
CGDebugInfo::CGDebugInfo(CGModule &CGM)
    : CGM(CGM), DBuilder(*CGM.getModule()) {
  llvm::SmallString<128> Path(
      CGM.getASTCtx().getFilename());
  llvm::sys::fs::make_absolute(Path);

  llvm::DIFile *File = DBuilder.createFile(
      llvm::sys::path::filename(Path),
      llvm::sys::path::parent_path(Path));

  bool IsOptimized = false;
  llvm::StringRef CUFlags;
  unsigned ObjCRunTimeVersion = 0;
  llvm::StringRef SplitName;
  llvm::DICompileUnit::DebugEmissionKind EmissionKind =
      llvm::DICompileUnit::DebugEmissionKind::FullDebug;
  CU = DBuilder.createCompileUnit(
      llvm::dwarf::DW_LANG_Modula2, File, "tinylang",
      IsOptimized, CUFlags, ObjCRunTimeVersion,
      SplitName, EmissionKind);
}
```

2. Often, we need to provide a line number. The line number can be derived from the source manager location, which is available in most AST nodes. The source manager can convert this into a line number:

```
unsigned CGDebugInfo::getLineNumber(SMLoc Loc) {
  return CGM.getASTCtx().getSourceMgr().FindLineNumber(
      Loc);
}
```

3. The information about a scope is held on a stack. We need methods to open and close a scope and retrieve the current scope. The compilation unit is the global scope, which we add automatically:

```
llvm::DIScope *CGDebugInfo::getScope() {
  if (ScopeStack.empty())
    openScope(CU->getFile());
  return ScopeStack.back();
}

void CGDebugInfo::openScope(llvm::DIScope *Scope) {
  ScopeStack.push_back(Scope);
}

void CGDebugInfo::closeScope() {
  ScopeStack.pop_back();
}
```

4. Next, we must create a method for each category of type we need to transform. The getPervasiveType() method creates the debug metadata for basic types. Note the use of the encoding parameter, which declares the INTEGER type as a signed type and the BOOLEAN type encoded as a Boolean:

```
llvm::DIType *
CGDebugInfo::getPervasiveType(TypeDeclaration *Ty) {
  if (Ty->getName() == "INTEGER") {
    return DBuilder.createBasicType(
        Ty->getName(), 64, llvm::dwarf::DW_ATE_signed);
  }
  if (Ty->getName() == "BOOLEAN") {
    return DBuilder.createBasicType(
        Ty->getName(), 1, llvm::dwarf::DW_ATE_boolean);
  }
  llvm::report_fatal_error(
      "Unsupported pervasive type");
}
```

5. If the type name is simply renamed, then we must map this to a type definition. Here, we need to make use of the scope and line number information:

```
llvm::DIType *
CGDebugInfo::getAliasType(AliasTypeDeclaration *Ty) {
  return DBuilder.createTypedef(
      getType(Ty->getType()), Ty->getName(),
      CU->getFile(), getLineNumber(Ty->getLocation()),
      getScope());
}
```

6. Creating the debug information for an array requires specifying the size and the alignment. We can retrieve this data from the DataLayout class. We also need to specify the index range of the array:

```
llvm::DIType *
CGDebugInfo::getArrayType(ArrayTypeDeclaration *Ty) {
  auto *ATy =
      llvm::cast<llvm::ArrayType>(CGM.convertType(Ty));
  const llvm::DataLayout &DL =
      CGM.getModule()->getDataLayout();

  Expr *Nums = Ty->getNums();
  uint64_t NumElements =
      llvm::cast<IntegerLiteral>(Nums)
          ->getValue()
          .getZExtValue();
  llvm::SmallVector<llvm::Metadata *, 4> Subscripts;
  Subscripts.push_back(
      DBuilder.getOrCreateSubrange(0, NumElements));
  return DBuilder.createArrayType(
      DL.getTypeSizeInBits(ATy) * 8,
      1 << Log2(DL.getABITypeAlign(ATy)),
      getType(Ty->getType()),
      DBuilder.getOrCreateArray(Subscripts));
}
```

7. Using all these single methods, we can create a central method to create the metadata for a type. This metadata is also responsible for caching the data:

```
llvm::DIType *
CGDebugInfo::getType(TypeDeclaration *Ty) {
  if (llvm::DIType *T = TypeCache[Ty])
    return T;
```

```
    if (llvm::isa<PervasiveTypeDeclaration>(Ty))
      return TypeCache[Ty] = getPervasiveType(Ty);
    else if (auto *AliasTy =
                 llvm::dyn_cast<AliasTypeDeclaration>(Ty))
      return TypeCache[Ty] = getAliasType(AliasTy);
    else if (auto *ArrayTy =
                 llvm::dyn_cast<ArrayTypeDeclaration>(Ty))
      return TypeCache[Ty] = getArrayType(ArrayTy);
    else if (auto *RecordTy =
                 llvm ::dyn_cast<RecordTypeDeclaration>(
                    Ty))
      return TypeCache[Ty] = getRecordType(RecordTy);
    llvm::report_fatal_error("Unsupported type");
    return nullptr;
}
```

8. We also need to add a method to emit metadata for global variables:

```
void CGDebugInfo::emitGlobalVariable(
    VariableDeclaration *Decl,
    llvm::GlobalVariable *V) {
  llvm::DIGlobalVariableExpression *GV =
      DBuilder.createGlobalVariableExpression(
          getScope(), Decl->getName(), V->getName(),
          CU->getFile(),
          getLineNumber(Decl->getLocation()),
          getType(Decl->getType()), false);
  V->addDebugInfo(GV);
}
```

9. To emit the debug information for procedures, we need to create the metadata for the procedure type. For this, we need a list of the types of the parameter, with the return type being the first entry. If the procedure has no return type, then we must use an unspecified type; this is called `void`, similar to how it is in C. If a parameter is a reference, then we need to add the reference type; otherwise, we must add the type to the list:

```
llvm::DISubroutineType *
CGDebugInfo::getType(ProcedureDeclaration *P) {
  llvm::SmallVector<llvm::Metadata *, 4> Types;
  const llvm::DataLayout &DL =
      CGM.getModule()->getDataLayout();
  // Return type at index 0
  if (P->getRetType())
    Types.push_back(getType(P->getRetType()));
```

```
    else
      Types.push_back(
          DBuilder.createUnspecifiedType("void"));
    for (const auto *FP : P->getFormalParams()) {
      llvm::DIType *PT = getType(FP->getType());
      if (FP->isVar()) {
        llvm::Type *PTy = CGM.convertType(FP->getType());
        PT = DBuilder.createReferenceType(
            llvm::dwarf::DW_TAG_reference_type, PT,
            DL.getTypeSizeInBits(PTy) * 8,
            1 << Log2(DL.getABITypeAlign(PTy)));
      }
      Types.push_back(PT);
    }
    return DBuilder.createSubroutineType(
        DBuilder.getOrCreateTypeArray(Types));
}
```

10. For the procedure itself, we can now create the debug information using the procedure type we created in the previous step. A procedure also opens a new scope, so we must push the procedure onto the scope stack. We must also associate the LLVM function object with the new debug information:

```
void CGDebugInfo::emitProcedure(
    ProcedureDeclaration *Decl, llvm::Function *Fn) {
  llvm::DISubroutineType *SubT = getType(Decl);
  llvm::DISubprogram *Sub = DBuilder.createFunction(
      getScope(), Decl->getName(), Fn->getName(),
      CU->getFile(), getLineNumber(Decl->getLocation()),
      SubT, getLineNumber(Decl->getLocation()),
      llvm::DINode::FlagPrototyped,
      llvm::DISubprogram::SPFlagDefinition);
  openScope(Sub);
  Fn->setSubprogram(Sub);
}
```

11. When the end of a procedure is reached, we must inform the builder to finish constructing the debug information for this procedure. We also need to remove the procedure from the scope stack:

```
void CGDebugInfo::emitProcedureEnd(
    ProcedureDeclaration *Decl, llvm::Function *Fn) {
  if (Fn && Fn->getSubprogram())
```

```
        DBuilder.finalizeSubprogram(Fn->getSubprogram());
    closeScope();
}
```

12. Lastly, when we've finished adding the debug information, we need to implement the `finalize()` method on the builder. The generated debug information is then validated. This is an important step during development as it helps you find wrongly generated metadata:

```
void CGDebugInfo::finalize() { DBuilder.finalize(); }
```

Debug information should only be generated if the user requested it. This means that we will need a new command-line switch for this. We will add this to the file of the `CGModule` class, and we will also use it inside this class:

```
static llvm::cl::opt<bool>
    Debug("g", llvm::cl::desc("Generate debug information"),
          llvm::cl::init(false));
```

The `-g` option can be used with the `tinylang` compiler to generate debug metadata.

Furthermore, the CGModule class holds an instance of the `std::unique_ptr<CGDebugInfo>` class. The pointer is initialized in the constructor for setting the command-line switch:

```
if (Debug)
    DebugInfo.reset(new CGDebugInfo(*this));
```

In the getter method defined in `CGModule.h`, we simply return the pointer:

```
CGDebugInfo *getDbgInfo() {
    return DebugInfo.get();
}
```

The common pattern to generate the debug metadata is to retrieve the pointer and check if it is valid. For example, after creating a global variable, we can add the debug information like so:

```
VariableDeclaration *Var = …;
llvm::GlobalVariable *V = …;
if (CGDebugInfo *Dbg = getDbgInfo())
    Dbg->emitGlobalVariable(Var, V);
```

To add line number information, we need a conversion method called `getDebugLoc()` in the CGDebugInfo class, which turns the location information from the AST into the debug metadata:

```
llvm::DebugLoc CGDebugInfo::getDebugLoc(SMLoc Loc) {
    std::pair<unsigned, unsigned> LineAndCol =
        CGM.getASTCtx().getSourceMgr().getLineAndColumn(Loc);
```

```
llvm::DILocation *DILoc = llvm::DILocation::get(
    CGM.getLLVMCtx(), LineAndCol.first, LineAndCol.second,
    getScope());
return llvm::DebugLoc(DILoc);
}
```

Additionally, a utility function in the `CGModule` class can be called to add the line number information to an instruction:

```
void CGModule::applyLocation(llvm::Instruction *Inst,
                            llvm::SMLoc Loc) {
  if (CGDebugInfo *Dbg = getDbgInfo())
    Inst->setDebugLoc(Dbg->getDebugLoc(Loc));
}
```

In this way, you can add the debug information for your compiler.

Summary

In this chapter, you learned how throwing and catching exceptions work in LLVM and the IR you can generate to exploit this feature. To enhance the scope of IR, you learned how you can attach various metadata to instructions. Metadata for type-based alias analysis provides additional information to the LLVM optimizer and helps with certain optimizations to produce better machine code. Users always appreciate the possibility of using a source-level debugger, and by adding debug information to the IR code, you can implement this important feature of a compiler.

Optimizing the IR code is the core task of LLVM. In the next chapter, we will learn how the pass manager works and how we can influence the optimization pipeline the pass manager governs.

7

Optimizing IR

LLVM uses a series of passes to optimize the IR. A pass operates on a unit of IR, such as a function or a module. The operation can be a transformation, which changes the IR in a defined way, or an analysis, which collects information such as dependencies. This series of passes is called the **pass pipeline**. The pass manager executes the pass pipeline on the IR, which our compiler produces. Therefore, you need to know what the pass manager does and how to construct a pass pipeline. The semantics of a programming language may require the development of new passes, and we must add these passes to the pipeline.

In this chapter, you will learn about the following:

- How to leverage the LLVM pass manager to implement passes within LLVM

- How to implement an instrumentation pass, as an example, within the LLVM project, as well as a separate plugin

- In using the ppprofiler pass with LLVM tools, you will learn how to use a pass plugin with `opt` and `clang`

- In adding an optimization pipeline to your compiler, you will extend the `tinylang` compiler with an optimization pipeline based on the new pass manager

By the end of this chapter, you will know how to develop a new pass and how you can add it to a pass pipeline. You will also be able to set up the pass pipeline in your compiler.

Technical requirements

The source code for this chapter is available at `https://github.com/PacktPublishing/Learn-LLVM-17/tree/main/Chapter07`.

The LLVM pass manager

The LLVM core libraries optimize the IR that your compiler creates and turn it into object code. This giant task is broken down into separate steps called **passes**. These passes need to be executed in the right order, which is the objective of the pass manager.

Why not hard-code the order of the passes? The user of your compiler usually expects your compiler to provide a different level of optimization. Developers prefer fast compilation speed over optimization during development time. The final application should run as fast as possible, and your compiler should be able to perform sophisticated optimizations, with longer a compilation time being accepted. A different level of optimization means a different number of optimization passes that need to be executed. Thus, as a compiler writer, you may want to provide your own passes to take advantage of your knowledge of your source language. For example, you may want to replace well-known library functions with inlined IR or even with the precomputed result. For C, such a pass is part of the LLVM libraries, but for other languages, you will need to provide it yourself. After introducing your own passes, you may need to re-order or add some passes. For example, if you know that the operation of your pass leaves some IR code unreachable, then you want to run the dead code removal pass additionally after your pass. The pass manager helps organize these requirements.

A pass is often categorized by the scope on which it works:

- A *module pass* takes a whole module as input. Such a pass performs its work on the given module and can be used for intra-procedure operations inside this module.

- A *call graph* pass operates on the **strongly connected components (SCCs)** of a call graph. It traverses the components in bottom-up order.

- A *function pass* takes a single function as input and performs its work on this function only.

- A *loop pass* works on a loop inside a function.

Besides the IR code, a pass may also require, update, or invalidate some analysis results. A lot of different analyses are performed, for example, alias analysis or the construction of a dominator tree. If a pass requires such analyses, then it can request it from an analyses manager. If the information is already computed, then the cached result will be returned. Otherwise, the information will be computed. If a pass changes the IR code, then it needs to announce which analysis results are preserved so that the cached analysis information can be invalidated if necessary.

Under the hood, the pass manager ensures the following:

- Analysis results are shared among passes. This requires keeping track of which pass requires which analysis and the state of each analysis. The goal is to avoid needless precomputation of analysis and to free up memory held by analysis results as soon as possible.

- The passes are executed in a pipeline fashion. For example, if several function passes should be executed in sequence, then the pass manager runs each of these function passes on the first function. Then, it will run all function passes on the second function, and so on. The underlying idea here is to improve the cache behavior as the compiler only performs transformations on a limited set of data (one IR function) and then moves on to the next limited set of data.

Let's implement a new IR transformation pass and explore how to add it to the optimization pipeline.

Implementing a new pass

A pass can perform arbitrary complex transformations on the LLVM IR. To illustrate the mechanics of adding a new pass, we add a pass that performs a simple instrumentation.

To investigate the performance of a program, it is interesting to know how often functions are called, and how long they run. One way to collect this data is to insert counters into each function. This process is called **instrumentation**. We will write a simple instrumentation pass that inserts a special function call at the entry of each function and each exit point. These functions collect the timing information and write it into a file. As a result, we can create a very basic profiler that we'll name the **poor person's profiler**, or in short, ppprofiler. We will develop the new pass so that it can be used as a standalone plugin or added as a plugin to the LLVM source tree. After that, we'll look at how the passes that come with LLVM are integrated into the framework.

Developing the ppprofiler pass as a plugin

In this section, we'll look at creating a new pass as a plugin out of the LLVM tree. The goal of the new pass is to insert a call to the __ppp_enter() function at the entry of a function, and a call to the __ppp_exit() function before each return instruction. Only the name of the current function is passed as a parameter. The implementation of these functions can then count the number of calls and measure the elapsed time. We will implement this runtime support at the end of this chapter. We'll examine how to develop the pass.

We'll store the source in the PPProfiler.cpp file. Follow these steps:

1. First, let's include some files:

```
#include "llvm/ADT/Statistic.h"
#include "llvm/IR/Function.h"
#include "llvm/IR/PassManager.h"
#include "llvm/Passes/PassBuilder.h"
#include "llvm/Passes/PassPlugin.h"
#include "llvm/Support/Debug.h"
```

2. To shorten the source, we'll tell the compiler that we're using the `llvm` namespace:

    ```
    using namespace llvm;
    ```

3. The built-in debug infrastructure of LLVM requires that we define a debug type, which is a string. This string is later shown in the printed statistic:

    ```
    #define DEBUG_TYPE "ppprofiler"
    ```

4. Next, we'll define one counter variable with the `ALWAYS_ENABLED_STATISTIC` macro. The first parameter is the name of the counter variable, while the second parameter is the text that will be printed in the statistic:

    ```
    ALWAYS_ENABLED_STATISTIC(
        NumOfFunc, "Number of instrumented functions.");
    ```

> **Note**
>
> Two macros can be used to define a counter variable. If you use the `STATISTIC` macro, then the statistic value will only be collected in a debug build if assertions are enabled, or if `LLVM_FORCE_ENABLE_STATS` is set to `ON` on the CMake command line. If you use the `ALWAYS_ENABLED_STATISTIC` macro instead, then the statistic value is always collected. However, printing the statistics using the `-stats` command-line option only works with the former methods. If needed, you can print the collected statistics by calling the `llvm::PrintStatistics(llvm::raw_ostream)` function.

5. Next, we must declare the pass class in an anonymous namespace. The class inherits from the `PassInfoMixin` template. This template only adds some boilerplate code, such as a `name()` method. It is not used to determine the type of the pass. The `run()` method is called by LLVM when the pass is executed. We also need a helper method called `instrument()`:

    ```
    namespace {
    class PPProfilerIRPass
        : public llvm::PassInfoMixin<PPProfilerIRPass> {
    public:
      llvm::PreservedAnalyses
      run(llvm::Module &M, llvm::ModuleAnalysisManager &AM);

    private:
      void instrument(llvm::Function &F,
                      llvm::Function *EnterFn,
                      llvm::Function *ExitFn);
    };
    }
    ```

6. Now, let's define how a function is instrumented. Besides the function to instrument, the functions to call are passed:

```
void PPProfilerIRPass::instrument(llvm::Function &F,
                                  Function *EnterFn,
                                  Function *ExitFn) {
```

7. Inside the function, we update the statistic counter:

```
++NumOfFunc;
```

8. To easily insert IR code, we need an instance of the `IRBuilder` class. We will set it to the first basic block, which is the entry block of the function:

```
IRBuilder<> Builder(&*F.getEntryBlock().begin());
```

9. Now that we have the builder, we can insert a global constant that holds the name of the function we wish to instrument:

```
GlobalVariable *FnName =
    Builder.CreateGlobalString(F.getName());
```

10. Next, we will insert a call to the `__ppp_enter()` function, passing the name as an argument:

```
Builder.CreateCall(EnterFn->getFunctionType(), EnterFn,
                   {FnName});
```

11. To call the `__ppp_exit()` function, we have to locate all return instructions. Conveniently, the insertion point that's set by the calling `SetInsertionPoint()` function is before the instruction that's passed as a parameter, so we can just insert the call at that point:

```
for (BasicBlock &BB : F) {
  for (Instruction &Inst : BB) {
    if (Inst.getOpcode() == Instruction::Ret) {
      Builder.SetInsertPoint(&Inst);
      Builder.CreateCall(ExitFn->getFunctionType(),
                         ExitFn, {FnName});
    }
  }
}
```

12. Next, we will implement the `run()` method. LLVM passes in the module our pass works on and an analysis manager from which we can request analysis results if needed:

```
PreservedAnalyses
```

```
PPProfilerIRPass::run(Module &M,
                      ModuleAnalysisManager &AM) {
```

13. There is a slight annoyance here: if the runtime module that contains the implementation of the __ppp_enter() and __ppp_exit() functions are instrumented, then we run into trouble because we create an infinite recursion. To avoid this, we must simply do nothing if one of those functions is defined:

```
if (M.getFunction("__ppp_enter") ||
    M.getFunction("__ppp_exit")) {
  return PreservedAnalyses::all();
}
```

14. Now, we are ready to declare the functions. There is nothing unusual here: first, the function type is created, followed by the functions:

```
Type *VoidTy = Type::getVoidTy(M.getContext());
PointerType *PtrTy =
    PointerType::getUnqual(M.getContext());
FunctionType *EnterExitFty =
    FunctionType::get(VoidTy, {PtrTy}, false);
Function *EnterFn = Function::Create(
    EnterExitFty, GlobalValue::ExternalLinkage,
    "__ppp_enter", M);
Function *ExitFn = Function::Create(
    EnterExitFty, GlobalValue::ExternalLinkage,
    "__ppp_exit", M);
```

15. All we need to do now is loop over all the functions of the module and instrument the found functions by calling our instrument() method. Of course, we need to ignore function declarations, which are just prototypes. There can also be functions without a name, which does not work well with our approach. We'll filter out those functions too:

```
for (auto &F : M.functions()) {
  if (!F.isDeclaration() && F.hasName())
    instrument(F, EnterFn, ExitFn);
}
```

16. Lastly, we must declare that we did not preserve any analysis. This is most likely too pessimistic but we are on the safe side by doing so:

```
  return PreservedAnalyses::none();
}
```

The functionality of our new pass is now implemented. To be able to use our pass, we need to register it with the `PassBuilder` object. This can happen in two ways: statically or dynamically. If the plugin is statically linked, then it needs to provide a function called `get<Plugin-Name>PluginInfo()`. To use dynamic linking, the `llvmGetPassPluginInfo()` function needs to be provided. In both cases, an instance of the `PassPluginLibraryInfo` struct is returned, which provides some basic information about a plugin. Most importantly, this structure contains a pointer to the function that registers the pass. Let's add this to our source.

17. In the `RegisterCB()` function, we register a Lambda function that is called when a pass pipeline string is parsed. If the name of the pass is `ppprofiler`, then we add our pass to the module pass manager. These callbacks will be expanded upon in the next section:

```
void RegisterCB(PassBuilder &PB) {
  PB.registerPipelineParsingCallback(
      [](StringRef Name, ModulePassManager &MPM,
         ArrayRef<PassBuilder::PipelineElement>) {
        if (Name == "ppprofiler") {
          MPM.addPass(PPProfilerIRPass());
          return true;
        }
        return false;
      });
}
```

18. The `getPPProfilerPluginInfo()` function is called when the plugin is statically linked. It returns some basic information about the plugin:

```
llvm::PassPluginLibraryInfo getPPProfilerPluginInfo() {
  return {LLVM_PLUGIN_API_VERSION, "PPProfiler", "v0.1",
          RegisterCB};
}
```

19. Finally, if the plugin is dynamically linked, then the `llvmGetPassPluginInfo()` function is called when the plugin is loaded. However, when linking this code statically into a tool, you might end up with linker errors because that function could be defined in several source files. The solution is to guard the function with a macro:

```
#ifndef LLVM_PPPROFILER_LINK_INTO_TOOLS
extern "C" LLVM_ATTRIBUTE_WEAK ::llvm::PassPluginLibraryInfo
llvmGetPassPluginInfo() {
  return getPPProfilerPluginInfo();
}
#endif
```

With that, we've implemented the pass plugin. Before we look at how to use the new plugin, let's examine what needs to be changed if we want to add the pass plugin to the LLVM source tree.

Adding the pass to the LLVM source tree

Implementing a new pass as a plugin is useful if you plan to use it with a precompiled clang, for example. On the other hand, if you write your own compiler, then there can be good reasons to add your new passes directly to the LLVM source tree. There are two different ways you can do this – as a plugin and as a fully integrated pass. The plugin approach requires fewer changes.

Utilizing the plugin mechanisms inside the LLVM source tree

The source of passes that perform transformations on LLVM IR is located in the `llvm-project/llvm/lib/Transforms` directory. Inside this directory, create a new directory called `PPProfiler` and copy the source file, `PPProfiler.cpp`, into it. You do not need to make any source changes!

To integrate the new plugin into the build system, create a file called `CMakeLists.txt` with the following content:

```
add_llvm_pass_plugin(PPProfiler PPProfiler.cpp)
```

Finally, in the `CmakeLists.txt` file in the parent directory, you need to include the new source directory by adding the following line:

```
add_subdirectory(PPProfiler)
```

You are now ready to build LLVM with `PPProfiler` added. Change into the build directory of LLVM and manually run Ninja:

```
$ ninja install
```

CMake will detect a change in the build description and rerun the configuration step. You will see an additional line:

```
-- Registering PPProfiler as a pass plugin (static build: OFF)
```

This tells you that the plugin was detected and has been built as a shared library. After the installation step, you will find that shared library, `PPProfiler.so`, in the `<install directory>/lib` directory.

So far, the only difference to the pass plugin from the previous section is that the shared library is installed as part of LLVM. But you can also statically link the new plugin to the LLVM tools. To do this, you need to rerun the CMake configuration and add the `-DLLVM_PPPROFILER_LINK_INTO_TOOLS=ON` option on the command line. Look for this information from CMake to confirm the changed build option:

```
-- Registering PPProfiler as a pass plugin (static build: ON)
```

After compiling and installing LLVM again, the following has changed:

- The plugin is compiled into the static library, `libPPProfiler.a`, and that library is installed in the `<install directory>/lib` directory.

- The LLVM tools, such as **opt**, are linked against that library.

- The plugin is registered as an extension. You can check that the `<install directory>/include/llvm/Support/Extension.def` file now contains the following line:

  ```
  HANDLE_EXTENSION(PPProfiler)
  ```

In addition, all tools that support this extension mechanism pick up the new pass. In the *Creating an optimization pipeline* section, you will learn how to do this in your compiler.

This approach works well because the new source files reside in a separate directory, and only one existing file was changed. This minimizes the probability of merge conflicts if you try to keep your modified LLVM source tree in sync with the main repository.

There are also situations where adding the new pass as a plugin is not the best way. The passes that LLVM provides use a different way for registration. If you develop a new pass and propose to add it to LLVM, and the LLVM community accepts your contribution, then you will want to use the same registration mechanism.

Fully integrating the pass into the pass registry

To fully integrate the new pass into LLVM, the source of the plugin needs to be structured slightly differently. The main reason for this is that the constructor of the pass class is called from the pass registry, which requires the class interface to be put into a header file.

Like before, you must put the new pass into the `Transforms` component of LLVM. Begin the implementation by creating the `llvm-project/llvm/include/llvm/Transforms/PPProfiler/PPProfiler.h` header file. The content of that file is the class definition; put it into the `llvm` namespace. No other changes are required:

```
#ifndef LLVM_TRANSFORMS_PPPROFILER_PPPROFILER_H
#define LLVM_TRANSFORMS_PPPROFILER_PPPROFILER_H

#include "llvm/IR/PassManager.h"

namespace llvm {
class PPProfilerIRPass
    : public llvm::PassInfoMixin<PPProfilerIRPass> {
public:
  llvm::PreservedAnalyses
  run(llvm::Module &M, llvm::ModuleAnalysisManager &AM);
```

```
private:
  void instrument(llvm::Function &F,
                  llvm::Function *EnterFn,
                  llvm::Function *ExitFn);
};
} // namespace llvm

#endif
```

Next, copy the source file of the pass plugin, PPProfiler.cpp, into the new directory, llvm-project/llvm/lib/Transforms/PPProfiler. This file needs to be updated in the following way:

1. Since the class definition is now in a header file, you must remove the class definition from this file. At the top, add the #include directive for the header file:

    ```
    #include "llvm/Transforms/PPProfiler/PPProfiler.h"
    ```

2. The llvmGetPassPluginInfo() function must be removed because the pass wasn't built into a shared library of its own.

As before, you also need to provide a CMakeLists.txt file for the build. You must declare the new pass as a new component:

```
add_llvm_component_library(LLVMPPProfiler
  PPProfiler.cpp

  LINK_COMPONENTS
  Core
  Support
)
```

After, like in the previous section, you need to include the new source directory by adding the following line to the CMakeLists.txt file in the parent directory:

```
add_subdirectory(PPProfiler)
```

Inside LLVM, the available passes are kept in the llvm/lib/Passes/ PassRegistry.def database file. You need to update this file. The new pass is a module pass, so we need to search inside the file for the section in which module passes are defined, for example, by searching for the MODULE_PASS macro. Inside this section, add the following line:

```
MODULE_PASS("ppprofiler", PPProfilerIRPass())
```

This database file is used in the `llvm/lib/Passes/PassBuilder.cpp` class. This file needs to include your new header file:

```
#include "llvm/Transforms/PPProfiler/PPProfiler.h"
```

These are all required source changes based on the plugin version of the new pass.

Since you created a new LLVM component, it is also necessary to add a link dependency in the `llvm/lib/Passes/CMakeLists.txt` file. Under the `LINK_COMPONENTS` keyword, you need to add a line with the name of the new component:

```
PPProfiler
```

Et voilà – you are ready to build and install LLVM. The new pass, `ppprofiler`, is now available to all LLVM tools. It has been compiled into the `libLLVMPPProfiler.a` library and available in the build system as the `PPProfiler` component.

So far, we have talked about how to create a new pass. In the next section, we will examine how to use the `ppprofiler` pass.

Using the ppprofiler pass with LLVM tools

Recall the ppprofiler pass that we developed as a plugin out of the LLVM tree in the *Developing the ppprofiler pass as a plugin* section. Here, we'll learn how to use this pass with LLVM tools, such as `opt` and `clang`, as they can load plugins.

Let's look at `opt` first.

Run the pass plugin in opt

To play around with the new plugin, you need a file containing LLVM IR. The easiest way to do this is to translate a C program, such as a basic "Hello World" style program:

```
#include <stdio.h>

int main(int argc, char *argv[]) {
  puts("Hello");
  return 0;
}
```

Compile this file, `hello.c`, with `clang`:

```
$ clang -S -emit-llvm -O1 hello.c
```

You will get a very simple IR file called `hello.ll` that contains the following code:

```
$ cat hello.ll
@.str = private unnamed_addr constant [6 x i8] c"Hello\00",
        align 1

define dso_local i32 @main(
            i32 noundef %0, ptr nocapture noundef readnone %1) {
  %3 = tail call i32 @puts(
                ptr noundef nonnull dereferenceable(1) @.str)
  ret i32 0
}
```

This is enough to test the pass.

To run the pass, you have to provide a couple of arguments. First, you need to tell `opt` to load the shared library via the `--load-pass-plugin` option. To run a single pass, you must specify the `--passes` option. Using the `hello.ll` file as input, you can run the following:

```
$ opt --load-pass-plugin=./PPProfile.so \
      --passes="ppprofiler" --stats hello.ll -o hello_inst.bc
```

If statistic generation is enabled, you will see the following output:

```
===-------------------------------------------------------------===
                    ... Statistics Collected ...
===-------------------------------------------------------------===

1 ppprofiler - Number of instrumented functions.
```

Otherwise, you will be informed that statistic collection is not enabled:

```
Statistics are disabled.  Build with asserts or with
-DLLVM_FORCE_ENABLE_STATS
```

The bitcode file, `hello_inst.bc`, is the result. You can turn this file into readable IR with the `llvm-dis` tool. As expected, you will see the calls to the `__ppp_enter()` and `__ppp_exit()` functions and a new constant for the name of the function:

```
$ llvm-dis hello_inst.bc -o -
@.str = private unnamed_addr constant [6 x i8] c"Hello\00",
        align 1
@0 = private unnamed_addr constant [5 x i8] c"main\00",
      align 1
```

```
define dso_local i32 @main(i32 noundef %0,
                           ptr nocapture noundef readnone %1) {
  call void @__ppp_enter(ptr @0)
  %3 = tail call i32 @puts(
                  ptr noundef nonnull dereferenceable(1) @.str)
  call void @__ppp_exit(ptr @0)
  ret i32 0
}
```

This already looks good! It would be even better if we could turn this IR into an executable and run it. For this, you need to provide implementations for the called functions.

Often, the runtime support for a feature is more complicated than adding that feature to the compiler itself. This is also true in this case. When the __ppp_enter() and __ppp_exit() functions are called, you can view this as an event. To analyze the data later, it is necessary to save the events. The basic data you would like to get is the event of the type, the name of the function and its address, and a timestamp. Without tricks, this is not as easy as it seems. Let's give it a try.

Create a file called runtime.c with the following content:

1. You need the file I/O, standard functions, and time support. This is provided by the following includes:

   ```
   #include <stdio.h>
   #include <stdlib.h>
   #include <time.h>
   ```

2. For the file, a file descriptor is needed. Moreover, when the program finishes, that file descriptor should be closed properly:

   ```
   static FILE *FileFD = NULL;

   static void cleanup() {
     if (FileFD == NULL) {
       fclose(FileFD);
       FileFD = NULL;
     }
   }
   ```

3. To simplify the runtime, only a fixed name for the output is used. If the file is not open, then open the file and register the cleanup function:

   ```
   static void init() {
     if (FileFD == NULL) {
       FileFD = fopen("ppprofile.csv", "w");
       atexit(&cleanup);
   ```

```
        }
    }
```

4. You can call the `clock_gettime()` function to get a timestamp. The `CLOCK_PROCESS_CPUTIME_ID` parameter returns the time consumed by this process. Please note that not all systems support this parameter. You can use one of the other clocks, such as `CLOCK_REALTIME`, if necessary:

```
typedef unsigned long long Time;

static Time get_time() {
    struct timespec ts;
    clock_gettime(CLOCK_PROCESS_CPUTIME_ID, &ts);
    return 1000000000L * ts.tv_sec + ts.tv_nsec;
}
```

5. Now, it is easy to define the `__ppp_enter()` function. Just make sure the file is open, get the timestamp, and write the event:

```
void __ppp_enter(const char *FnName) {
    init();
    Time T = get_time();
    void *Frame = __builtin_frame_address(1);
    fprintf(FileFD,
            // "enter|name|clock|frame"
            „enter|%s|%llu|%p\n", FnName, T, Frame);
}
```

6. The `__ppp_exit()` function only differs in terms of the event type:

```
void __ppp_exit(const char *FnName) {
    init();
    Time T = get_time();
    void *Frame = __builtin_frame_address(1);
    fprintf(FileFD,
            // "exit|name|clock|frame"
            „exit|%s|%llu|%p\n", FnName, T, Frame);
}
```

That concludes a very simple implementation for runtime support. Before we try it, some remarks should be made about the implementation as it should be obvious that there are several problematic parts.

First of all, the implementation is not thread-safe since there is only one file descriptor, and access to it is not protected. Trying to use this runtime implementation with a multithreaded program will most likely lead to disturbed data in the output file.

In addition, we omitted checking the return value of the I/O-related functions, which can result in data loss.

But most importantly, the timestamp of the event is not precise. Calling a function already adds overhead, but performing I/O operations in that function makes it even worse. In principle, you can match the enter and exit events for a function and calculate the runtime of the function. However, this value is inherently flawed because it may include the time required for I/O. In summary, do not trust the times recorded here.

Despite all the flaws, this small runtime file allows us to produce some output. Compile the bitcode of the instrumented file together with the file containing the runtime code and run the resulting executable:

```
$ clang hello_inst.bc runtime.c
$ ./a.out
```

This results in a new file called ppprofile.csv in the directory that contains the following content:

```
$ cat ppprofile.csv
enter|main|3300868|0x1
exit|main|3760638|0x1
```

Cool – the new pass and the runtime seem to work!

> **Specifying a pass pipeline**
>
> With the --passes option, you can not only name a single pass but you can also describe a whole pipeline. For example, the default pipeline for optimization level 2 is named default<O2>. You can run the ppprofile pass before the default pipeline with the --passes="ppprofile,default<O2>" argument. Please note that the pass names in such a pipeline description must be of the same type.

Now, let's turn to using the new pass with clang.

Plugging the new pass into clang

In the previous section, you learned how you can run a single pass using opt. This is useful if you need to debug a pass but for a real compiler, the steps should not be that involved.

To achieve the best result, a compiler needs to run the optimization passes in a certain order. The LLVM pass manager has a default order for pass execution. This is also called the **default pass pipeline**. Using opt, you can specify a different pass pipeline with the -passes option. This is flexible but also complicated for the user. It also turns out that most of the time, you just want to add a new pass at very specific points, such as before optimization passes are run or at the end of the loop optimization processes. These points are called **extension points**. The PassBuilder class allows you to register a pass at an extension point. For example, you can call the registerPipelineStartEPCallback()

method to add a pass to the beginning of the optimization pipeline. This is exactly the place we need for the ppprofiler pass. During optimization, functions may be inlined, and the pass will miss those inline functions. Instead, running the pass before the optimization passes guarantees that all functions are instrumented.

To use this approach, you need to extend the RegisterCB() function in the pass plugin. Add the following code to the function:

```
PB.registerPipelineStartEPCallback(
    [](ModulePassManager &PM, OptimizationLevel Level) {
      PM.addPass(PPProfilerIRPass());
    });
```

Whenever the pass manager populates the default pass pipeline, it calls all the callbacks for the extension points. We simply add the new pass here.

To load the plugin into clang, you can use the -fpass-plugin option. Creating the instrumented executable of the hello.c file now becomes almost trivial:

```
$ clang -fpass-plugin=./PPProfiler.so hello.c runtime.c
```

Please run the executable and verify that the run creates the ppprofiler.csv file.

> **Note**
> The runtime.c file is not instrumented because the pass checks that the special functions are not yet declared in a module.

This already looks better, but does it scale to larger programs? Let's assume you want to build an instrumented binary of the tinylang compiler for *Chapter 5*. How would you do this?

You can pass compiler and linker flags on the CMake command line, which is exactly what we need. The flags for the C++ compiler are given in the CMAKE_CXX_FLAGS variable. Thus, specifying the following on the CMake command line adds the new pass to all compiler runs:

```
-DCMAKE_CXX_FLAGS="-fpass-plugin=<PluginPath>/PPProfiler.so"
```

Please replace <PluginPath> with the absolute path to the shared library.

Similarly, specifying the following adds the runtime.o file to each linker invocation. Again, please replace <RuntimePath> with the absolute path to a compiled version of runtime.c:

```
-DCMAKE_EXE_LINKER_FLAGS="<RuntimePath>/runtime.o"
```

Of course, this requires clang as the build compiler. The fastest way to make sure clang is used as the build compiler is to set the CC and CXX environment variables accordingly:

```
export CC=clang
export CXX=clang++
```

With these additional options, the CMake configuration from *Chapter 5* should run as usual.

After building the tinylang executable, you can run it with the example Gcd.mod file. The ppprofile.csv file will also be written, this time with more than 44,000 lines!

Of course, having such a dataset raises the question of if you can get something useful out of it. For example, getting a list of the 10 most often called functions, together with the call count and the time spent in the function, would be useful information. Luckily, on a Unix system, you have a couple of tools that can help. Let's build a short pipeline that matches enter events with exit events, counts the functions, and displays the top 10 functions. The awk Unix tool helps with most of these steps.

To match an enter event with an exit event, the enter event must be stored in the record associative map. When an exit event is matched, the stored enter event is looked up, and the new record is written. The emitted line contains the timestamp from the enter event, the timestamp from the exit event, and the difference between both. We must put this into the join.awk file:

```
BEGIN { FS = "|"; OFS = "|" }
/enter/ { record[$2] = $0 }
/exit/ { split(record[$2],val,"|")
         print val[2], val[3], $3, $3-val[3], val[4] }
```

To count the function calls and the execution, two associative maps, count and sum, are used. In count, the function calls are counted, while in sum, the execution time is added. In the end, the maps are dumped. You can put this into the avg.awk file:

```
BEGIN { FS = "|"; count[""] = 0; sum[""] = 0 }
{ count[$1]++; sum[$1] += $4 }
END { for (i in count) {
        if (i != "") {
            print count[i], sum[i], sum[i]/count[i], I }
} }
```

After running these two scripts, the result can be sorted in descending order, and then the top 10 lines can be taken from the file. However, we can still improve the function names, __ppp_enter() and __ppp_exit(), which are mangled and are therefore difficult to read. Using the llvm-cxxfilt tool, the names can be demangled. The demangle.awk script is as follows:

```
{ cmd = "llvm-cxxfilt " $4
  (cmd) | getline name
  close(cmd); $4 = name; print }
```

To get the top 10 function calls, you can run the following:

```
$ cat ppprofile.csv | awk -f join.awk | awk -f avg.awk |\
  sort -nr | head -15 | awk -f demangle.awk
```

Here are some sample lines from the output:

```
446 1545581 3465.43 charinfo::isASCII(char)
409 826261 2020.2 llvm::StringRef::StringRef()
382 899471 2354.64
        tinylang::Token::is(tinylang::tok::TokenKind) const
171 1561532 9131.77 charinfo::isIdentifierHead(char)
```

The first number is the call count of the function, the second is the cumulated execution time, and the third number is the average execution time. As explained previously, do not trust the time values, though the call counts should be accurate.

So far, we've implemented a new instrumentation pass, either as a plugin or as an addition to LLVM, and we used it in some real-world scenarios. In the next section, we'll explore how to set up an optimization pipeline in our compiler.

Adding an optimization pipeline to your compiler

The tinylang compiler we developed in the previous chapters performs no optimizations on the IR code. In the next few subsections, we'll add an optimization pipeline to the compiler to achieve this accordingly.

Creating an optimization pipeline

The PassBuilder class is central to setting up the optimization pipeline. This class knows about all registered passes and can construct a pass pipeline from a textual description. We can use this class to either create the pass pipeline from a description given on the command line or use a default pipeline based on the requested optimization level. We also support the use of pass plugins, such as the ppprofiler pass plugin we discussed in the previous section. With this, we can mimic part of the functionality of the **opt** tool and also use similar names for the command-line options.

The PassBuilder class populates an instance of a ModulePassManager class, which is the pass manager that holds the constructed pass pipeline and runs it. The code generation passes still use the old pass manager. Therefore, we have to retain the old pass manager for this purpose.

For the implementation, we will extend the `tools/driver/Driver.cpp` file from our `tinylang` compiler:

1. We'll use new classes, so we'll begin with adding new include files. The `llvm/Passes/PassBuilder.h` file defines the `PassBuilder` class. The `llvm/Passes/PassPlugin.h` file is required for plugin support. Finally, the `llvm/Analysis/TargetTransformInfo.h` file provides a pass that connects IR-level transformations with target-specific information:

```
#include "llvm/Passes/PassBuilder.h"
#include "llvm/Passes/PassPlugin.h"
#include "llvm/Analysis/TargetTransformInfo.h"
```

2. To use certain features of the new pass manager, we must add three command-line options, using the same names as the `opt` tool does. The `--passes` option allows the textual specification of the pass pipeline, while the `--load-pass-plugin` option allows the use of pass plugins. If the `--debug-pass-manager` option is given, then the pass manager prints out information about the executed passes:

```
static cl::opt<bool>
    DebugPM("debug-pass-manager", cl::Hidden,
            cl::desc("Print PM debugging information"));
static cl::opt<std::string> PassPipeline(
    "passes",
    cl::desc("A description of the pass pipeline"));
static cl::list<std::string> PassPlugins(
    "load-pass-plugin",
    cl::desc("Load passes from plugin library"));
```

3. The user influences the construction of the pass pipeline with the optimization level. The `PassBuilder` class supports six different optimization levels: no optimization, three levels for optimizing speed, and two levels for reducing size. We can capture all levels in one command-line option:

```
static cl::opt<signed char> OptLevel(
    cl::desc("Setting the optimization level:"),
    cl::ZeroOrMore,
    cl::values(
        clEnumValN(3, "O", "Equivalent to -O3"),
        clEnumValN(0, "O0", "Optimization level 0"),
        clEnumValN(1, "O1", "Optimization level 1"),
        clEnumValN(2, "O2", "Optimization level 2"),
        clEnumValN(3, "O3", "Optimization level 3"),
        clEnumValN(-1, "Os",
```

```
                              "Like -O2 with extra optimizations "
                              "for size"),
                clEnumValN(
                    -2, "Oz",
                    "Like -Os but reduces code size further")),
            cl::init(0));
```

4. The plugin mechanism of LLVM supports a plugin registry for statically linked plugins, which is created during the configuration of the project. To make use of this registry, we must include the `llvm/Support/Extension.def` database file to create the prototype for the functions that return the plugin information:

```
#define HANDLE_EXTENSION(Ext)                          \
    llvm::PassPluginLibraryInfo get##Ext##PluginInfo();
#include "llvm/Support/Extension.def"
```

5. Now, we must replace the existing `emit()` function with a new version. Additionally, we must declare the required `PassBuilder` instance at the top of the function:

```
bool emit(StringRef Argv0, llvm::Module *M,
          llvm::TargetMachine *TM,
          StringRef InputFilename) {
    PassBuilder PB(TM);
```

6. To implement the support for pass plugins given on the command line, we must loop through the list of plugin libraries given by the user and try to load the plugin. We'll emit an error message if this fails; otherwise, we'll register the passes:

```
for (auto &PluginFN : PassPlugins) {
    auto PassPlugin = PassPlugin::Load(PluginFN);
    if (!PassPlugin) {
        WithColor::error(errs(), Argv0)
            << "Failed to load passes from '" << PluginFN
            << "'. Request ignored.\n";
        continue;
    }

    PassPlugin->registerPassBuilderCallbacks(PB);
}
```

7. The information from the static plugin registry is used in a similar way to register those plugins with our `PassBuilder` instance:

```
#define HANDLE_EXTENSION(Ext)                               \
    get##Ext##PluginInfo().RegisterPassBuilderCallbacks( \
```

```
        PB);
#include "llvm/Support/Extension.def"
```

8. Now, we need to declare variables for the different analysis managers. The only parameter is the debug flag:

    ```
    LoopAnalysisManager LAM(DebugPM);
    FunctionAnalysisManager FAM(DebugPM);
    CGSCCAnalysisManager CGAM(DebugPM);
    ModuleAnalysisManager MAM(DebugPM);
    ```

9. Next, we must populate the analysis managers with calls to the respective `register` method on the `PassBuilder` instance. Through this call, the analysis manager is populated with the default analysis passes and also runs registration callbacks. We must also make sure that the function analysis manager uses the default alias-analysis pipeline and that all analysis managers know about each other:

    ```
    FAM.registerPass(
        [&] { return PB.buildDefaultAAPipeline(); });
    PB.registerModuleAnalyses(MAM);
    PB.registerCGSCCAnalyses(CGAM);
    PB.registerFunctionAnalyses(FAM);
    PB.registerLoopAnalyses(LAM);
    PB.crossRegisterProxies(LAM, FAM, CGAM, MAM);
    ```

10. The MPM module pass manager holds the pass pipeline that we constructed. The instance is initialized with the debug flag:

    ```
    ModulePassManager MPM(DebugPM);
    ```

11. Now, we need to implement two different ways to populate the module pass manager with the pass pipeline. If the user provided a pass pipeline on the command line – that is, they have used the --passes option – then we use this as the pass pipeline:

    ```
    if (!PassPipeline.empty()) {
      if (auto Err = PB.parsePassPipeline(
             MPM, PassPipeline)) {
        WithColor::error(errs(), Argv0)
            << toString(std::move(Err)) << "\n";
        return false;
      }
    }
    ```

12. Otherwise, we use the chosen optimization level to determine the pass pipeline to construct. The name of the default pass pipeline is `default`, and it takes the optimization level as a parameter:

```
else {
  StringRef DefaultPass;
  switch (OptLevel) {
  case 0: DefaultPass = "default<O0>"; break;
  case 1: DefaultPass = "default<O1>"; break;
  case 2: DefaultPass = "default<O2>"; break;
  case 3: DefaultPass = "default<O3>"; break;
  case -1: DefaultPass = "default<Os>"; break;
  case -2: DefaultPass = "default<Oz>"; break;
  }
  if (auto Err = PB.parsePassPipeline(
          MPM, DefaultPass)) {
    WithColor::error(errs(), Argv0)
        << toString(std::move(Err)) << "\n";
    return false;
  }
}
```

13. With that, the pass pipeline to run transformations on the IR code has been set up. After this step, we need an open file to write the result to. The system assembler and LLVM IR output are text-based, so we should set the `OF_Text` flag for them:

```
std::error_code EC;
sys::fs::OpenFlags OpenFlags = sys::fs::OF_None;
CodeGenFileType FileType = codegen::getFileType();
if (FileType == CGFT_AssemblyFile)
  OpenFlags |= sys::fs::OF_Text;
auto Out = std::make_unique<llvm::ToolOutputFile>(
    outputFilename(InputFilename), EC, OpenFlags);
if (EC) {
  WithColor::error(errs(), Argv0)
      << EC.message() << '\n';
  return false;
}
```

14. For the code generation process, we have to use the old pass manager. We must simply declare the `CodeGenPM` instances and add the pass, which makes target-specific information available at the IR transformation level:

```
legacy::PassManager CodeGenPM;
CodeGenPM.add(createTargetTransformInfoWrapperPass(
    TM->getTargetIRAnalysis()));
```

15. To output LLVM IR, we must add a pass that prints the IR into a stream:

```
if (FileType == CGFT_AssemblyFile && EmitLLVM) {
  CodeGenPM.add(createPrintModulePass(Out->os()));
}
```

16. Otherwise, we must let the `TargetMachine` instance add the required code generation passes, directed by the `FileType` value we pass as an argument:

```
else {
  if (TM->addPassesToEmitFile(CodeGenPM, Out->os(),
                              nullptr, FileType)) {
    WithColor::error()
        << "No support for file type\n";
    return false;
  }
}
```

17. After all this preparation, we are now ready to execute the passes. First, we must run the optimization pipeline on the IR module. Next, the code generation passes are run. Of course, after all this work, we want to keep the output file:

```
MPM.run(*M, MAM);
CodeGenPM.run(*M);
Out->keep();
return true;
}
```

18. That was a lot of code, but the process was straightforward. Of course, we have to update the dependencies in the `tools/driver/CMakeLists.txt` build file too. Besides adding the target components, we must add all the transformation and code generation components from LLVM. The names roughly resemble the directory names where the source is located. The component name is translated into the link library name during the configuration process:

```
set(LLVM_LINK_COMPONENTS ${LLVM_TARGETS_TO_BUILD}
  AggressiveInstCombine Analysis AsmParser
  BitWriter CodeGen Core Coroutines IPO IRReader
  InstCombine Instrumentation MC ObjCARCOpts Remarks
  ScalarOpts Support Target TransformUtils Vectorize
  Passes)
```

19. Our compiler driver supports plugins, and we must announce this support:

```
add_tinylang_tool(tinylang Driver.cpp SUPPORT_PLUGINS)
```

20. As before, we have to link against our own libraries:

```
target_link_libraries(tinylang
    PRIVATE tinylangBasic tinylangCodeGen
    tinylangLexer tinylangParser tinylangSema)
```

These are necessary additions to the source code and the build system.

21. To build the extended compiler, you must change into your `build` directory and type the following:

```
$ ninja
```

Changes to the files of the build system are automatically detected, and `cmake` is run before compiling and linking our changed source. If you need to re-run the configuration step, please follow the instructions in *Chapter 1*, *Installing LLVM*, the *Compiling the tinylang application* section.

As we have used the options for the `opt` tool as a blueprint, you should try running `tinylang` with the options to load a pass plugin and run the pass, as we did in the previous sections.

With the current implementation, we can either run a default pass pipeline or we can construct one ourselves. The latter is very flexible, but in almost all cases, it would be overkill. The default pipeline runs very well for C-like languages. However, what is missing is a way to extend the pass pipeline. We'll look at how to implement this in the next section.

Extending the pass pipeline

In the previous section, we used the `PassBuilder` class to create a pass pipeline, either from a user-provided description or a predefined name. Now, let's look at another way to customize the pass pipeline: using extension points.

During the construction of the pass pipeline, the pass builder allows passes contributed by the user to be added. These places are called **extension points**. A couple of extension points exist, as follows:

- The pipeline start extension point, which allows us to add passes at the beginning of the pipeline

- The peephole extension point, which allows us to add passes after each instance of the instruction combiner pass

Other extension points exist too. To employ an extension point, you must register a callback. During the construction of the pass pipeline, your callback is run at the defined extension point and can add passes to the given pass manager.

To register a callback for the pipeline start extension point, you must call the `registerPipelineStartEPCallback()` method of the `PassBuilder` class. For example, to add our `PPProfiler` pass to the beginning of the pipeline, you would adapt the pass to be used as a module pass with a call to the `createModuleToFunctionPassAdaptor()` template function and then add the pass to the module pass manager:

```
PB.registerPipelineStartEPCallback(
    [](ModulePassManager &MPM) {
        MPM.addPass(PPProfilerIRPass());
    });
```

You can add this snippet in the pass pipeline setup code anywhere before the pipeline is created – that is, before the `parsePassPipeline()` method is called.

A very natural extension to what we did in the previous section is to let the user pass a pipeline description for an extension point on the command line. The `opt` tool allows this too. Let's do this for the pipeline start extension point. Add the following code to the `tools/driver/Driver.cpp` file:

1. First, we must a new command line for the user to specify the pipeline description. Again, we take the option name from the `opt` tool:

    ```
    static cl::opt<std::string> PipelineStartEPPipeline(
        "passes-ep-pipeline-start",
        cl::desc("Pipeline start extension point));
    ```

2. Using a Lambda function as a callback is the most convenient way to do this. To parse the pipeline description, we must call the `parsePassPipeline()` method of the `PassBuilder` instance. The passes are added to the PM pass manager and given as an argument to the Lambda function. If an error occurs, we only print an error message without stopping the application. You can add this snippet after the call to the `crossRegisterProxies()` method:

    ```
    PB.registerPipelineStartEPCallback(
        [&PB, Argv0](ModulePassManager &PM) {
            if (auto Err = PB.parsePassPipeline(
                    PM, PipelineStartEPPipeline)) {
                WithColor::error(errs(), Argv0)
                    << "Could not parse pipeline "
                    << PipelineStartEPPipeline.ArgStr << ": "
                    << toString(std::move(Err)) << "\n";
            }
        });
    ```

> **Tip**
> To allow the user to add passes at every extension point, you need to add the preceding code snippet for each extension point.

3. Now is a good time to try out the different `pass manager` options. With the `--debug-pass-manager` option, you can follow which passes are executed in which order. You can also print the IR before or after each pass, which is invoked with the `--print-before-all` and `--print-after-all` options. If you created your own pass pipeline, then you can insert the `print` pass in points of interest. For example, try the `--passes="print,inline,print"` option. Furthermore, to identify which pass changes the IR code, you can use the `--print-changed` option, which will only print the IR code if it has changed compared to the result from the pass before. The greatly reduced output makes it much easier to follow IR transformations.

    The `PassBuilder` class has a nested `OptimizationLevel` class to represent the six different optimization levels. Instead of using the `"default<O?>"` pipeline description as an argument to the `parsePassPipeline()` method, we can also call the `buildPerModuleDefaultPipeline()` method, which builds the default optimization pipeline for the request level – except for level O0. This optimization level means that no optimization is performed.

    Consequently, no passes are added to the pass manager. If we still want to run a certain pass, then we can add it to the pass manager manually. A simple pass to run at this level is the `AlwaysInliner` pass, which inlines a function marked with the `always_inline` attribute into the caller. After translating the command-line option value for the optimization level into the corresponding member of the `OptimizationLevel` class, we can implement this as follows:

    ```
    PassBuilder::OptimizationLevel Olevel = …;
    if (OLevel == PassBuilder::OptimizationLevel::O0)
      MPM.addPass(AlwaysInlinerPass());
    else
      MPM = PB.buildPerModuleDefaultPipeline(OLevel, DebugPM);
    ```

    Of course, it is possible to add more than one pass to the pass manager in this fashion. `PassBuilder` also uses the `addPass()` method when constructing the pass pipeline.

> **Running extension point callbacks**
> Because the pass pipeline is not populated for optimization level O0, the registered extension points are not called. If you use the extension points to register passes that should also run at O0 level, this is problematic. You can call the `runRegisteredEPCallbacks()` method to run the registered extension point callbacks, resulting in a pass manager populated only with the passes that were registered through the extension points.

By adding the optimization pipeline to tinylang, you created an optimizing compiler similar to clang. The LLVM community works on improving the optimizations and the optimization pipeline with each release. Due to this, it is very seldom that the default pipeline is not used. Most often, new passes are added to implement certain semantics of the programming language.

Summary

In this chapter, you learned how to create a new pass for LLVM. You ran the pass using a pass pipeline description and an extension point. You extended your compiler with the construction and execution of a pass pipeline similar to clang, turning tinylang into an optimizing compiler. The pass pipeline allows the addition of passes at extension points, and you learned how you can register passes at these points. This allows you to extend the optimization pipeline with your developed passes or existing passes.

In the next chapter, you will learn the basics of the **TableGen** language, which is used extensively in LLVM and clang to significantly reduce manual programming.

Part 3:
Taking LLVM to the Next Level

In this section, you will delve into various low-level details of LLVM. Firstly, you will explore the TableGen language, which is LLVM's domain-specific language, and understand how it can be utilized in the backend. LLVM also has a **just-in-time** (**JIT**) compiler, and you will explore how to use it and tailor it to your needs. Furthermore, you will also try out various tools and libraries designed to identify bugs in applications. With all of this in mind, this knowledge will empower you to take advantage of new architectures not yet supported by LLVM.

This section comprises the following chapters:

- *Chapter 8, The TableGen Language*
- *Chapter 9, JIT Compilation*
- *Chapter 10, Debugging Using LLVM Tools*

The TableGen Language

Large parts of backends in LLVM are written in the TableGen language, a special language used to generate fragments of C++ source code to avoid implementing code similar for each backend and to shorten the amount of source code. Having knowledge of TableGen is therefore important.

In this chapter, you will learn the following:

- In *Understanding the TableGen language*, you will learn about the main idea behind TableGen
- In *Experimenting with the TableGen language*, you will define your own TableGen classes and records, and learn the syntax of the TableGen language
- In *Generating C++ code from a TableGen file*, you will develop your own TableGen backend
- Drawbacks of TableGen

By the end of the chapter, you will be able to use existing TableGen classes to define your own records. You will also acquire knowledge of how to create TableGen classes and records from scratch, and how to develop a TableGen backend to emit source code.

Technical requirements

You can find the source code used in this chapter on GitHub: https://github.com/PacktPublishing/Learn-LLVM-17/tree/main/Chapter08.

Understanding the TableGen language

LLVM comes with its own **domain-specific language** (**DSL**) called **TableGen**. It is used to generate C++ code for a wide range of use cases, thus reducing the amount of code a developer has to produce. The TableGen language is not a full-fledged programming language. It is only used to define records, which is a fancy word for a collection of names and values. To understand why such a restricted language is useful, let's examine two examples.

Typical data you need to define one machine instruction of a CPU is:

- The mnemonic of the instruction

- The bit pattern

- The number and types of operands

- Possible restrictions or side effects

It is easy to see that this data can be represented as a record. For example, a field named `asmstring` could hold the value of the mnemonic; say, `"add"`. Also, a field named `opcode` could hold the binary representation of the instruction. Together, the record would describe an additional instruction. Each LLVM backend describes the instruction set in this way.

Records are such a general concept that you can describe a wide variety of data with them. Another example is the definition of command-line options. A command-line option:

- Has a name

- May have an optional argument

- Has a help text

- May belong to a group of options

Again, this data can be easily seen as a record. Clang uses this approach for the command-line options of the Clang driver.

> **The TableGen language**
>
> In LLVM, the TableGen language is used for a variety of tasks. Large parts of a backend are written in the TableGen language; for example, the definition of a register file, all instructions with mnemonic and binary encoding, calling conventions, patterns for instruction selection, and scheduling models for instruction scheduling. Other uses of LLVM are the definition of intrinsic functions, the definition of attributes, and the definition of command-line options.
>
> You'll find the *Programmer's Reference* at `https://llvm.org/docs/TableGen/ProgRef.html` and the *Backend Developer's Guide* at `https://llvm.org/docs/TableGen/BackGuide.html`.

To achieve this flexibility, the parsing and the semantics of the TableGen language are implemented in a library. To generate C++ code from the records, you need to create a tool that takes the parsed records and generates C++ code from it. In LLVM, that tool is called `llvm-tblgen`, and in Clang, it is called `clang-tblgen`. Those tools contain the code generators required by the project. But they can also be used to learn more about the TableGen language, which is what we'll do in the next section.

Experimenting with the TableGen language

Very often, beginners feel overwhelmed by the TableGen language. But as soon as you start experimenting with the language, it becomes much easier.

Defining records and classes

Let's define a simple record for an instruction:

```
def ADD {
    string Mnemonic = "add";
    int Opcode = 0xA0;
}
```

The def keyword signals that you define a record. It is followed by the name of the record. The record body is surrounded by curly braces, and the body consists of field definitions, similar to a structure in C++.

You can use the llvm-tblgen tool to see the generated records. Save the preceding source code in an inst.td file and run the following:

```
$ llvm-tblgen --print-records inst.td
------------ Classes -----------------
------------ Defs ----------------
def ADD {
    string Mnemonic = "add";
    int Opcode = 160;
}
```

This is not yet exciting; it only shows the defined record was parsed correctly.

Defining instructions using single records is not very comfortable. A modern CPU has hundreds of instructions, and with this amount of records, it is very easy to introduce typing errors in the field names. And if you decide to rename a field or add a new field, then the number of records to change becomes a challenge. Therefore, a blueprint is needed. In C++, classes have a similar purpose, and in TableGen, it is also called a **class**. Here is the definition of an Inst class and two records based on that class:

```
class Inst<string mnemonic, int opcode> {
    string Mnemonic = mnemonic;
    int Opcode = opcode;
}

def ADD : Inst<"add", 0xA0>;
def SUB : Inst<"sub", 0xB0>;
```

The syntax for classes is similar to that of records. The `class` keyword signals that a class is defined, followed by the name of the class. A class can have a parameter list. Here, the `Inst` class has two parameters, `mnemonic` and `opcode`, which are used to initialize the records' fields. The values for those fields are given when the class is instantiated. The ADD and SUB records show two instantiations of the class. Again, let's use `llvm-tblgen` to look at the records:

```
$ llvm-tblgen --print-records inst.td
------------ Classes -----------------
class Inst<string Inst:mnemonic = ?, int Inst:opcode = ?> {
  string Mnemonic = Inst:mnemonic;
  int Opcode = Inst:opcode;
}
------------ Defs -----------------
def ADD {        // Inst
  string Mnemonic = "add";
  int Opcode = 160;
}
def SUB {        // Inst
  string Mnemonic = "sub";
  int Opcode = 176;
}
```

Now, you have one class definition and two records. The name of the class used to define the records is shown as a comment. Please note that the arguments of the class have the default value ?, which indicates `int` is uninitialized.

> **Tip for debugging**
>
> To get a more detailed dump of the records, you can use the `--print-detailed-records` option. The output includes the line numbers of record and class definitions, and where record fields are initialized. They can be very helpful if you try to track down why a record field was assigned a certain value.

In general, the ADD and SUB instructions have a lot in common, but there is also a difference: addition is a commutative operation but subtraction is not. Let's capture that fact in the record, too. A small challenge is that TableGen only supports a limited set of data types. You already used `string` and `int` in the examples. The other available data types are `bit`, `bits<n>`, `list<type>`, and `dag`. The `bit` type represents a single bit; that is, 0 or 1. If you need a fixed number of bits, then you use the `bits<n>` type. For example, `bits<5>` is an integer type 5 bits wide. To define a list based on another type, you use the `list<type>` type. For example, `list<int>` is a list of integers, and `list<Inst>` is a list of records of the `Inst` class from the example. The `dag` type represents **directed acyclic graph (DAG)** nodes. This type is useful for defining patterns and operations and is used extensively in LLVM backends.

To represent a flag, a single bit is sufficient, so you can use one to mark an instruction as commutable. The majority of instructions are not commutable, so you can take advantage of default values:

```
class Inst<string mnemonic, int opcode, bit commutable = 0> {
  string Mnemonic = mnemonic;
  int Opcode = opcode;
  bit Commutable = commutable;
}


def ADD : Inst<"add", 0xA0, 1>;
def SUB : Inst<"sub", 0xB0>;
```

You should run llvm-tblgen to verify that the records are defined as expected.

There is no requirement for a class to have parameters. It is also possible to assign values later. For example, you can define that all instructions are not commutable:

```
class Inst<string mnemonic, int opcode> {
  string Mnemonic = mnemonic;
  int Opcode = opcode;
  bit Commutable = 0;
}


def SUB : Inst<"sub", 0xB0>;
```

Using a let statement, you can overwrite that value:

```
let Commutable = 1 in
  def ADD : Inst<"add", 0xA0>;
```

Alternatively, you can open a record body to overwrite the value:

```
def ADD : Inst<"add", 0xA0> {
  let Commutable = 1;
}
```

Again, please use llvm-tblgen to verify that the Commutable flag is set to 1 in both cases.

Classes and records can be inherited from multiple classes, and it is always possible to add new fields or overwrite the value of existing fields. You can use inheritance to introduce a new CommutableInst class:

```
class Inst<string mnemonic, int opcode> {
  string Mnemonic = mnemonic;
  int Opcode = opcode;
  bit Commutable = 0;
```

```
}

class CommutableInst<string mnemonic, int opcode>
   : Inst<mnemonic, opcode> {
   let Commutable = 1;
}

def SUB : Inst<"sub", 0xB0>;
def ADD : CommutableInst<"add", 0xA0>;
```

The resulting records are always the same, but the language allows you to define records in different ways. Please note that, in the latter example, the Commutable flag may be superfluous: the code generator can query a record for the classes it is based on, and if that list contains the CommutableInst class, then it can set the flag internally.

Creating multiple records at once with multiclasses

Another often-used statement is multiclass. A multiclass allows you to define multiple records at once. Let's expand the example to show why this can be useful.

The definition of an add instruction is very simplistic. In reality, a CPU often has several add instructions. A common variant is that one instruction has two register operands while another instruction has one register operand and an immediate operand, which is a small number. Assume that for the instruction having an immediate operand, the designer of the instruction set decided to mark them with i as a suffix. So, we end up with the add and addi instructions. Further, assume that the opcodes differ by 1. Many arithmetic and logical instructions follow this scheme; therefore, you want the definition to be as compact as possible.

The first challenge is that you need to manipulate values. There is a limited number of operators that you can use to modify a value. For example, to produce the sum of 1 and the value of the field opcode, you write:

```
!add(opcode, 1)
```

Such an expression is best used as an argument for a class. Testing a field value and then changing it based on the found value is generally not possible because it requires dynamic statements that are not available. Always remember that all calculations are done while the records are constructed!

In a similar way, strings can be concatenated:

```
!strconcat(mnemonic,"i")
```

Because all operators begin with an exclamation mark (!), they are also called **bang operators**. You find a full list of bang operators in the *Programmer's Reference*: https://llvm.org/docs/TableGen/ProgRef.html#appendix-a-bang-operators.

Now, you can define a multiclass. The `Inst` class serves again as the base:

```
class Inst<string mnemonic, int opcode> {
  string Mnemonic = mnemonic;
  int Opcode = opcode;
}
```

The definition of a multiclass is a bit more involved, so let's do it in steps:

1. The definition of a multiclass uses a similar syntax to classes. The new multiclass is named `InstWithImm` and has two parameters, `mnemonic` and `opcode`:

    ```
    multiclass InstWithImm<string mnemonic, int opcode> {
    ```

2. First, you define an instruction with two register operands. As in a normal record definition, you use the `def` keyword to define the record, and you use the `Inst` class to create the record content. You also need to define an empty name. We will explain later why this is necessary:

    ```
    def "": Inst<mnemonic, opcode>;
    ```

3. Next, you define an instruction with the immediate operand. You derive the values for the mnemonic and the opcode from the parameters of the multiclass, using bang operators. The record is named `I`:

    ```
    def I: Inst<!strconcat(mnemonic,"i"), !add(opcode, 1)>;
    ```

4. That is all; the class body can be closed, like so:

    ```
    }
    ```

To instantiate the records, you must use the `defm` keyword:

```
defm ADD : InstWithImm<"add", 0xA0>;
```

These statements result in the following:

1. The `Inst<"add", 0xA0>` record is instantiated. The name of the record is the concatenation of the name following the `defm` keyword and of the name following `def` inside the multiclass statement, which results in the name ADD.

2. The `Inst<"addi", 0xA1>` record is instantiated and, following the same scheme, is given the name `ADDI`.

Let's verify this claim with `llvm-tblgen`:

```
$ llvm-tblgen -print-records inst.td
------------- Classes ------------------
class Inst<string Inst:mnemonic = ?, int Inst:opcode = ?> {
```

```
    string Mnemonic = Inst:mnemonic;
    int Opcode = Inst:opcode;
}
------------ Defs ----------------
def ADD {        // Inst
    string Mnemonic = "add";
    int Opcode = 160;
}
def ADDI {        // Inst
    string Mnemonic = "addi";
    int Opcode = 161;
}
```

Using a multiclass, it is very easy to generate multiple records at once. This feature is used very often!

A record does not need to have a name. Anonymous records are perfectly fine. Omitting the name is all you need to do to define an anonymous record. The name of a record generated by a multiclass is made up of two names, and both names must be given to create a named record. If you omit the name after defm, then only anonymous records are created. Similarly, if the def inside the multiclass is not followed by a name, an anonymous record is created. This is the reason why the first definition in the multiclass example used the empty name " ": without it, the record would be anonymous.

Simulating function calls

In some cases, using a multiclass like in the previous example can lead to repetitions. Assume that the CPU also supports memory operands, in a way similar to immediate operands. You can support this by adding a new record definition to the multiclass:

```
multiclass InstWithOps<string mnemonic, int opcode> {
    def "": Inst<mnemonic, opcode>;
    def "I": Inst<!strconcat(mnemonic,"i"), !add(opcode, 1)>;
    def "M": Inst<!strconcat(mnemonic,"m"), !add(opcode, 2)>;
}
```

This is perfectly fine. But now, imagine you do not have 3 but 16 records to define, and you need to do this multiple times. A typical scenario where such a situation can arise is when the CPU supports many vector types, and the vector instructions vary slightly based on the used type.

Please note that all three lines with the def statement have the same structure. The variation is only in the suffix of the name and of the mnemonic, and the delta value is added to the opcode. In C, you could put the data into an array and implement a function that returns the data based on an index value. Then, you could create a loop over the data instead of manually repeating statements.

Amazingly, you can do something similar in the TableGen language! Here is how to transform the example:

1. To store the data, you define a class with all required fields. The class is called `InstDesc`, because it describes some properties of an instruction:

    ```
    class InstDesc<string name, string suffix, int delta> {
       string Name = name;
       string Suffix = suffix;
       int Delta = delta;
    }
    ```

2. Now, you can define records for each operand type. Note that it exactly captures the differences observed in the data:

    ```
    def RegOp : InstDesc<"", "", 0>;
    def ImmOp : InstDesc<"I", """, 1>;
    def MemOp : InstDesc"""","""", 2>;
    ```

3. Imagine you have a loop enumerating the numbers 0, 1, and 2, and you want to select one of the previously defined records based on the index. How can you do this? The solution is to create a `getDesc` class that takes the index as a parameter. It has a single field, `ret`, that you can interpret as a return value. To assign the correct value to this field, the `!cond` operator is used:

    ```
    class getDesc<int n> {
       InstDesc ret = !cond(!eq(n, 0) : RegOp,
                            !eq(n, 1) : ImmOp,
                            !eq(n, 2) : MemOp);
    }
    ```

    This operator works similarly to a `switch`/`case` statement in C.

4. Now, you are ready to define the multiclass. The TableGen language has a `loop` statement, and it also allows us to define variables. But remember that there is no dynamic execution! As a consequence, the loop range is statically defined, and you can assign a value to a variable, but you cannot change that value later. However, this is enough to retrieve the data. Please note how the use of the `getDesc` class resembles a function call. But there is no function call! Instead, an anonymous record is created, and the values are taken from that record. Lastly, the past operator (#) performs a string concatenation, similar to the `!strconcat` operator used earlier:

    ```
    multiclass InstWithOps<string mnemonic, int opcode> {
       foreach I = 0-2 in {
          defvar Name = getDesc<I>.ret.Name;
          defvar Suffix = getDesc<I>.ret.Suffix;
          defvar Delta = getDesc<I>.ret.Delta;
          def Name: Inst<mnemonic # Suffix,
    ```

```
                              !add(opcode, Delta)>;
            }
        }
```

5. Now, you use the multiclass as before to define records:

    ```
    defm ADD : InstWithOps<"add", 0xA0>;
    ```

Please run `llvm-tblgen` and examine the records. Besides the various ADD records, you will also see a couple of anonymous records generated by the use of the `getDesc` class.

This technique is used in the instruction definition of several LLVM backends. With the knowledge you have acquired, you should have no problem understanding those files.

The `foreach` statement used the syntax `0-2` to denote the bounds of the range. This is called a **range piece**. An alternative syntax is to use three dots (`0...3`), which is useful if the numbers are negative. Lastly, you are not restricted to numerical ranges; you can also loop over a list of elements, which allows you to use strings or previously defined records. For example, you may like the use of the `foreach` statement, but you think that using the `getDesc` class is too complicated. In this case, looping over the `InstDesc` records is the solution:

```
multiclass InstWithOps<string mnemonic, int opcode> {
  foreach I = [RegOp, ImmOp, MemOp] in {
    defvar Name = I.Name;
    defvar Suffix = I.Suffix;
    defvar Delta = I.Delta;
    def Name: Inst<mnemonic # Suffix, !add(opcode, Delta)>;
  }
}
```

So far, you only defined records in the TableGen language, using the most commonly used statements. In the next section, you'll learn how to generate C++ source code from records defined in the TableGen language.

Generating C++ code from a TableGen file

In the previous section, you defined records in the TableGen language. To make use of those records, you need to write your own TableGen backend that can produce C++ source code or do other things using the records as input.

In *Chapter 3*, *Turning the Source File into an Abstract Syntax Tree*, the implementation of the `Lexer` class uses a database file to define tokens and keywords. Various query functions make use of that database file. Besides that, the database file is used to implement a keyword filter. The keyword filter is a hash map, implemented using the `llvm::StringMap` class. Whenever an identifier is found, the keyword filter is called to find out if the identifier is actually a keyword. If you take a closer look at the implementation using the `ppprofiler` pass from *Chapter 6*, *Advanced IR Generation*, then you will see that this function is called quite often. Therefore, it may be useful to experiment with different implementations to make that functionality as fast as possible.

However, this is not as easy as it seems. For example, you can try to replace the lookup in the hash map with a binary search. This requires that the keywords in the database file are sorted. Currently, this seems to be the case, but during development, a new keyword might be added in the wrong place undetected. The only way to make sure that the keywords are in the right order is to add some code that checks the order at runtime.

You can speed up the standard binary search by changing the memory layout. For example, instead of sorting the keywords, you can use the Eytzinger layout, which enumerates the search tree in breadth-first order. This layout increases the cache locality of the data and therefore speeds up the search. Personally speaking, maintaining the keywords in breadth-first order manually in the database file is not possible.

Another popular approach for searching is the generation of minimal perfect hash functions. If you insert a new key into a dynamic hash table such as `llvm::StringMap`, then that key might be mapped to an already occupied slot. This is called a **key collision**. Key collisions are unavoidable, and many strategies have been developed to mitigate that problem. However, if you know all the keys, then you can construct hash functions without key collisions. Such hash functions are called **perfect**. In case they do not require more slots than keys, then they are called minimal. Perfect hash functions can be generated efficiently – for example, with the `gperf` GNU tool.

In summary, there is some incentive to be able to generate a lookup function from keywords. So, let's move the database file to TableGen!

Defining data in the TableGen language

The `TokenKinds.def` database file defines three different macros. The `TOK` macro is used for tokens that do not have a fixed spelling – for example, for integer literals. The `PUNCTUATOR` macro is used for all kinds of punctuation marks and includes a preferred spelling. Lastly, the `KEYWORD` macro defines a keyword that is made up of a literal and a flag, which is used to indicate at which language level this literal is a keyword. For example, the `thread_local` keyword was added to C++11.

One way to express this in the TableGen language is to create a Token class that holds all the data. You can then add subclasses of that class to make the usage more comfortable. You also need a Flag class for flags defined together with a keyword. And last, you need a class to define a keyword filter. These classes define the basic data structure and can be potentially reused in other projects. Therefore, you create a Keyword.td file for it. Here are the steps:

1. A flag is modeled as a name and an associated value. This makes it easy to generate an enumeration from this data:

    ```
    class Flag<string name, int val> {
        string Name = name;
        int Val = val;
    }
    ```

2. The Token class is used as the base class. It just carries a name. Please note that this class has no parameters:

    ```
    class Token {
        string Name;
    }
    ```

3. The Tok class has the same function as the corresponding TOK macro from the database file. it represents a token without fixed spellings. It derives from the base class, Token, and just adds initialization for the name:

    ```
    class Tok<string name> : Token {
        let Name = name;
    }
    ```

4. In the same way, the Punctuator class resembles the PUNCTUATOR macro. It adds a field for the spelling of the token:

    ```
    class Punctuator<string name, string spelling> : Token {
        let Name = name;
        string Spelling = spelling;
    }
    ```

5. And last, the Keyword class needs a list of flags:

    ```
    class Keyword<string name, list<Flag> flags> : Token {
        let Name = name;
        list<Flag> Flags = flags;
    }
    ```

6. With these definitions in place, you can now define a class for the keyword filter, called `TokenFilter`. It takes a list of tokens as a parameter:

```
class TokenFilter<list<Token> tokens> {
    string FunctionName;
    list<Token> Tokens = tokens;
}
```

With these class definitions, you are certainly able to capture all the data from the `TokenKinds`. `def` database file. The TinyLang language does not utilize the flags, since there is only this version of the language. Real-world languages such as C and C++ have undergone a couple of revisions, and they usually require flags. Therefore, we use keywords from C and C++ as an example. Let's create a `KeywordC.td` file, as follows:

1. First, you include the class definitions created earlier:

```
Include "Keyword.td"
```

2. Next, you define flags. The value is the binary value of the flag. Note how the `!or` operator is used to create a value for the `KEYALL` flag:

```
def KEYC99  : Flag<"KEYC99", 0x1>;
def KEYCXX  : Flag<"KEYCXX", 0x2>;
def KEYCXX11: Flag<"KEYCXX11", 0x4>;
def KEYGNU  : Flag<"KEYGNU", 0x8>;
def KEYALL  : Flag<"KEYALL",
                   !or(KEYC99.Val, KEYCXX.Val,
                       KEYCXX11.Val , KEYGNU.Val)>;
```

3. There are tokens without a fixed spelling – for example, a comment:

```
def : Tok<"comment">;
```

4. Operators are defined using the `Punctuator` class, as in this example:

```
def : Punctuator<"plus", "+">;
def : Punctuator<"minus", "-">;
```

5. Keywords need to use different flags:

```
def kw_auto: Keyword<"auto", [KEYALL]>;
def kw_inline: Keyword<"inline", [KEYC99,KEYCXX,KEYGNU]>;
def kw_restrict: Keyword<"restrict", [KEYC99]>;
```

6.   And last, here's the definition of the keyword filter:

```
def : TokenFilter<[kw_auto, kw_inline, kw_restrict]>;
```

Of course, this file does not include all tokens from C and C++. However, it demonstrates all possible usages of the defined TableGen classes.

Based on these TableGen files, you'll implement a TableGen backend in the next section.

Implementing a TableGen backend

Since parsing and creation of records are done through an LLVM library, you only need to care about the backend implementation, which consists mostly of generating C++ source code fragments based on the information in the records. First, you need to be clear about what source code to generate before you can put it into the backend.

Sketching the source code to be generated

The output of the TableGen tool is a single file containing C++ fragments. The fragments are guarded by macros. The goal is to replace the `TokenKinds.def` database file. Based on the information in the TableGen file, you can generate the following:

1.   The enumeration members used to define flags. The developer is free to name the type; however, it should be based on the `unsigned` type. If the generated file is named `TokenKinds.inc`, then the intended use is this:

```
enum Flags : unsigned {
#define GET_TOKEN_FLAGS
#include "TokenKinds.inc"
}
```

2.   The `TokenKind` enumeration, and the prototypes and definitions of the `getTokenName()`, `getPunctuatorSpelling()`, and `getKeywordSpelling()` functions. This code replaces the `TokenKinds.def` database file, most of the `TokenKinds.h` include file and the `TokenKinds.cpp.` source file.

3.   A new `lookupKeyword()` function that can be used instead of the current implementation using the `llvm::StringMap.` type. This is the function you want to optimize.

Knowing what you want to generate, you can now turn to implementing the backend.

Creating a new TableGen tool

A simple structure for your new tool is to have a driver that evaluates the command-line options and calls the generation functions and the actual generator functions in a different file. Let's call the driver file `TableGen.cpp` and the file containing the generator `TokenEmitter.cpp`. You also need a `TableGenBackends.h` header file. Let's begin the implementation with the generation of the C++ code in the `TokenEmitter.cpp` file:

1. As usual, the file begins with including the required headers. The most important one is `llvm/TableGen/Record.h`, which defines a `Record` class, used to hold records generated by parsing the `.td` file:

    ```
    #include "TableGenBackends.h"
    #include "llvm/Support/Format.h"
    #include "llvm/TableGen/Record.h"
    #include "llvm/TableGen/TableGenBackend.h"
    #include <algorithm>
    ```

2. To simplify coding, the `llvm` namespace is imported:

    ```
    using namespace llvm;
    ```

3. The `TokenAndKeywordFilterEmitter` class is responsible for generating the C++ source code. The `emitFlagsFragment()`, `emitTokenKind()`, and `emitKeywordFilter()` methods emit the source code, as described in the previous section, *Sketching the source code to be generated*. The only public method, `run()`, calls all the code-emitting methods. The records are held in an instance of `RecordKeeper`, which is passed as a parameter to the constructor. The class is inside an anonymous namespace:

    ```
    namespace {
    class TokenAndKeywordFilterEmitter {
      RecordKeeper &Records;

    public:
      explicit TokenAndKeywordFilterEmitter(RecordKeeper &R)
          : Records(R) {}

      void run(raw_ostream &OS);

    private:
      void emitFlagsFragment(raw_ostream &OS);
      void emitTokenKind(raw_ostream &OS);
      void emitKeywordFilter(raw_ostream &OS);
    };
    } // End anonymous namespace
    ```

4. The `run()` method calls all the emitting methods. It also times the length of each phase. You specify the `--time-phases` option, and then the timing is shown after all code is generated:

```
void TokenAndKeywordFilterEmitter::run(raw_ostream &OS) {
  // Emit Flag fragments.
  Records.startTimer("Emit flags");
  emitFlagsFragment(OS);

  // Emit token kind enum and functions.
  Records.startTimer("Emit token kind");
  emitTokenKind(OS);

  // Emit keyword filter code.
  Records.startTimer("Emit keyword filter");
  emitKeywordFilter(OS);
  Records.stopTimer();
}
```

5. The `emitFlagsFragment()` method shows the typical structure of a function emitting C++ source code. The generated code is guarded by the GET_TOKEN_FLAGS macro. To emit the C++ source fragment, you loop over all records that are derived from the `Flag` class in the TableGen file. Having such a record, it is easy to query the record for the name and the value. Please note that the names `Flag`, `Name`, and `Val` must be written exactly as in the TableGen file. If you rename `Val` to `Value` in the TableGen file, then you also need to change the string in this function. All the generated source code is written to the provided stream, OS:

```
void TokenAndKeywordFilterEmitter::emitFlagsFragment(
    raw_ostream &OS) {
  OS << "#ifdef GET_TOKEN_FLAGS\n";
  OS << "#undef GET_TOKEN_FLAGS\n";
  for (Record *CC :
       Records.getAllDerivedDefinitions("Flag")) {
    StringRef Name = CC->getValueAsString("Name");
    int64_t Val = CC->getValueAsInt("Val");
    OS << Name << " = " << format_hex(Val, 2) << ",\n";
  }
  OS << "#endif\n";
}
```

6. The `emitTokenKind()` method emits a declaration and definition of token classification functions. Let's have a look at emitting the declarations first. The overall structure is the same as the previous method – only more C++ source code is emitted. The generated source fragment is guarded by the GET_TOKEN_KIND_DECLARATION macro. Please note that this method tries to generate nicely formatted C++ code, using new lines and indentation as a human

developer would do. In case the emitted source code is not correct, and you need to examine it to find the error, this will be tremendously helpful. It is also easy to make such errors: after all, you are writing a C++ function that emits C++ source code.

First, the `TokenKind` enumeration is emitted. The name for a keyword should be prefixed with a `kw_` string. The loop goes over all records of the `Token` class, and you can query the records if they are also a subclass of the `Keyword` class, which enables you to emit the prefix:

```
OS << "#ifdef GET_TOKEN_KIND_DECLARATION\n"
   << "#undef GET_TOKEN_KIND_DECLARATION\n"
   << "namespace tok {\n"
   << "  enum TokenKind : unsigned short {\n";
for (Record *CC :
       Records.getAllDerivedDefinitions("Token")) {
  StringRef Name = CC->getValueAsString("Name");
  OS << "    ";
  if (CC->isSubClassOf("Keyword"))
    OS << "kw_";
  OS << Name << ",\n";
}
OS << "    NUM_TOKENS\n"
   << "  };\n";
```

7. Next, the function declarations are emitted. This is only a constant string, so nothing exciting happens. This finishes emitting the declarations:

```
OS << "  const char *getTokenName(TokenKind Kind) "
      "LLVM_READNONE;\n"
   << "  const char *getPunctuatorSpelling(TokenKind "
      "Kind) LLVM_READNONE;\n"
   << "  const char *getKeywordSpelling(TokenKind "
      "Kind) "
      "LLVM_READNONE;\n"
   << "}\n"
   << "#endif\n";
```

8. Now, let's turn to emitting the definitions. Again, this generated code is guarded by a macro called `GET_TOKEN_KIND_DEFINITION`. First, the token names are emitted into a `TokNames` array, and the `getTokenName()` function uses that array to retrieve the name. Please note that the quote symbol must be escaped as `\ "` when used inside a string:

```
OS << "#ifdef GET_TOKEN_KIND_DEFINITION\n";
OS << "#undef GET_TOKEN_KIND_DEFINITION\n";
OS << "static const char * const TokNames[] = {\n";
for (Record *CC :
       Records.getAllDerivedDefinitions("Token")) {
```

```
      OS << "  \"" << CC->getValueAsString("Name")
         << "\",\n";
   }
   OS << "};\n\n";
   OS << "const char *tok::getTokenName(TokenKind Kind) "
         "{\n"
      << "  if (Kind <= tok::NUM_TOKENS)\n"
      << "    return TokNames[Kind];\n"
      << "  llvm_unreachable(\"unknown TokenKind\");\n"
      << "  return nullptr;\n"
      << "};\n\n";
```

9. Next, the `getPunctuatorSpelling()` function is emitted. The only notable difference to the other parts is that the loop goes over all records derived from the `Punctuator` class. Also, a `switch` statement is generated instead of an array:

```
   OS << "const char "
         "*tok::getPunctuatorSpelling(TokenKind "
         "Kind) {\n"
      << "  switch (Kind) {\n";
   for (Record *CC :
         Records.getAllDerivedDefinitions("Punctuator")) {
     OS << "    " << CC->getValueAsString("Name")
         << ": return \""
         << CC->getValueAsString("Spelling") << "\";\n";
   }
   OS << "    default: break;\n"
      << "  }\n"
      << "  return nullptr;\n"
      << "};\n\n";
```

10. And finally, the `getKeywordSpelling()` function is emitted. The coding is similar to emitting `getPunctuatorSpelling()`. This time, the loop goes over all records of the `Keyword` class, and the name is again prefixed with kw_:

```
   OS << "const char *tok::getKeywordSpelling(TokenKind "
         "Kind) {\n"
      << "  switch (Kind) {\n";
   for (Record *CC :
         Records.getAllDerivedDefinitions("Keyword")) {
     OS << "    kw_" << CC->getValueAsString("Name")
         << ": return \"" << CC->getValueAsString("Name")
         << "\";\n";
   }
```

```
    OS << "    default: break;\n"
       << "  }\n"
       << "  return nullptr;\n"
       << "};\n\n";
    OS << "#endif\n";
}
```

11. The emitKeywordFilter() method is more complex than the previous methods since emitting the filter requires collecting some data from the records. The generated source code uses the std::lower_bound() function, thus implementing a binary search.

    Now, let's make a shortcut. There can be several records of the TokenFilter class defined in the TableGen file. For demonstration purposes, just emit at most one token filter method:

    ```
    std::vector<Record *> AllTokenFilter =
        Records.getAllDerivedDefinitionsIfDefined(
            "TokenFilter");
    if (AllTokenFilter.empty())
      return;
    ```

12. The keywords used for the filter are in the list named Tokens. To get access to that list, you first need to look up the Tokens field in the record. This returns a pointer to an instance of the RecordVal class, from which you can retrieve the Initializer instance via the calling method, getValue(). The Tokens field is defined as a list, so you cast the initializer instance to ListInit. If this fails, then exit the function:

    ```
    ListInit *TokenFilter = dyn_cast_or_null<ListInit>(
        AllTokenFilter[0]
            ->getValue("Tokens")
            ->getValue());
    if (!TokenFilter)
      return;
    ```

13. Now, you are ready to construct a filter table. For each keyword stored in the TokenFilter, list, you need the name and the value of the Flag field. That field is again defined as a list, so you need to loop over those elements to calculate the final value. The resulting name/flag value pair is stored in a Table vector:

    ```
    using KeyFlag = std::pair<StringRef, uint64_t>;
    std::vector<KeyFlag> Table;
    for (size_t I = 0, E = TokenFilter->size(); I < E;
         ++I) {
     Record *CC = TokenFilter->getElementAsRecord(I);
     StringRef Name = CC->getValueAsString("Name");
     uint64_t Val = 0;
     ListInit *Flags = nullptr;
    ```

```
      if (RecordVal *F = CC->getValue("Flags"))
        Flags = dyn_cast_or_null<ListInit>(F->getValue());
      if (Flags) {
        for (size_t I = 0, E = Flags->size(); I < E; ++I) {
          Val |=
              Flags->getElementAsRecord(I)->getValueAsInt(
                  "Val");
        }
      }
      Table.emplace_back(Name, Val);
    }
```

14. To be able to perform a binary search, the table needs to be sorted. The comparison function is provided by a lambda function:

```
    llvm::sort(Table.begin(), Table.end(),
               [](const KeyFlag A, const KeyFlag B) {
                 return A.first < B.first;
               });
```

15. Now, you can emit the C++ source code. First, you emit the sorted table containing the name of the keyword and the associated flag value:

```
    OS << "#ifdef GET_KEYWORD_FILTER\n"
       << "#undef GET_KEYWORD_FILTER\n";
    OS << "bool lookupKeyword(llvm::StringRef Keyword, "
          "unsigned &Value) {\n";
    OS << "  struct Entry {\n"
       << "    unsigned Value;\n"
       << "    llvm::StringRef Keyword;\n"
       << "  };\n"
       << "static const Entry Table[" << Table.size()
       << "] = {\n";
    for (const auto &[Keyword, Value] : Table) {
     OS << "    { " << Value << ", llvm::StringRef(\""
        << Keyword << "\", " << Keyword.size()
        << ") },\n";
    }
    OS << "  };\n\n";
```

16. Next, you look up the keyword in the sorted table, using the `std::lower_bound()` standard C++ function. If the keyword is in the table, then the `Value` parameter receives the value of the flags associated with the keyword, and the function returns `true`. Otherwise, the function simply returns `false`:

```
    OS << "  const Entry *E = "
         "std::lower_bound(&Table[0], "
         "&Table["
       << Table.size()
       << "], Keyword, [](const Entry &A, const "
         "StringRef "
         "&B) {\n";
    OS << "    return A.Keyword < B;\n";
    OS << "  });\n";
    OS << "  if (E != &Table[" << Table.size()
       << "]) {\n";
    OS << "    Value = E->Value;\n";
    OS << "    return true;\n";
    OS << "  }\n";
    OS << "  return false;\n";
    OS << "}\n";
    OS << "#endif\n";
}
```

17. The only missing part now is a way to call this implementation, for which you define a global function, `EmitTokensAndKeywordFilter()`. The `emitSourceFileHeader()` function declared in the `llvm/TableGen/TableGenBackend.h` header emits a comment at the top of the generated file:

```
void EmitTokensAndKeywordFilter(RecordKeeper &RK,
                                raw_ostream &OS) {
  emitSourceFileHeader("Token Kind and Keyword Filter "
                       "Implementation Fragment",
                       OS);
  TokenAndKeywordFilterEmitter(RK).run(OS);
}
```

With that, you finished the implementation of the source emitter in the `TokenEmitter.cpp` file. Overall, the coding is not too complicated.

The TableGenBackends.h header file only contains the declaration of the EmitTokensAndKeywordFilter() function. To avoid including other files, you use forward declarations for the raw_ostream and RecordKeeper classes:

```
#ifndef TABLEGENBACKENDS_H
#define TABLEGENBACKENDS_H

namespace llvm {
class raw_ostream;
class RecordKeeper;
} // namespace llvm

void EmitTokensAndKeywordFilter(llvm::RecordKeeper &RK,
                                llvm::raw_ostream &OS);

#endif
```

The missing part is the implementation of the driver. Its task is to parse the TableGen file and emit the records according to the command-line options. The implementation is in the TableGen.cpp file:

1. As usual, the implementation begins with including the required headers. The most important one is llvm/TableGen/Main.h because this header declares the frontend of TableGen:

   ```
   #include "TableGenBackends.h"
   #include "llvm/Support/CommandLine.h"
   #include "llvm/Support/PrettyStackTrace.h"
   #include "llvm/Support/Signals.h"
   #include "llvm/TableGen/Main.h"
   #include "llvm/TableGen/Record.h"
   ```

2. To simplify coding, the llvm namespace is imported:

   ```
   using namespace llvm;
   ```

3. The user can choose one action. The ActionType enumeration contains all possible actions:

   ```
   enum ActionType {
     PrintRecords,
     DumpJSON,
     GenTokens,
   };
   ```

4. A single command-line option object called `Action` is used. The user needs to specify the `--gen-tokens` option to emit the token filter you implemented. The other two options, `--print-records` and `--dump-json`, are standard options to dump read records. Note that the object is in an anonymous namespace:

```
namespace {
cl::opt<ActionType> Action(
    cl::desc("Action to perform:"),
    cl::values(
        clEnumValN(
            PrintRecords, "print-records",
            "Print all records to stdout (default)"),
        clEnumValN(DumpJSON, "dump-json",
                   "Dump all records as "
                   "machine-readable JSON"),
        clEnumValN(GenTokens, "gen-tokens",
                   "Generate token kinds and keyword "
                   "filter")));
```

5. The `Main()` function performs the requested action based on the value of `Action`. Most importantly, your `EmitTokensAndKeywordFilter()` function is called if `--gen-tokens` was specified on the command line. After the end of the function, the anonymous namespace is closed:

```
bool Main(raw_ostream &OS, RecordKeeper &Records) {
  switch (Action) {
  case PrintRecords:
    OS << Records; // No argument, dump all contents
    break;
  case DumpJSON:
    EmitJSON(Records, OS);
    break;
  case GenTokens:
    EmitTokensAndKeywordFilter(Records, OS);
    break;
  }

  return false;
}
} // namespace
```

6. And lastly, you define a `main()` function. After setting up the stack trace handler and parsing the command-line options, the `TableGenMain()` function is called to parse the TableGen file and create records. That function also calls your `Main()` function if there are no errors:

```
int main(int argc, char **argv) {
  sys::PrintStackTraceOnErrorSignal(argv[0]);
  PrettyStackTraceProgram X(argc, argv);
  cl::ParseCommandLineOptions(argc, argv);

  llvm_shutdown_obj Y;

  return TableGenMain(argv[0], &Main);
}
```

Your own TableGen tool is now implemented. After compiling, you can run it with the `KeywordC.td` sample input file as follows:

```
$ tinylang-tblgen --gen-tokens -o TokenFilter.inc KeywordC.td
```

The generated C++ source code is written to the `TokenFilter.inc` file.

Performance of the token filter

Using a plain binary search for the keyword filter does not give a better performance than the implementation based on the `llvm::StringMap` type. To beat the performance of the current implementation, you need to generate a perfect hash function.

The classic algorithm from Czech, Havas, and Majewski can be easily implemented, and it gives you a very good performance. It is described in *An optimal algorithm for generating minimal perfect hash functions, Information Processing Letters, Volume 43, Issue 5, 1992*. See `https://www.sciencedirect.com/science/article/abs/pii/002001909290220P`.

A state-of-the-art algorithm is PTHash from Pibiri and Trani, described in *PTHash: Revisiting FCH Minimal Perfect Hashing, SIGIR '21*. See `https://arxiv.org/pdf/2104.10402.pdf`.

Both algorithms are good candidates for generating a token filter that is actually faster than `llvm::StringMap`.

Drawbacks of TableGen

Here are a few drawbacks of TableGen:

- The TableGen language is built on a simple concept. As a consequence, it does not have the same computing capabilities as other DSLs. Obviously, some programmers would like to replace TableGen with a different, more powerful language, and this topic comes up from time to time in the LLVM discussion forum.

- With the possibility of implementing your own backends, the TableGen language is very flexible. However, it also means that the semantics of a given definition are hidden inside the backend. Thus, you can create TableGen files that are basically not understandable by other developers.

- And last, the backend implementation can be very complex if you try to solve a non-trivial task. It is reasonable to expect that this effort would be lower if the TableGen language were more powerful.

Even if not all developers are happy with the capabilities of TableGen, the tool is used widely in LLVM, and for a developer, it is important to understand it.

Summary

In this chapter, you first learned the main idea behind TableGen. Then, you defined your first classes and records in the TableGen language, and you acquired knowledge of the syntax of TableGen. Finally, you developed a TableGen backend emitting fragments of C++ source code, based on the TableGen classes you defined.

In the next chapter, we examine another unique feature of LLVM: generating and executing code in one step, also known as **Just-In-Time (JIT)** compilation.

9

JIT Compilation

The LLVM core libraries come with the **ExecutionEngine** component that allows the compilation and execution of **intermediate representation** (**IR**) code in memory. Using this component, we can build **just-in-time** (**JIT**) compilers, which allows for direct execution of IR code. A JIT compiler works more like an interpreter because no object code needs to be stored on secondary storage.

In this chapter, you will learn about applications for JIT compilers, and how the LLVM JIT compiler works in principle. You will explore the LLVM dynamic compiler and interpreter and learn how to implement JIT compiler tools on your own. Furthermore, you will also learn how to use a JIT compiler as part of a static compiler, and the associated challenges.

This chapter will cover the following topics:

- Getting an overview of LLVM's JIT implementation and use cases
- Using JIT compilation for direct execution
- Implementing your own JIT compiler from existing classes
- Implementing your own JIT compiler from scratch

By the end of the chapter, you will understand and know how to develop a JIT compiler, either using a preconfigured class or a customized version fitting for your needs.

Technical requirements

You can find the code used in this chapter at `https://github.com/PacktPublishing/Learn-LLVM-17/tree/main/Chapter09` .

LLVM's overall JIT implementation and use cases

So far, we have only looked at **ahead-of-time** (**AOT**) compilers. These compilers compile the whole application. The application can only run after the compilation is finished. If the compilation is performed at the runtime of the application, then the compiler is a JIT compiler. A JIT compiler has interesting use cases:

- **Implementation of a virtual machine**: A programming language can be translated to byte code with an AOT compiler. At runtime, a JIT compiler is used to compile the byte code to machine code. The advantage of this approach is that the byte code is hardware-independent, and thanks to the JIT compiler, there is no performance penalty compared to an AOT compiler. Java and C# use this model today, but this is not a new idea: the USCD Pascal compiler from 1977 already used a similar approach.

- **Expression evaluation**: A spreadsheet application can compile often-executed expressions with a JIT compiler. For example, this can speed up financial simulations. The `lldb` LLVM debugger uses this approach to evaluate source expressions at debug time.

- **Database queries**: A database creates an execution plan from a database query. The execution plan describes operations on tables and columns, which leads to a query answer when executed. A JIT compiler can be used to translate the execution plan into machine code, which speeds up the execution of the query.

The static compilation model of LLVM is not as far away from the JIT model as one may think. The `llc` LLVM static compiler compiles LLVM IR into machine code and saves the result as an object file on disk. If the object file is not stored on disk but in memory, would the code be executable? Not directly, as references to global functions and global data use relocations instead of absolute addresses. Conceptually, a **relocation** describes how to calculate the address – for example, as an offset to a known address. If we resolve relocations into addresses, as the linker and the dynamic loader do, then we can execute the object code. Running the static compiler to compile IR code into an object file in memory, performing a link step on the in-memory object file, and running the code gives us a JIT compiler. The JIT implementation in the LLVM core libraries is based on this idea.

During the development history of LLVM, there were several JIT implementations, with different feature sets. The latest JIT API is the **On-Request Compilation** (**ORC**) engine. In case you were curious about the acronym, it was the lead developer's intention to invent yet another acronym based on Tolkien's universe, after **Executable and Linking Format** (**ELF**) and **Debugging Standard** (**DWARF**) were already present.

The ORC engine builds on and extends the idea of using the static compiler and a dynamic linker on the in-memory object file. The implementation uses a layered approach. The two basic levels are the compile layer and the link layer. On top of this sits a layer providing support for lazy compilation. A transformation layer can be stacked on top or below the lazy compilation layer, allowing the developer to add arbitrary transformations or simply to be notified of certain events. Moreover, this layered approach has the advantage that the JIT engine is customizable for diverse requirements. For example, a high-performance virtual machine may choose to compile everything upfront and make no use of the lazy compilation layer. On the other hand, other virtual machines will emphasize startup time and responsiveness to the user and will achieve this with the help of the lazy compilation layer.

The older MCJIT engine is still available, and its API is derived from an even older, already-removed JIT engine. Over time, this API gradually became bloated, and it lacks the flexibility of the ORC API. The goal is to remove this implementation, as the ORC engine now provides all the functionality of the MCJIT engine, and new developments should use the ORC API.

In the next section, we look at lli, the LLVM interpreter, and the dynamic compiler, before we dive into implementing a JIT compiler.

Using JIT compilation for direct execution

Running LLVM IR directly is the first idea that comes to mind when thinking about a JIT compiler. This is what the lli tool, the LLVM interpreter, and the dynamic compiler do. We will explore the lli tool in the next section.

Exploring the lli tool

Let's try the lli tool with a very simple example. The following LLVM IR can be stored as a file called hello.ll, which is the equivalent of a C hello world application. This file declares a prototype for the printf() function from the C library. The hellostr constant contains the message to be printed. Inside the main() function, a call to the printf() function is generated, and this function contains a hellostr message that will be printed. The application always returns 0.

The complete source code is as follows:

```
declare i32 @printf(ptr, ...)

@hellostr = private unnamed_addr constant [13 x i8] c"Hello
world\0A\00"

define dso_local i32 @main(i32 %argc, ptr %argv) {
  %res = call i32 (ptr, ...) @printf(ptr @hellostr)
  ret i32 0
}
```

This LLVM IR file is generic enough that it is valid for all platforms. We can directly execute the IR using the lli tool with the following command:

```
$ lli hello.ll
Hello world
```

The interesting point here is how the `printf()` function is found. The IR code is compiled to machine code, and a lookup for the `printf` symbol is triggered. This symbol is not found in the IR, so the current process is searched for it. The `lli` tool dynamically links against the C library, and the symbol is found there.

Of course, the `lli` tool does not link against the libraries you created. To enable the use of such functions, the `lli` tool supports the loading of shared libraries and objects. The following C source just prints a friendly message:

```
#include <stdio.h>

void greetings() {
    puts("Hi!");
}
```

Stored in `greetings.c`, we use this to explore loading objects with `lli`. The following command will compile this source into a shared library. The `-fPIC` option instructs `clang` to generate position-independent code, which is required for shared libraries. Moreover, the compiler creates a `greetings.so` shared library with   `-shared`:

```
$ clang greetings.c -fPIC -shared -o greetings.so
```

We also compile the file into the `greetings.o` object file:

```
$ clang greetings.c -c -o greetings.o
```

We now have two files, the `greetings.so` shared library and the `greetings.o` object file, which we will load into the `lli` tool.

We also need an LLVM IR file that calls the `greetings()` function. For this, create a `main.ll` file that contains a single call to the function:

```
declare void @greetings(...)

define dso_local i32 @main(i32 %argc, i8** %argv) {
    call void (...) @greetings()
    ret i32 0
}
```

Notice that on executing, the previous IR crashes, as `lli` cannot locate the greetings symbol:

```
$ lli main.ll
JIT session error: Symbols not found: [ _greetings ]
lli: Failed to materialize symbols: { (main, { _main }) }
```

The greetings() function is defined in an external file, and to fix the crash, we have to tell the lli tool which additional file needs to be loaded. In order to use the shared library, you must use the -load option, which takes the path to the shared library as an argument:

```
$ lli -load ./greetings.so main.ll
Hi!
```

It is important to specify the path to the shared library if the directory containing the shared library is not in the search path for the dynamic loader. If omitted, then the library will not be found.

Alternatively, we can instruct lli to load the object file with -extra-object:

```
$ lli -extra-object greetings.o main.ll
Hi!
```

Other supported options are -extra-archive, which loads an archive, and   -extra-module, which loads another bitcode file. Both options require the path to the file as an argument.

You now know how you can use the lli tool to directly execute LLVM IR. In the next section, we will implement our own JIT tool.

Implementing our own JIT compiler with LLJIT

The lli tool is nothing more than a thin wrapper around LLVM APIs. In the first section, we learned that the ORC engine uses a layered approach. The ExecutionSession class represents a running JIT program. Besides other items, this class holds information such as used JITDylib instances. A JITDylib instance is a symbol table that maps symbol names to addresses. For example, these can be symbols defined in an LLVM IR file or the symbols of a loaded shared library.

For executing LLVM IR, we do not need to create a JIT stack on our own, as the LLJIT class provides this functionality. You can also make use of this class when migrating from the older MCJIT implementation, as this class essentially provides the same functionality.

To illustrate the functions of the LLJIT utility, we will be creating an interactive calculator application while incorporating JIT functionality. The main source code of our JIT calculator will be extended from the calc example from *Chapter 2, The Structure of a Compiler*.

The primary idea behind our interactive JIT calculator will be as follows:

1. Allow the user to input a function definition, such as def f(x) = x*2.

2. The function inputted by the user is then compiled by the LLJIT utility into a function – in this case, f.

3. Allow the user to call the function they have defined with a numerical value: f(3).

4. Evaluate the function with the provided argument, and print the result to the console: 6.

Before we discuss incorporating JIT functionality into the calculator source code, there are a few main differences to point out with respect to the original calculator example:

- Firstly, we previously only input and parsed functions beginning with the `with` keyword, rather than the `def` keyword described previously. For this chapter, we instead only accept function definitions beginning with `def`, and this is represented as a particular node in our **abstract syntax tree (AST)** class, known as `DefDecl`. The `DefDecl` class is aware of the arguments and their names it is defined with, and the function name is also stored within this class.

- Secondly, we also need our AST to be aware of function calls, to represent the functions that the `LLJIT` utility has consumed or JIT'ted. Whenever a user inputs the name of a function, followed by arguments enclosed in parentheses, the AST recognizes these as `FuncCallFromDef` nodes. This class essentially is aware of the same information as the `DefDecl` class.

Due to the addition of these two AST classes, it is obvious to expect that the semantic analysis, parser, and code generation classes will be adapted accordingly to handle the changes in our AST. One additional thing to note is the addition of a new data structure, called `JITtedFunctions`, which all these classes are aware of. This data structure is a map with the defined function names as keys, and the number of arguments a function is defined with is stored as values within the map. We will see later how this data structure will be utilized in our JIT calculator.

For more details on the changes we have made to the `calc` example, the full source containing the changes from `calc` and this section's JIT implementation can be found within the `lljit` source directory.

Integrating the LLJIT engine into the calculator

Firstly, let's discuss how to set up the JIT engine in our interactive calculator. All of the implementation pertaining to the JIT engine exists within `Calc.cpp`, and this file has one `main()` loop for the execution of the program:

1. We must include several header files, aside from the headers including our code generation, semantic analyzer, and parser implementation. The `LLJIT.h` header defines the `LLJIT` class and the core classes of the ORC API. Next, the `InitLLVM.h` header is needed for the basic initialization of the tool, and the `TargetSelect.h` header is needed for the initialization of the native target. Finally, we also include the `<iostream>` C++ header to allow for user input into our calculator application:

```
#include "CodeGen.h"
#include "Parser.h"
#include "Sema.h"
#include "llvm/ExecutionEngine/Orc/LLJIT.h"
#include "llvm/Support/InitLLVM.h"
#include "llvm/Support/TargetSelect.h"
#include <iostream>
```

2. Next, we add the `llvm` and `llvm::orc` namespaces to the current scope:

    ```
    using namespace llvm;
    using namespace llvm::orc;
    ```

3. Many of the calls from our `LLJIT` instance that we will be creating return an error type,
    `Error`. The `ExitOnError` class allows us to discard `Error` values that are returned by
    the calls from the `LLJIT` instance while logging to `stderr` and exiting the application. We
    declare a global `ExitOnError` variable as follows:

    ```
    ExitOnError ExitOnErr;
    ```

4. Then, we add the `main()` function, which initializes the tool and the native target:

    ```
    int main(int argc, const char **argv{
      InitLLVM X(argc, argv);

      InitializeNativeTarget();
      InitializeNativeTargetAsmPrinter();
      InitializeNativeTargetAsmParser();
    ```

5. We use the `LLJITBuilder` class to create an `LLJIT` instance, wrapped in the previously
    declared `ExitOnErr` variable in case an error occurs. A possible source of error would be
    that the platform does not yet support JIT compilation:

    ```
    auto JIT = ExitOnErr(LLJITBuilder().create());
    ```

6. Next, we declare our `JITtedFunctions` map that keeps track of the function definitions,
    as we have previously described:

    ```
    StringMap<size_t> JITtedFunctions;
    ```

7. To facilitate an environment that waits for user input, we add a `while()` loop and allow the
    user to type in an expression, saving the line that the user typed within a string called `calcExp`:

    ```
    while (true) {
      outs() << "JIT calc > ";
      std::string calcExp;
      std::getline(std::cin, calcExp);
    ```

8. Afterward, the LLVM context class is initialized, along with a new LLVM module. The module's
    data layout is also set accordingly, and we also declare a code generator, which will be used to
    generate IR for the function that the user has defined on the command line:

    ```
    std::unique_ptr<LLVMContext> Ctx = std::make_
    unique<LLVMContext>();
    std::unique_ptr<Module> M = std::make_unique<Module>("JIT
    calc.expr", *Ctx);
    ```

```
M->setDataLayout(JIT->getDataLayout());
CodeGen CodeGenerator;
```

9. We must interpret the line that was entered by the user to determine if the user is defining a new function or calling a previous function that they have defined with an argument. A `Lexer` class is defined while taking in the line of input that the user has given. We will see that there are two main cases that the lexer cares about:

```
Lexer Lex(calcExp);
Token::TokenKind CalcTok = Lex.peek();
```

10. The lexer can check the first token of the user input. If the user is defining a new function (represented by the `def` keyword, or the `Token::KW_def` token), then we parse it and check its semantics. If the parser or the semantic analyzer detects any issues with the user-defined function, errors will be emitted accordingly, and the calculator program will halt. If no errors are detected from either the parser or the semantic analyzer, this means we have a valid AST data structure, `DefDecl`:

```
if (CalcTok == Token::KW_def) {
    Parser Parser(Lex);
    AST *Tree = Parser.parse();
    if (!Tree || Parser.hasError()) {
      llvm::errs() << "Syntax errors occured\n";
      return 1;
    }
    Sema Semantic;
    if (Semantic.semantic(Tree, JITtedFunctions)) {
      llvm::errs() << "Semantic errors occured\n";
      return 1;
    }
}
```

11. We then can pass our newly constructed AST into our code generator to compile the IR for the function that the user has defined. The specifics of IR generation will be discussed afterward, but this function that compiles to the IR needs to be aware of the module and our `JITtedFunctions` map. After generating the IR, we can add this information to our `LLJIT` instance by calling `addIRModule()` and wrapping our module and context in a `ThreadSafeModule` class, to prevent these from being accessed by other concurrent threads:

```
CodeGenerator.compileToIR(Tree, M.get(), JITtedFunctions);
ExitOnErr(
      JIT->addIRModule(ThreadSafeModule(std::move(M),
      std::move(Ctx))));
```

12. Instead, if the user is calling a function with parameters, which is represented by the Token::ident token, we also need to parse and semantically check if the user input is valid prior to converting the input into a valid AST. The parsing and checking here are slightly different compared to before, as it can include checks such as ensuring the number of parameters that the user has supplied to the function call matches the number of parameters that the function was originally defined with:

```
} else if (CalcTok == Token::ident) {
    outs() << "Attempting to evaluate expression:\n";
    Parser Parser(Lex);
    AST *Tree = Parser.parse();
    if (!Tree || Parser.hasError()) {
      llvm::errs() << "Syntax errors occured\n";
      return 1;
    }
    Sema Semantic;
    if (Semantic.semantic(Tree, JITtedFunctions)) {
      llvm::errs() << "Semantic errors occured\n";
      return 1;
    }
```

13. Once a valid AST is constructed for a function call, FuncCallFromDef, we get the name of the function from the AST, and then the code generator prepares to generate the call to the function that was previously added to the LLJIT instance. What occurs under the cover is that the user-defined function is regenerated as an LLVM call within a separate function that will be created that does the actual evaluation of the original function. This step requires the AST, the module, the function call name, and our map of function definitions:

```
llvm::StringRef FuncCallName = Tree->getFnName();
CodeGenerator.prepareCalculationCallFunc(Tree, M.get(),
FuncCallName, JITtedFunctions);
```

14. After the code generator has completed its work to regenerate the original function and to create a separate evaluation function, we must add this information to the LLJIT instance. We create a ResourceTracker instance to track the memory that is allocated to the functions that have been added to LLJIT, as well as another ThreadSafeModule instance of the module and context. These two instances are then added to the JIT as an IR module:

```
auto RT = JIT->getMainJITDylib().createResourceTracker();
auto TSM = ThreadSafeModule(std::move(M), std::move(Ctx));
ExitOnErr(JIT->addIRModule(RT, std::move(TSM)));
```

15. The separate evaluation function is then queried for within our LLJIT instance through the lookup() method, by supplying the name of our evaluation function, calc_expr_func, into the function. If the query is successful, the address for the calc_expr_func function

is cast to the appropriate type, which is a function that takes no arguments and returns a single integer. Once the function's address is acquired, we call the function to generate the result of the user-defined function with the parameters they have supplied and then print the result to the console:

```
auto CalcExprCall = ExitOnErr(JIT->lookup("calc_expr_
func"));
int (*UserFnCall)() = CalcExprCall.toPtr<int (*)()>();
outs() << "User defined function evaluated to:
" << UserFnCall() << "\n";
```

16. After the function call is completed, the memory that was previously associated with our functions is then freed by `ResourceTracker`:

```
ExitOnErr(RT->remove());
```

Code generation changes to support JIT compilation via LLJIT

Now, let's take a brief look at some of the changes we have made within `CodeGen.cpp` to support our JIT-based calculator:

1. As previously mentioned, the code generation class has two important methods: one to compile the user-defined function into LLVM IR and print the IR to the console, and another to prepare the calculation evaluation function, `calc_expr_func`, which contains a call to the original user-defined function for evaluation. This second function also prints the resulting IR to the user:

```
void CodeGen::compileToIR(AST *Tree, Module *M,
                    StringMap<size_t> &JITtedFunctions) {
  ToIRVisitor ToIR(M, JITtedFunctions);
  ToIR.run(Tree);
  M->print(outs(), nullptr);
}
void CodeGen::prepareCalculationCallFunc(AST *FuncCall,
        Module *M, llvm::StringRef FnName,
        StringMap<size_t> &JITtedFunctions) {
  ToIRVisitor ToIR(M, JITtedFunctions);
  ToIR.genFuncEvaluationCall(FuncCall);
  M->print(outs(), nullptr);
}
```

2. As noted in the preceding source, these code generation functions define a `ToIRVisitor` instance that takes in our module and a `JITtedFunctions` map to be used in its constructor upon initialization:

```
class ToIRVisitor : public ASTVisitor {
  Module *M;
```

```
    IRBuilder<> Builder;
    StringMap<size_t> &JITtedFunctionsMap;
    . . .


public:
    ToIRVisitor(Module *M,
                StringMap<size_t> &JITtedFunctions)
        : M(M), Builder(M->getContext()),
        JITtedFunctionsMap(JITtedFunctions) {
```

3. Ultimately, this information is used to either generate IR or evaluate the function that the IR was previously generated for. When generating the IR, the code generator expects to see a `DefDecl` node, which represents defining a new function. The function name, along with the number of arguments it is defined with, is stored within the function definitions map:

```
virtual void visit(DefDecl &Node) override {
    llvm::StringRef FnName = Node.getFnName();
    llvm::SmallVector<llvm::StringRef, 8> FunctionVars =
    Node.getVars();
    (JITtedFunctionsMap)[FnName] = FunctionVars.size();
```

4. Afterward, the actual function definition is created by the `genUserDefinedFunction()` call:

```
    Function *DefFunc = genUserDefinedFunction(FnName);
```

5. Within `genUserDefinedFunction()`, the first step is to check if the function exists within the module. If it does not, we ensure that the function prototype exists within our map data structure. Then, we use the name and the number of arguments to construct a function that has the number of arguments that were defined by the user, and make the function return a single integer value:

```
Function *genUserDefinedFunction(llvm::StringRef Name) {
    if (Function *F = M->getFunction(Name))
      return F;

    Function *UserDefinedFunction = nullptr;
    auto FnNameToArgCount = JITtedFunctionsMap.find(Name);
    if (FnNameToArgCount != JITtedFunctionsMap.end()) {
        std::vector<Type *> IntArgs(FnNameToArgCount->second,
        Int32Ty);
        FunctionType *FuncType = FunctionType::get(Int32Ty,
        IntArgs, false);
        UserDefinedFunction =
            Function::Create(FuncType,
            GlobalValue::ExternalLinkage, Name, M);
    }
```

```
        return UserDefinedFunction;
    }
```

6. After generating the user-defined function, a new basic block is created, and we insert our function into the basic block. Each function argument is also associated with a name that is defined by the user, so we also set the names for all function arguments accordingly, as well as generate mathematical operations that operate on the arguments within the function:

```
BasicBlock *BB = BasicBlock::Create(M->getContext(),
"entry", DefFunc);
Builder.SetInsertPoint(BB);
unsigned FIdx = 0;
for (auto &FArg : DefFunc->args()) {
  nameMap[FunctionVars[FIdx]] = &FArg;
  FArg.setName(FunctionVars[FIdx++]);
}
Node.getExpr()->accept(*this);
};
```

7. When evaluating the user-defined function, the AST that is expected in our example is called a FuncCallFromDef node. First, we define the evaluation function and name it calc_expr_func (taking in zero arguments and returning one result):

```
virtual void visit(FuncCallFromDef &Node) override {
  llvm::StringRef CalcExprFunName = "calc_expr_func";
  FunctionType *CalcExprFunTy = FunctionType::get(Int32Ty, {},
  false);
  Function *CalcExprFun = Function::Create(
      CalcExprFunTy, GlobalValue::ExternalLinkage,
      CalcExprFunName, M);
```

8. Next, we create a new basic block to insert calc_expr_func into:

```
BasicBlock *BB = BasicBlock::Create(M->getContext(),
"entry", CalcExprFun);
Builder.SetInsertPoint(BB);
```

9. Similar to before, the user-defined function is retrieved by genUserDefinedFunction(), and we pass the numerical parameters of the function call into the original function that we have just regenerated:

```
llvm::StringRef CalleeFnName = Node.getFnName();
Function *CalleeFn = genUserDefinedFunction(CalleeFnName);
```

10. Once we have the actual `llvm::Function` instance available, we utilize `IRBuilder` to create a call to the defined function and also return the result so that it is accessible when the result is printed to the user in the end:

```
auto CalleeFnVars = Node.getArgs();
llvm::SmallVector<Value *> IntParams;
for (unsigned i = 0, end = CalleeFnVars.size(); i != end;
++i) {
  int ArgsToIntType;
  CalleeFnVars[i].getAsInteger(10, ArgsToIntType);
  Value *IntParam = ConstantInt::get(Int32Ty, ArgsToIntType,
  true);
  IntParams.push_back(IntParam);
}
Builder.CreateRet(Builder.CreateCall(CalleeFn, IntParams,
"calc_expr_res"));
};
```

Building an LLJIT-based calculator

Finally, to compile our JIT calculator source, we also need to create a `CMakeLists.txt` file with the build description, saved beside `Calc.cpp` and our other source files:

1. We set the minimal required CMake version to the number required by LLVM and give the project a name:

```
cmake_minimum_required (VERSION 3.20.0)
project ("jit")
```

2. The LLVM package needs to be loaded, and we add the directory of the CMake modules provided by LLVM to the search path. Then, we include the `DetermineGCCCompatible` and `ChooseMSVCCRT` modules, which check if the compiler has GCC-compatible command-line syntax and ensure that the same C runtime is used as by LLVM, respectively:

```
find_package(LLVM REQUIRED CONFIG)
list(APPEND CMAKE_MODULE_PATH ${LLVM_DIR})
include(DetermineGCCCompatible)
include(ChooseMSVCCRT)
```

3. We also need to add definitions and the `include` path from LLVM. The used LLVM components are mapped to the library names with a function call:

```
add_definitions(${LLVM_DEFINITIONS})
include_directories(SYSTEM ${LLVM_INCLUDE_DIRS})
llvm_map_components_to_libnames(llvm_libs Core OrcJIT
                                         Support native)
```

4. Afterward, if it is determined that the compiler has GCC-compatible command-line syntax, we also check if runtime type information and exception handling are enabled. If they are not enabled, C++ flags to turn off these features are added to our compilation accordingly:

```
if(LLVM_COMPILER_IS_GCC_COMPATIBLE)
  if(NOT LLVM_ENABLE_RTTI)
    set(CMAKE_CXX_FLAGS "${CMAKE_CXX_FLAGS} -fno-rtti")
  endif()
  if(NOT LLVM_ENABLE_EH)
    set(CMAKE_CXX_FLAGS "${CMAKE_CXX_FLAGS} -fno-exceptions")
  endif()
endif()
```

5. Lastly, we define the name of the executable, the source files to compile, and the library to link against:

```
add_executable (calc
  Calc.cpp CodeGen.cpp Lexer.cpp Parser.cpp Sema.cpp)
target_link_libraries(calc PRIVATE ${llvm_libs})
```

The preceding steps are all that is required for our JIT-based interactive calculator tool. Next, create and change into a build directory, and then run the following command to create and compile the application:

```
$ cmake -G Ninja <path to source directory>
$ ninja
```

This compiles the `calc` tool. We can then launch the calculator, start defining functions, and see how our calculator is able to evaluate the functions that we define.

The following example invocations show the IR of the function that is first defined, and then the `calc_expr_func` function that is created to generate a call to our originally defined function in order to evaluate the function with whichever parameter passed into it:

```
$ ./calc
JIT calc > def f(x) = x*2
define i32 @f(i32 %x) {
entry:
  %0 = mul nsw i32 %x, 2
  ret i32 %0
}

JIT calc > f(20)
Attempting to evaluate expression:
define i32 @calc_expr_func() {
entry:
```

```
  %calc_expr_res = call i32 @f(i32 20)
  ret i32 %calc_expr_res
}

declare i32 @f(i32)
User defined function evaluated to: 40

JIT calc > def g(x,y) = x*y+100
define i32 @g(i32 %x, i32 %y) {
entry:
  %0 = mul nsw i32 %x, %y
  %1 = add nsw i32 %0, 100
  ret i32 %1
}

JIT calc > g(8,9)
Attempting to evaluate expression:
define i32 @calc_expr_func() {
entry:
  %calc_expr_res = call i32 @g(i32 8, i32 9)
  ret i32 %calc_expr_res
}

declare i32 @g(i32, i32)
User defined function evaluated to: 172
```

That's it! We have just created a JIT-based calculator application!

As our JIT calculator is meant to be a simple example that describes how to incorporate `LLJIT` into our projects, it is worth noting that there are some limitations:

- This calculator does not accept negatives of decimal values

- We cannot redefine the same function more than once

For the second limitation, this occurs by design, and so is expected and enforced by the ORC API itself:

```
$ ./calc
JIT calc > def f(x) = x*2
define i32 @f(i32 %x) {
entry:
  %0 = mul nsw i32 %x, 2
  ret i32 %0
}
JIT calc > def f(x,y) = x+y
```

```
define i32 @f(i32 %x, i32 %y) {
entry:
  %0 = add nsw i32 %x, %y
  ret i32 %0
}
Duplicate definition of symbol '_f'
```

Keep in mind that there are numerous other possibilities to expose names, besides exposing the symbols for the current process or from a shared library. For example, the `StaticLibraryDefinitionGenerator` class exposes the symbols found in a static archive and can be used in the `DynamicLibrarySearchGenerator` class.

Furthermore, the `LLJIT` class has also an `addObjectFile()` method to expose the symbols of an object file. You can also provide your own `DefinitionGenerator` implementation if the existing implementations do not fit your needs.

As we can see, using the predefined `LLJIT` class is convenient, but it can limit our flexibility. In the next section, we'll look at how to implement a JIT compiler using the layers provided by the ORC API.

Building a JIT compiler class from scratch

Using the layered approach of ORC, it is very easy to build a JIT compiler customized for the requirements. There is no one-size-fits-all JIT compiler, and the first section of this chapter gave some examples. Let's have a look at how to set up a JIT compiler from scratch.

The ORC API uses layers that are stacked together. The lowest level is the object-linking layer, represented by the `llvm::orc::RTDyldObjectLinkingLayer` class. It is responsible for linking in-memory objects and turning them into executable code. The memory required for this task is managed by an instance of the `MemoryManager` interface. There is a default implementation, but we can also use a custom version if we need.

Above the object-linking layer is the compile layer, which is responsible for creating an in-memory object file. The `llvm::orc::IRCompileLayer` class takes an IR module as input and compiles it to an object file. The `IRCompileLayer` class is a subclass of the `IRLayer` class, which is a generic class for layer implementations accepting LLVM IR.

Both of these layers already form the core of a JIT compiler: they add an LLVM IR module as input, which is compiled and linked in memory. To add extra functionality, we can incorporate more layers on top of both layers.

For example, the `CompileOnDemandLayer` class splits a module so that only the requested functions are compiled. This can be used to implement lazy compilation. Moreover, the `CompileOnDemandLayer` class is also a subclass of the `IRLayer` class. In a very generic way, the `IRTransformLayer` class, also a subclass of the `IRLayer` class, allows us to apply a transformation to the module.

Another important class is the `ExecutionSession` class. This class represents a running JIT program. Essentially, this means that the class manages `JITDylib` symbol tables, provides lookup functionality for symbols, and keeps track of used resource managers.

The generic recipe for a JIT compiler is as follows:

1. Initialize an instance of the `ExecutionSession` class.

2. Initialize the layer, at least consisting of an `RTDyldObjectLinkingLayer` class and an `IRCompileLayer` class.

3. Create the first `JITDylib` symbol table, usually with `main` or a similar name.

The general usage of the JIT compiler is also very straightforward:

1. Add an IR module to the symbol table.

2. Look up a symbol, triggering the compilation of the associated function, and possibly the whole module.

3. Execute the function.

In the next subsection, we implement a JIT compiler class following the generic recipe.

Creating a JIT compiler class

To keep the implementation of the JIT compiler class simple, everything is placed in `JIT.h`, within a source directory we can create called `jit`. However, the initialization of the class is a bit more complex compared to using `LLJIT`. Due to handling possible errors, we need a factory method to create some objects upfront before we can call the constructor. The steps to create the class are as follows:

1. We begin with guarding the header file against multiple inclusion with the `JIT_H` preprocessor definition:

   ```
   #ifndef JIT_H
   #define JIT_H
   ```

2. Firstly, a number of `include` files are required. Most of them provide a class with the same name as the header file. The `Core.h` header provides a couple of basic classes, including the `ExecutionSession` class. Additionally, the `ExecutionUtils.h` header provides the `DynamicLibrarySearchGenerator` class to search libraries for symbols. Furthermore, the `CompileUtils.h` header provides the `ConcurrentIRCompiler` class:

   ```
   #include "llvm/Analysis/AliasAnalysis.h"
   #include "llvm/ExecutionEngine/JITSymbol.h"
   #include "llvm/ExecutionEngine/Orc/CompileUtils.h"
   #include "llvm/ExecutionEngine/Orc/Core.h"
   #include "llvm/ExecutionEngine/Orc/ExecutionUtils.h"
   ```

```
#include "llvm/ExecutionEngine/Orc/IRCompileLayer.h"
#include "llvm/ExecutionEngine/Orc/IRTransformLayer.h"
#include
    "llvm/ExecutionEngine/Orc/JITTargetMachineBuilder.h"
#include "llvm/ExecutionEngine/Orc/Mangling.h"
#include
    "llvm/ExecutionEngine/Orc/RTDyldObjectLinkingLayer.h"
#include
        "llvm/ExecutionEngine/Orc/TargetProcessControl.h"
#include "llvm/ExecutionEngine/SectionMemoryManager.h"
#include "llvm/Passes/PassBuilder.h"
#include "llvm/Support/Error.h"
```

3. Declare a new class. Our new class will be called JIT:

    ```
    class JIT {
    ```

4. The private data members reflect the ORC layers and some helper classes. The
    ExecutionSession, ObjectLinkingLayer, CompileLayer, OptIRLayer, and
    MainJITDylib instances represent the running JIT program, the layers, and the symbol
    table, as already described. Moreover, the TargetProcessControl instance is used for
    interaction with the JIT target process. This can be the same process, another process on the
    same machine, or a remote process on a different machine, possibly with a different architecture.
    The DataLayout and MangleAndInterner classes are required to mangle symbols'
    names in the correct way. Additionally, the symbol names are internalized, which means that
    all equal names have the same address. This means that to check if two symbol names are equal,
    it is then sufficient to compare the addresses, which is a very fast operation:

    ```
    std::unique_ptr<llvm::orc::TargetProcessControl> TPC;
    std::unique_ptr<llvm::orc::ExecutionSession> ES;
    llvm::DataLayout DL;
    llvm::orc::MangleAndInterner Mangle;
    std::unique_ptr<llvm::orc::RTDyldObjectLinkingLayer>
        ObjectLinkingLayer;
    std::unique_ptr<llvm::orc::IRCompileLayer>
        CompileLayer;
    std::unique_ptr<llvm::orc::IRTransformLayer>
        OptIRLayer;
    llvm::orc::JITDylib &MainJITDylib;
    ```

5. The initialization is split into three parts. In C++, a constructor cannot return an error. The simple and recommended solution is to create a static factory method that can do the error handling before constructing an object. The initialization of the layers is more complex, so we introduce factory methods for them, too.

    In the `create()` factory method, we first create a `SymbolStringPool` instance, which is used to implement string internalization and is shared by several classes. To take control of the current process, we create a `SelfTargetProcessControl` instance. If we want to target a different process, then we need to change this instance.

    Next, we construct a `JITTargetMachineBuilder` instance, for which we need to know the target triple of the JIT process. Afterward, we query the target machine builder for the data layout. This step can fail if the builder is not able to instantiate the target machine based on the provided triple – for example, because support for this target is not compiled into the LLVM libraries:

    ```
    public:
      static llvm::Expected<std::unique_ptr<JIT>> create() {
        auto SSP =
            std::make_shared<llvm::orc::SymbolStringPool>();
        auto TPC =
            llvm::orc::SelfTargetProcessControl::Create(SSP);
        if (!TPC)
          return TPC.takeError();
        llvm::orc::JITTargetMachineBuilder JTMB(
            (*TPC)->getTargetTriple());
        auto DL = JTMB.getDefaultDataLayoutForTarget();
        if (!DL)
          return DL.takeError();
    ```

6. At this point, we have handled all calls that could potentially fail. We are now able to initialize the `ExecutionSession` instance. Finally, the constructor of the `JIT` class is called with all instantiated objects, and the result is returned to the caller:

    ```
        auto ES =
            std::make_unique<llvm::orc::ExecutionSession>(
                std::move(SSP));

        return std::make_unique<JIT>(
            std::move(*TPC), std::move(ES), std::move(*DL),
            std::move(JTMB));
      }
    ```

7. The constructor of the `JIT` class moves the passed parameters to the private data members. Layer objects are constructed with a call to static factory names with the `create` prefix. Furthermore, each layer factory method requires a reference to the `ExecutionSession` instance, which connects the layer to the running JIT session. Except for the object-linking layer, which is at the bottom of the layer stack, each layer requires a reference to the previous layer, illustrating the stacking order:

```
JIT(std::unique_ptr<llvm::orc::ExecutorProcessControl>
        EPCtrl,
    std::unique_ptr<llvm::orc::ExecutionSession>
        ExeS,
    llvm::DataLayout DataL,
    llvm::orc::JITTargetMachineBuilder JTMB)
    : EPC(std::move(EPCtrl)), ES(std::move(ExeS)),
      DL(std::move(DataL)), Mangle(*ES, DL),
      ObjectLinkingLayer(std::move(
          createObjectLinkingLayer(*ES, JTMB))),
      CompileLayer(std::move(createCompileLayer(
          *ES, *ObjectLinkingLayer,
          std::move(JTMB)))),
      OptIRLayer(std::move(
          createOptIRLayer(*ES, *CompileLayer))),
      MainJITDylib(
          ES->createBareJITDylib("<main>")) {
```

8. In the body of the constructor, we add a generator to search the current process for symbols. The `GetForCurrentProcess()` method is special, as the return value is wrapped in an `Expected<>` template, indicating that an `Error` object can also be returned. However, since we know that no error can occur, the current process will eventually run! Thus, we unwrap the result with the `cantFail()` function, which terminates the application if an error occurred anyway:

```
MainJITDylib.addGenerator(llvm::cantFail(
    llvm::orc::DynamicLibrarySearchGenerator::
        GetForCurrentProcess(DL.getGlobalPrefix())));
}
```

9. To create an object-linking layer, we need to provide a memory manager. Here, we stick to the default `SectionMemoryManager` class, but we could also provide a different implementation if needed:

```
static std::unique_ptr<
    llvm::orc::RTDyldObjectLinkingLayer>
createObjectLinkingLayer(
    llvm::orc::ExecutionSession &ES,
```

```
llvm::orc::JITTargetMachineBuilder &JTMB) {
auto GetMemoryManager = [] () {
  return std::make_unique<
      llvm::SectionMemoryManager>();
};
auto OLLayer = std::make_unique<
    llvm::orc::RTDyldObjectLinkingLayer>(
    ES, GetMemoryManager);
```

10. A slight complication exists for the **Common Object File Format (COFF)** object file format, which is used on Windows. This file format does not allow functions to be marked as exported. This subsequently leads to failures in checks inside the object-linking layer: flags stored in the symbol are compared with the flags from IR, which leads to a mismatch because of the missing export marker. The solution is to override the flags only for this file format. This finishes the construction of the object layer, and the object is returned to the caller:

```
if (JTMB.getTargetTriple().isOSBinFormatCOFF()) {
  OLLayer
      ->setOverrideObjectFlagsWithResponsibilityFlags(
          true);
  OLLayer
      ->setAutoClaimResponsibilityForObjectSymbols(
          true);
}
return OLLayer;
}
```

11. To initialize the compiler layer, an IRCompiler instance is required. The IRCompiler instance is responsible for compiling an IR module into an object file. If our JIT compiler does not use threads, then we can use the SimpleCompiler class, which compiles the IR module using a given target machine. The TargetMachine class is not threadsafe, and therefore the SimpleCompiler class is not, either. To support compilation with multiple threads, we use the ConcurrentIRCompiler class, which creates a new TargetMachine instance for each module to compile. This approach solves the problem with multiple threads:

```
static std::unique_ptr<llvm::orc::IRCompileLayer>
createCompileLayer(
    llvm::orc::ExecutionSession &ES,
    llvm::orc::RTDyldObjectLinkingLayer &OLLayer,
    llvm::orc::JITTargetMachineBuilder JTMB) {
  auto IRCompiler = std::make_unique<
      llvm::orc::ConcurrentIRCompiler>(
      std::move(JTMB));
  auto IRCLayer =
```

```
        std::make_unique<llvm::orc::IRCompileLayer>(
            ES, OLLayer, std::move(IRCompiler));
    return IRCLayer;
}
```

12. Instead of compiling the IR module directly to machine code, we install a layer that optimizes the IR first. This is a deliberate design decision: we turn our JIT compiler into an optimizing JIT compiler, which produces faster code that takes longer to produce, meaning a delay for the user. We do not add lazy compilation, so whole modules are compiled when just a symbol is looked up. This can add up to a significant amount of time before the user sees the code executing.

> **Note**
>
> Introducing lazy compilation is not a proper solution in all circumstances. Lazy compilation is realized by moving each function into a module of its own, which is compiled when the function name is looked up. This prevents inter-procedural optimizations such as *inlining* because the inliner pass needs access to the body of called functions to inline them. As a result, users see a faster startup with lazy compilation, but the produced code is not as optimal as it can be. These design decisions depend on the intended use. Here, we decide on fast code, accepting a slower startup time. Furthermore, this means that the optimization layer is essentially a transformation layer.

The `IRTransformLayer` class delegates the transformation to a function – in our case, to the `optimizeModule` function:

```
static std::unique_ptr<llvm::orc::IRTransformLayer>
createOptIRLayer(
    llvm::orc::ExecutionSession &ES,
    llvm::orc::IRCompileLayer &CompileLayer) {
  auto OptIRLayer =
      std::make_unique<llvm::orc::IRTransformLayer>(
          ES, CompileLayer,
          optimizeModule);
  return OptIRLayer;
}
```

13. The `optimizeModule()` function is an example of a transformation on an IR module. The function gets the module to transform as a parameter and returns the transformed version of the IR module. Since the JIT compiler can potentially run with multiple threads, the IR module is wrapped in a `ThreadSafeModule` instance:

```
static llvm::Expected<llvm::orc::ThreadSafeModule>
optimizeModule(
    llvm::orc::ThreadSafeModule TSM,
```

```
const llvm::orc::MaterializationResponsibility
    &R) {
```

14. To optimize the IR, we recall some information from *Chapter 7, Optimizing IR*, in the *Adding an optimization pipeline to your compiler* section. We need a `PassBuilder` instance to create an optimization pipeline. First, we define a couple of analysis managers and register them afterward at the pass builder. Afterward, we populate a `ModulePassManager` instance with the default optimization pipeline for the O2 level. This is again a design decision: the O2 level produces already fast machine code, but it produces even faster code at the O3 level. Next, we run the pipeline on the module, and finally, the optimized module is returned to the caller:

```
TSM.withModuleDo([](llvm::Module &M) {
  bool DebugPM = false;
  llvm::PassBuilder PB(DebugPM);
  llvm::LoopAnalysisManager LAM(DebugPM);
  llvm::FunctionAnalysisManager FAM(DebugPM);
  llvm::CGSCCAnalysisManager CGAM(DebugPM);
  llvm::ModuleAnalysisManager MAM(DebugPM);
  FAM.registerPass(
      [&] { return PB.buildDefaultAAPipeline(); });
  PB.registerModuleAnalyses(MAM);
  PB.registerCGSCCAnalyses(CGAM);
  PB.registerFunctionAnalyses(FAM);
  PB.registerLoopAnalyses(LAM);
  PB.crossRegisterProxies(LAM, FAM, CGAM, MAM);
  llvm::ModulePassManager MPM =
      PB.buildPerModuleDefaultPipeline(
          llvm::PassBuilder::OptimizationLevel::O2,
          DebugPM);
  MPM.run(M, MAM);
});

return TSM;
}
```

15. The client of the `JIT` class needs a way to add an IR module, which we provide with the `addIRModule()` function. Recall the layer stack we created: we must add the IR module to the top layer; otherwise, we would accidentally bypass some of the layers. This would be a programming error that is not easily spotted: if the `OptIRLayer` member is replaced by the `CompileLayer` member, then our `JIT` class still works, but not as an optimizing JIT because

we have bypassed this layer. This is no concern for this small implementation, but in a large JIT optimization, we would introduce a function to return the top-level layer:

```
llvm::Error addIRModule(
    llvm::orc::ThreadSafeModule TSM,
    llvm::orc::ResourceTrackerSP RT = nullptr) {
  if (!RT)
    RT = MainJITDylib.getDefaultResourceTracker();
  return OptIRLayer->add(RT, std::move(TSM));
}
```

16. Likewise, a client of our JIT class needs a way to look up a symbol. We delegate this to the ExecutionSession instance, passing in a reference to the main symbol table and the mangled and internalized name of the requested symbol:

```
llvm::Expected<llvm::orc::ExecutorSymbolDef>
lookup(llvm::StringRef Name) {
  return ES->lookup({&MainJITDylib},
                    Mangle(Name.str()));
}
```

As we can see, the initialization of this JIT class can be tricky, as it involves a factory method and a constructor call for the JIT class, and factory methods for each layer. Although this distribution is caused by limitations in C++, the code itself is straightforward.

Next, we are going to use the new JIT compiler class to implement a simple command-line utility that takes an LLVM IR file as input.

Using our new JIT compiler class

We start off by creating a file called JIT.cpp, in the same directory as the JIT.h file, and add the following to this source file:

1. Firstly, several header files are included. We must include JIT.h to use our new class, and the IRReader.h header because it defines a function to read LLVM IR files. The CommandLine.h header allows us to parse the command-line options in the LLVM style. Next, InitLLVM.h is needed for the basic initialization of the tool. Finally, TargetSelect.h is needed for the initialization of the native target:

```
#include "JIT.h"
#include "llvm/IRReader/IRReader.h"
#include "llvm/Support/CommandLine.h"
#include "llvm/Support/InitLLVM.h"
#include "llvm/Support/TargetSelect.h"
```

2. Next, we add the `llvm` namespace to the current scope:

    ```
    using namespace llvm;
    ```

3. Our JIT tool expects exactly one input file on the command line, which we declare with the `cl::opt<>` class:

    ```
    static cl::opt<std::string>
        InputFile(cl::Positional, cl::Required,
                  cl::desc("<input-file>"));
    ```

4. To read the IR file, we call the `parseIRFile()` function. The file can be a textual IR representation or a bitcode file. The function returns a pointer to the created module. Additionally, the error handling is a bit different, because a textual IR file can be parsed, which is not necessarily syntactically correct. Finally, the `SMDiagnostic` instance holds the error information in case of a syntax error. In the event of an error, an error message is printed, and the application is exited:

    ```
    std::unique_ptr<Module>
    loadModule(StringRef Filename, LLVMContext &Ctx,
               const char *ProgName) {
      SMDiagnostic Err;
      std::unique_ptr<Module> Mod =
          parseIRFile(Filename, Err, Ctx);
      if (!Mod.get()) {
        Err.print(ProgName, errs());
        exit(-1);
      }
      return Mod;
    }
    ```

5. The `jitmain()` function is placed after the `loadModule()` method. This function sets up our JIT engine and compiles an LLVM IR module. The function needs the LLVM module with the IR to execute. The LLVM context class is also required for this module because the context class contains important type information. The goal is to call the `main()` function, so we also pass the usual `argc` and `argv` parameters:

    ```
    Error jitmain(std::unique_ptr<Module> M,
                  std::unique_ptr<LLVMContext> Ctx,
                  int argc, char *argv[]) {
    ```

6. Next, we create an instance of our JIT class that we constructed earlier. If an error occurs, then we return an error message accordingly:

    ```
    auto JIT = JIT::create();
    if (!JIT)
      return JIT.takeError();
    ```

7. Then, we add the module to the main `JITDylib` instance, wrapping the module and a context in a `ThreadSafeModule` instance yet again. If an error occurs, then we return an error message:

```
if (auto Err = (*JIT)->addIRModule(
        orc::ThreadSafeModule(std::move(M),
                              std::move(Ctx))))
    return Err;
```

8. Following this, we look up the `main` symbol. This symbol must be in the IR module given on the command line. The lookup triggers the compilation of that IR module. If other symbols are referenced inside the IR module, then they are resolved using the generator added in the previous step. The result is of the `ExecutorAddr` class, where it represents the address of the executor process:

```
llvm::orc::ExecutorAddr MainExecutorAddr = MainSym-
>getAddress();
auto *Main = MainExecutorAddr.toPtr<int(int, char**)>();
```

9. Now, we can call the `main()` function in the IR module, and pass the `argc` and `argv` parameters that the function expects. We ignore the return value:

```
(void)Main(argc, argv);
```

10. We report success after the execution of the function:

```
    return Error::success();
}
```

11. After implementing a `jitmain()` function, we add a `main()` function, which initializes the tool and the native target and parses the command line:

```
int main(int argc, char *argv[]) {
    InitLLVM X(argc, argv);
    InitializeNativeTarget();
    InitializeNativeTargetAsmPrinter();
    InitializeNativeTargetAsmParser();

    cl::ParseCommandLineOptions(argc, argv, "JIT\n");
```

12. Afterward, the LLVM context class is initialized, and we load the IR module named on the command line:

```
auto Ctx = std::make_unique<LLVMContext>();
std::unique_ptr<Module> M =
    loadModule(InputFile, *Ctx, argv[0]);
```

13. After loading the IR module, we can call the `jitmain()` function. To handle errors, we use the `ExitOnError` utility class to print an error message and exit the application when an error is encountered. We also set a banner with the name of the application, which is printed before the error message:

```
ExitOnError ExitOnErr(std::string(argv[0]) + ": ");
ExitOnErr(jitmain(std::move(M), std::move(Ctx),
                  argc, argv));
```

14. If the control flow reaches this point, then the IR was successfully executed. We return 0 to indicate success:

```
    return 0;
}
```

We can now test our newly implemented JIT compiler by compiling a simple example that prints `Hello World!` to the console. Under the hood, the new class uses a fixed optimization level, so with large enough modules, we can note differences in the startup and runtime.

To build our JIT compiler, we can follow the same CMake steps as we did near the end of the *Implementing our own JIT compiler with LLJIT* section, and we just need to ensure that the `JIT.cpp` source file is being compiled with the correct libraries to link against:

```
add_executable(JIT JIT.cpp)
include_directories(${CMAKE_SOURCE_DIR})
target_link_libraries(JIT ${llvm_libs})
```

We then change into the `build` directory and compile the application:

```
$ cmake -G Ninja <path to jit source directory>
$ ninja
```

Our `JIT` tool is now ready to be used. A simple `Hello World!` program can be written in C, like the following:

```
$ cat main.c
#include <stdio.h>

int main(int argc, char** argv) {
  printf("Hello world!\n");
  return 0;
}
```

Next, we can compile the Hello World C source into LLVM IR with the following command:

```
$ clang -S -emit-llvm main.c
```

Remember – we compile the C source into LLVM IR because our JIT compiler accepts an IR file as input. Finally, we can invoke our JIT compiler with our IR example, as follows:

```
$ JIT main.ll
Hello world!
```

Summary

In this chapter, you learned how to develop a JIT compiler. You began with learning about the possible applications of JIT compilers, and you explored lli, the LLVM dynamic compiler and interpreter. Using the predefined LLJIT class, you built an interactive JIT-based calculator tool and learned about looking up symbols and adding IR modules to LLJIT. To be able to take advantage of the layered structure of the ORC API, you also implemented an optimizing JIT class.

In the next chapter, you will learn how to utilize LLVM tools for debugging purposes.

10

Debugging Using LLVM Tools

LLVM comes with a set of tools that helps you identify certain errors in your application. All these tools make use of the LLVM and **clang** libraries.

In this chapter, you will learn how to instrument an application with **sanitizers**, as well as how to use the most common sanitizer to identify a wide range of bugs, after which you'll implement fuzz testing for your application. This will help you identify bugs that are usually not found with unit testing. You will also learn how to identify performance bottlenecks in your application, run the **static analyzer** to identify problems normally not found by the compiler, and create your own clang-based tool, in which you can extend clang with new functionality.

This chapter will cover the following topics:

- Instrumenting an application with sanitizers

- Finding bugs with **libFuzzer**

- Performance profiling with **XRay**

- Checking the source with the **Clang Static Analyzer**

- Creating your own clang-based tool

By the end of this chapter, you will know how to use the various LLVM and clang tools to identify a large category of errors in an application. You will also acquire the knowledge to extend clang with new functionality, for example, to enforce naming conventions or to add new source analysis.

Technical requirements

To create the **flame graph** in the *Performance profiling with XRay* section, you need to install the scripts from `https://github.com/brendangregg/FlameGraph`. Some systems, such as **Fedora** and **FreeBSD**, provide a package for these scripts, which you can also use.

To view the **Chrome visualization** in the same section, you need to have the **Chrome** browser installed. You can download the browser from `https://www.google.com/chrome/` or use the package manager of your system to install **Chrome** browser.

Additionally, to run the static analyzer via the `scan-build` script, you need to have the `perl-core` package installed on **Fedora** and **Ubuntu**.

Instrumenting an application with sanitizers

LLVM comes with a couple of **sanitizers**. These are passes that instrument the **intermediate representation** (**IR**) to check for certain misbehavior of an application. Usually, they require library support, which is part of the `compiler-rt` project. The sanitizers can be enabled in clang, which makes them very comfortable to use. To build the `compiler-rt` project, we can simply add the `-DLLVM_ENABLE_RUNTIMES=compiler-rt` CMake variable to the initial CMake configuration step when building LLVM.

In the following sections, we will look at the `address`, `memory`, and `thread` sanitizers. First, we'll look at the `address` sanitizer.

Detecting memory access problems with the address sanitizer

You can use the `address` sanitizer to detect different types of memory access bugs within an application. This includes common errors such as using dynamically allocated memory after freeing it or writing to dynamically allocated memory outside the boundaries of the allocated memory.

When enabled, the `address` sanitizer replaces calls to the `malloc()` and `free()` functions with its own version and instruments all memory accesses with a checking guard. Of course, this adds a lot of overhead to the application, and you will only use the `address` sanitizer during the testing phase of the application. If you are interested in the implementation details, then you can find the source of the pass in the `llvm/lib/Transforms/Instrumentation/AddressSanitzer.cpp` file and a description of the implemented algorithm at `https://github.com/google/sanitizers/wiki/AddressSanitizerAlgorithm`.

Let's run a short example to show the capabilities of the `address` sanitizer!

The following example application, `outofbounds.c`, allocates `12` bytes of memory, but initializes `14` bytes:

```
#include <stdlib.h>
#include <string.h>

int main(int argc, char *argv[]) {
  char *p = malloc(12);
  memset(p, 0, 14);
```

```
    return (int)*p;
}
```

You can compile and run this application without noticing a problem as this behavior is typical for this kind of error. Even in larger applications, such kinds of bugs can go unnoticed for a long time. However, if you enable the `address` sanitizer with the `-fsanitize=address` option, then the application stops after detecting the error.

It is also useful to enable debug symbols with the `-g` options because it helps identify the location of the error in the source. The following code is an example of how to compile the source file with the `address` sanitizer and debug symbols enabled:

```
$ clang -fsanitize=address -g outofbounds.c -o outofbounds
```

Now, you get a lengthy error report when running the application:

```
$ ./outofbounds
=================================================================
==1067==ERROR: AddressSanitizer: heap-buffer-overflow on address
0x60200000001c at pc 0x00000023a6ef bp 0x7fffffffeb10 sp
0x7fffffffe2d8
WRITE of size 14 at 0x60200000001c thread T0
    #0 0x23a6ee in __asan_memset /usr/src/contrib/llvm-project/
compiler-rt/lib/asan/asan_interceptors_memintrinsics.cpp:26:3
    #1 0x2b2a03 in main /home/kai/sanitizers/outofbounds.c:6:3
    #2 0x23331f in _start /usr/src/lib/csu/amd64/crt1.c:76:7
```

The report also contains detailed information about the memory content. The important information is the type of the error – **heap buffer overflow**, in this case – and the offending source line. To find the source line, you must look at the stack trace at location #1, which is the last location before the address sanitizer intercepts the execution of the application. It shows *line 6* in the outofbounds.c file, which is the line containing the call to memset(). This is the exact place where the buffer overflow happens.

If you replace the line containing memset(p, 0, 14); in the outofbounds.c file with the following code, then you can introduce access to memory once you've freed the memory. You'll need to store the source in the useafterfree.c file:

```
    memset(p, 0, 12);
    free(p);
```

Again, if you compile and run it, the sanitizer detects the use of the pointer after the memory is freed:

```
$ clang -fsanitize=address -g useafterfree.c -o useafterfree
$ ./useafterfree
=================================================================
==1118==ERROR: AddressSanitizer: heap-use-after-free on address
0x602000000010 at pc 0x0000002b2a5c bp 0x7fffffffeb00 sp
```

```
0x7fffffffeaf8
READ of size 1 at 0x602000000010 thread T0
    #0 0x2b2a5b in main /home/kai/sanitizers/useafterfree.c:8:15
    #1 0x23331f in _start /usr/src/lib/csu/amd64/crt1.c:76:7
```

This time, the report points to *line 8*, which contains dereferencing of the p pointer.

On **x86_64 Linux** and **macOS**, you can also enable a leak detector. If you set the ASAN_OPTIONS environment variable to detect_leaks=1 before running the application, then you also get a report about memory leaks.

On the command line, you can do this as follows:

```
$ ASAN_OPTIONS=detect_leaks=1 ./useafterfree
```

The address sanitizer is very useful because it catches a category of bugs that are otherwise difficult to detect. The memory sanitizer does a similar task. We'll examine its use cases in the next section.

Finding uninitialized memory accesses with the memory sanitizer

Using uninitialized memory is another category of bugs that are hard to find. In **C** and **C++**, the general memory allocation routines do not initialize the memory buffer with a default value. The same is true for automatic variables on the stack.

There are lots of opportunities for errors, and the memory sanitizer helps find these bugs. If you are interested in the implementation details, you can find the source for the memory sanitizer pass in the llvm/lib/Transforms/Instrumentation/MemorySanitizer.cpp file. The comment at the top of the file explains the ideas behind the implementation.

Let's run a small example and save the following source as the memory.c file. Note that the x variable is not initialized and is used as a return value:

```
int main(int argc, char *argv[]) {
  int x;
  return x;
}
```

Without the sanitizer, the application will run just fine. However, you will get an error report if you use the -fsanitize=memory option:

```
$ clang -fsanitize=memory -g memory.c -o memory
$ ./memory
==1206==WARNING: MemorySanitizer: use-of-uninitialized-value
    #0 0x10a8f49 in main /home/kai/sanitizers/memory.c:3:3
    #1 0x1053481 in _start /usr/src/lib/csu/amd64/crt1.c:76:7
```

```
SUMMARY: MemorySanitizer: use-of-uninitialized-value /home/kai/
sanitizers/memory.c:3:3 in main
Exiting
```

Like the `address` sanitizer, the memory sanitizer stops the application at the first error that's found. As shown here the memory sanitizer provides a **use of initialized value** warning.

Finally, in the next section, we'll look at how we can use the `thread` sanitizer to detect data races in multi-threaded applications.

Pointing out data races with the thread sanitizer

To leverage the power of modern CPUs, applications now use multiple threads. This is a powerful technique, but it also introduces new sources of errors. A very common problem in multi-threaded applications is that the access to global data is not protected, for example, with a **mutex** or **semaphore**. This is called a **data race**. The `thread` sanitizer can detect data races in **Pthreads**-based applications and in applications using the LLVM **libc++** implementation. You can find the implementation in the `llvm/lib/Transforms/Instrumentation/ThreadSanitizer.cpp` file.

To demonstrate the functionality of the `thread` sanitizer, we will create a very simple producer-consumer-style application. The producer thread increments a global variable, while the consumer thread decrements the same variable. Access to the global variable is not protected, so this is a data race.

You'll need to save the following source in the `thread.c` file:

```c
#include <pthread.h>

int data = 0;

void *producer(void *x) {
  for (int i = 0; i < 10000; ++i) ++data;
  return x;
}

void *consumer(void *x) {
  for (int i = 0; i < 10000; ++i) --data;
  return x;
}

int main() {
  pthread_t t1, t2;
  pthread_create(&t1, NULL, producer, NULL);
  pthread_create(&t2, NULL, consumer, NULL);
  pthread_join(t1, NULL);
  pthread_join(t2, NULL);
```

```
        return data;
    }
```

In the preceding code, the data variable is shared between two threads. Here, it is of the int type to make the example simple since often, a data structure such as the std::vector class or similar would be used. Furthermore, these two threads run the producer() and consumer() functions.

The producer() function only increments the data variable, while the consumer() function decrements it. No access protection is implemented, so this constitutes a data race. The main() function starts both threads with the pthread_create() function, waits for the end of the threads with the pthread_join() function, and returns the current value of the data variable.

If you compile and run this application, then you will note no error – that is, the return value is always zero. An error – in this case, a return value not equal to zero – will appear if the number of loops that are performed is increased by a factor of 100. At this point, you will begin to notice other values appear.

We can use the thread sanitizer to identify the data race within our program. To compile with the thread sanitizer enabled, you'll need to pass the -fsanitize=thread option to clang. Adding debug symbols with the −g options gives you line numbers in the report, which is also helpful. Note that you also need to link the pthread library:

```
$ clang -fsanitize=thread -g thread.c -o thread -lpthread
$ ./thread
==================
WARNING: ThreadSanitizer: data race (pid=1474)
  Write of size 4 at 0x000000cdf8f8 by thread T2:
    #0 consumer /home/kai/sanitizers/thread.c:11:35 (thread+0x2b0fb2)

  Previous write of size 4 at 0x000000cdf8f8 by thread T1:
    #0 producer /home/kai/sanitizers/thread.c:6:35 (thread+0x2b0f22)

  Location is global 'data' of size 4 at 0x000000cdf8f8
(thread+0x000000cdf8f8)

  Thread T2 (tid=100437, running) created by main thread at:
    #0 pthread_create /usr/src/contrib/llvm-project/compiler-rt/lib/
tsan/rtl/tsan_interceptors_posix.cpp:962:3 (thread+0x271703)
    #1 main /home/kai/sanitizers/thread.c:18:3 (thread+0x2b1040)

  Thread T1 (tid=100436, finished) created by main thread at:
    #0 pthread_create /usr/src/contrib/llvm-project/compiler-rt/lib/
tsan/rtl/tsan_interceptors_posix.cpp:962:3 (thread+0x271703)
    #1 main /home/kai/sanitizers/thread.c:17:3 (thread+0x2b1021)

SUMMARY: ThreadSanitizer: data race /home/kai/sanitizers/
thread.c:11:35 in consumer
```

```
===================
ThreadSanitizer: reported 1 warnings
```

The report points us to *lines 6* and *11* of the source file, where the global variable is accessed. It also shows that two threads named *T1* and *T2* accessed the variable and the file and line number of the respective calls to the `pthread_create()` function.

With that, we've learned how to use three different types of sanitizers to identify common problems in applications. The `address` sanitizer helps us identify common memory access errors, such as out-of-bounds accesses or using memory after it's been freed. Using the `memory` sanitizer, we can find access to uninitialized memory, and the `thread` sanitizer helps us identify data races.

In the next section, we'll try to trigger the sanitizers by running our application on random data. This process is known as **fuzz testing**.

Finding bugs with libFuzzer

To test your application, you'll need to write **unit tests**. This is a great way to make sure your software behaves correctly and as you might expect. However, because of the exponential number of possible inputs, you'll probably miss certain weird inputs, and a few bugs as well.

Fuzz testing can help here. The idea is to present your application with randomly generated data, or data based on valid input but with random changes. This is done repeatedly, so your application is tested with a large number of inputs, which is why fuzz testing can be a powerful testing approach. It has been noted that fuzz testing has assisted in finding hundreds of bugs within web browsers and other software.

Interestingly, LLVM comes with its own fuzz testing library. Originally part of the LLVM core libraries, the **libFuzzer** implementation was finally moved to `compiler-rt`. The library is designed to test small and fast functions.

Let's run a small example to see how libFuzzer works. First, you will need to provide the `LLVMFuzzerTestOneInput()` function. This function is called by the **fuzzer driver** and provides you with some input. The following function counts consecutive ASCII digits in the input. Once it's done that, we'll feed random input to it.

You'll need to save the example in the `fuzzer.c` file:

```c
#include <stdint.h>
#include <stdlib.h>

int count(const uint8_t *Data, size_t Size) {
  int cnt = 0;
  if (Size)
    while (Data[cnt] >= '0' && Data[cnt] <= '9') ++cnt;
```

```
        return cnt;
}

int LLVMFuzzerTestOneInput(const uint8_t *Data, size_t Size) {
    count(Data, Size);
    return 0;
}
```

In the preceding code, the count() function counts the number of digits in the memory pointed to by the Data variable. The size of the data is only checked to determine if there are any bytes available. Inside the while loop, the size is not checked.

Used with normal **C strings**, there will be no error because C strings are always terminated by a 0 byte. The LLVMFuzzerTestOneInput() function is the so-called **fuzz target**, and it is the function called by libFuzzer. It calls the function we want to test and returns 0, which is currently the only allowed value.

To compile the file with libFuzzer, you must add the -fsanitize=fuzzer option. The recommendation is to also enable the address sanitizer and the generation of debug symbols. We can use the following command to compile the fuzzer.c file:

```
$ clang -fsanitize=fuzzer,address -g fuzzer.c -o fuzzer
```

When you run the test, it emits a lengthy report. The report contains more information than a stack trace, so let's have a closer look at it:

1. The first line tells you the seed that was used to initialize the random number generator. You can use the –seed= option to repeat this execution:

    ```
    INFO: Seed: 1297394926
    ```

2. By default, libFuzzer limits inputs to, at most, 4096 bytes. You can change the default by using the –max_len= option:

    ```
    INFO: -max_len is not provided; libFuzzer will not generate
    inputs larger than 4096 bytes
    ```

3. Now, we can run the test without providing sample input. The set of all sample inputs is called corpus, and it is empty for this run:

    ```
    INFO: A corpus is not provided, starting from an empty corpus
    ```

4. Some information about the generated test data will follow. It shows you that 28 inputs were tried and 6 inputs, which together have a length of 19 bytes, were found, which together cover 6 coverage points or basic blocks:

```
#28      NEW    cov: 6 ft: 9 corp: 6/19b lim: 4 exec/s: 0 rss:
29Mb L: 4/4 MS: 4 CopyPart-PersAutoDict-CopyPart-ChangeByte- DE:
"1\x00"-
```

5. After this, a buffer overflow was detected, and it followed the information from the address sanitizer. Lastly, the report tells you where the input causing the buffer overflow is saved:

```
artifact_prefix='./'; Test unit written to ./crash-17ba0791499db
908433b80f37c5fbc89b870084b
```

With the saved input, the test case can be executed with the same crashing input again:

```
$ ./fuzzer crash-17ba0791499db908433b80f37c5fbc89b870084b
```

This helps identify the problem as we can use the saved input as a direct reproducer to fix whatever problems that may arise. However, only using random data is often not very helpful in every situation. If you try to fuzz test the tinylang lexer or parser, then pure random data leads to immediate rejection of the input because no valid token can be found.

In such cases, it is more useful to provide a small set of valid input, called the corpus. In this situation, the files of the corpus are randomly mutated and used as input. You can think of the input as mostly valid, with just a few bits flipped. This also works great with other input, which must have a certain format. For example, for a library that processes **JPEG** and **PNG** files, you will provide some small **JPEG** and **PNG** files as corpus.

An example of providing the corpus looks like this. You can save the corpus files in one or more directories and you can create a simple corpus for our fuzz test with the help of the printf command:

```
$ mkdir corpus
$ printf "012345\0" >corpus/12345.txt
$ printf "987\0" >corpus/987.txt
```

When running the test, you must provide the directory on the command line:

```
$ ./fuzzer corpus/
```

The corpus is then used as the base for generating random input, as the report tells you:

```
INFO: seed corpus: files: 2 min: 4b max: 7b total: 11b rss: 29Mb
```

Furthermore, if you are testing a function that works on tokens or other magic values, such as a programming language, then you can speed up the process by providing a dictionary with the tokens. For a programming language, the dictionary would contain all the keywords and special symbols used in the language. Moreover, the dictionary definitions follow a simple key-value style. For example, to define the `if` keyword in the dictionary, you can add the following:

```
kw1="if"
```

However, the key is optional, and you can leave it out. Now, you can specify the dictionary file on the command line with the `-dict=` option.

Now that we've covered using libFuzzer to find bugs, let's look at the limitations and alternatives for the libFuzzer implementation.

Limitations and alternatives

The libFuzzer implementation is fast but poses several restrictions on the test target. They are as follows:

- The function under `test` must accept the input as an array in memory. Some library functions require a file path to the data instead, and they cannot be tested with libFuzzer.
- The `exit()` function should not be called.
- The global state should not be altered.
- Hardware random number generators should not be used.

The first two restrictions are an implication of the implementation of libFuzzer as a library. The latter two restrictions are needed to avoid confusion in the evaluation algorithm. If one of these restrictions is not met, then two identical calls to the fuzz target can yield different results.

The best-known alternative tool for fuzz testing is **AFL**, which can be found at `https://github.com/google/AFL`. AFL requires an instrumented binary (an LLVM plugin for instrumentation is provided) and requires the application to take the input as a file path on the command line. AFL and libFuzzer can share the same corpus and the same dictionary files. Thus, it is possible to test an application with both tools. Furthermore, where libFuzzer is not applicable, AFL may be a good alternative.

There are many more ways of influencing the way libFuzzer works. You can read the reference page at `https://llvm.org/docs/LibFuzzer.html` for more details.

In the next section, we look at a different problem an application can have – we'll try to identify performance bottlenecks while using the XRay tool.

Performance profiling with XRay

If your application seems to run slow, then you might want to know where the time is spent in the code. Here, instrumenting the code with **XRay** can assist with this task. Basically, at each function entry and exit, a special call is inserted into the runtime library. This allows you to count how often a function is called, and also how much time is spent in the function. You can find the implementation for the instrumentation pass in the `llvm/lib/XRay/` directory. The runtime portion is part of `compiler-rt`.

In the following example source, real work is simulated by calling the `usleep()` function. The `func1()` function sleeps for 10 µs. The `func2()` function calls `func1()` or sleeps for 100 µs, depending on if the n parameter is odd or even. Inside the `main()` function, both functions are called inside a loop. This is already enough to get interesting information. You'll need to save the following source code in the `xraydemo.c` file:

```
#include <unistd.h>

void func1() { usleep(10); }

void func2(int n) {
  if (n % 2) func1();
  else usleep(100);
}

int main(int argc, char *argv[]) {
  for (int i = 0; i < 100; i++) { func1(); func2(i); }
  return 0;
}
```

To enable the XRay instrumentation during compilation, you will need to specify the `-fxray-instrument` option. It is worth noting that functions with less than 200 instructions are not instrumented. This is because this is an arbitrary threshold defined by the developers, and in our case, the functions would not be instrumented. The threshold can be specified with the `-fxray-instruction-threshold=` option.

Alternatively, we can add a function attribute to control if a function should be instrumented. For example, adding the following prototype would result in us always instrumenting the function:

```
void func1() __attribute__((xray_always_instrument));
```

Likewise, by using the `xray_never_instrument` attribute, you can turn off instrumentation for a function.

We will now use the command-line option and compile the `xraydemo.c` file, as follows:

```
$ clang -fxray-instrument -fxray-instruction-threshold=1 -g\
  xraydemo.c -o xraydemo
```

In the resulting binary, instrumentation is turned off by default. If you run the binary, you will note no difference compared to a non-instrumented binary. The `XRAY_OPTIONS` environment variable is used to control the recording of runtime data. To enable data collection, you can run the application as follows:

```
$ XRAY_OPTIONS="patch_premain=true xray_mode=xray-basic"\
  ./xraydemo
```

The `xray_mode=xray-basic` option tells the runtime that we want to use basic mode. In this mode, all runtime data is collected, which can result in large log files. When the `patch_premain=true` option is given, functions that are run before the `main()` function are also instrumented.

After running this command, a new file will be created in the directory, in which the collected data is stored. You will need to use the **llvm-xray** tool to extract any readable information from this file.

The llvm-xray tool supports various sub-commands. First of all, you can use the `account` sub-command to extract some basic statistics. For example, to get the top 10 most called functions, you can add the `-top=10` option to limit the output, and the `-sort=count` option to specify the function call count as the sort criteria. You can also influence the sort order with the `-sortorder=` option.

The following commands can be run to get the statistics from our program:

```
$ llvm-xray account xray-log.xraydemo.xVsWiE --sort=count\
  --sortorder=dsc --instr_map ./xraydemo
Functions with latencies: 3
   funcid      count        sum   function
        1        150   0.166002   demo.c:4:0:  func1
        2        100   0.543103   demo.c:9:0:  func2
        3          1   0.655643   demo.c:17:0:  main
```

As you can see, the `func1()` function is called most often; you can also see the accumulated time spent in this function. This example only has three functions, so the `-top=` option has no visible effect here, but for real applications, it is very useful.

From the collected data, it is possible to reconstruct all the stack frames that occurred during runtime. You use the `stack` sub-command to view the top 10 stacks. The output shown here has been reduced for brevity:

```
$ llvm-xray stack xray-log.xraydemo.xVsWiE -instr_map\
  ./xraydemo
Unique Stacks: 3
Top 10 Stacks by leaf sum:

Sum: 1325516912
lvl    function            count              sum
#0     main                    1       1777862705
```

```
#1      func2                   50          1325516912


Top 10 Stacks by leaf count:

Count: 100
lvl     function            count                sum
#0      main                    1          1777862705
#1      func1                 100           303596276
```

A **stack frame** is the sequence of how a function is called. The func2 () function is called by the main () function, and this is the stack frame with the largest accumulated time. The depth depends on how many functions are called, and the stack frames are usually large.

This sub-command can also be used to create a **flame graph** from the stack frames. With a flame graph, you can easily identify which functions have a large, accumulated runtime. The output is the stack frames with count and runtime information. Using the flamegraph.pl script, you can convert the data into a **scalable vector graphics (SVG)** file that you can view in your browser.

With the following command, you instruct llvm-xray to output all stack frames with the –all-stacks option. Using the –stack-format=flame option, the output is in the format expected by the flamegraph.pl script. Moreover, with the –aggregation-type option, you can choose if stack frames are aggregated by total time or by the number of invocations. The output of llvm-xray is piped into the flamegraph.pl script, and the resulting output is saved in the flame.svg file:

```
$ llvm-xray stack xray-log.xraydemo.xVsWiE --all-stacks\
  --stack-format=flame --aggregation-type=time\
  --instr_map ./xraydemo | flamegraph.pl >flame.svg
```

After running the command and generating the new flame graph, you can open the generated flame. svg file in your browser. The graphic looks as follows:

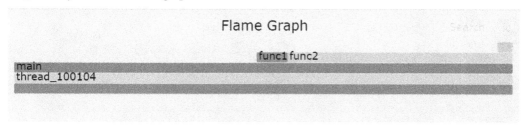

Figure 10.1 – Flame graph produced by llvm-xray

Flame graphs can be confusing at first glance because the *X*-axis does not have the usual meaning of elapsed time. Instead, the functions are simply sorted alphabetically by name. Furthermore, the *Y*-axis of the flame graph shows the stack depth, where the bottom begins counting from zero. The colors are chosen to have a good contrast and have no other meaning. From the preceding graph, you can easily determine the call hierarchy and the time spent in a function.

Information about a stack frame is displayed only after you move the mouse cursor over the rectangle representing the frame. By clicking on the frame, you can zoom into this stack frame. Flame graphs are of great help if you want to identify functions worth optimizing. To find out more about flame graphs, please visit the website of Brendan Gregg, the inventor of flame graphs: `http://www.brendangregg.com/flamegraphs.html`.

Additionally, you can use the `convert` subcommand to convert the data into `.yaml` format or the format used by the **Chrome Trace Viewer Visualization**. The latter is another nice way to create a graphic from the data. To save the data in the `xray.evt` file, you can run the following command:

```
$ llvm-xray convert --output-format=trace_event\
    --output=xray.evt --symbolize --sort\
    --instr_map=./xraydemo xray-log.xraydemo.xVsWiE
```

If you do not specify the `-symbolize` option, then no function names are shown in the resulting graph.

Once you've done that, open Chrome and type `chrome:///tracing`. Next, click on the **Load** button to load the `xray.evt` file. You will see the following visualization of the data:

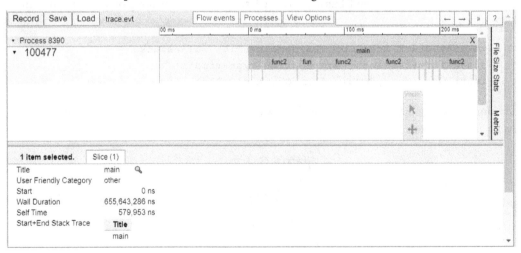

Figure 10.2 – Chrome Trace Viewer Visualization generated by llvm-xray

In this view, the stack frames are sorted by the time the function call occurs. For further interpretation of the visualization, please read the tutorial at `https://www.chromium.org/developers/how-tos/trace-event-profiling-tool`.

> **Tip**
>
> The llvm-xray tool has more functionality that is applicable for performance profiling. You can read about it on the LLVM website at `https://llvm.org/docs/XRay.html` and `https://llvm.org/docs/XRayExample.html`.

In this section, we learned how to instrument an application with XRay, how to collect runtime information, and how to visualize that data. We can use this knowledge to find performance bottlenecks in applications.

Another approach to identifying errors in an application is to analyze the source code, which is done with the clang static analyzer.

Checking the source with the clang static analyzer

The **clang static analyzer** is a tool that performs additional checks on C, C++, and **Objective C** source code. The checks that are performed by the static analyzer are more thorough than the checks the compiler performs. They are also more costly in terms of time and required resources. The static analyzer has a set of checkers, which check for certain bugs.

The tool performs a symbolic interpretation of the source code, which looks at all code paths through an application and derives constraints on the values used in the application from it. **Symbolic interpretation** is a common technique that's used in compilers, for example, to identify constant values. In the context of the static analyzer, the checkers are applied to the derived values.

For example, if the divisor of a division is zero, then the static analyzer warns us about it. We can check this with the following example stored in the `div.c` file:

```
int divbyzero(int a, int b) { return a / b; }

int bug() { return divbyzero(5, 0); }
```

The static analyzer will warn about a division by 0 in this example. However, when compiling, the file, when compiled with the `clang -Wall -c div.c` command, will show no warning.

There are two ways to invoke the static analyzer from the command line. The older tool is **scan-build**, which is included in LLVM and can be used for simple scenarios. The newer tool is **CodeChecker**, available at `https://github.com/Ericsson/codechecker/`. To check a single file, the `scan-build` tool is the easiest solution. You simply pass the `compile` command to the tool; everything else is done automatically:

```
$ scan-build clang -c div.c
scan-build: Using '/usr/home/kai/LLVM/llvm-17/bin/clang-17' for static
analysis
div.c:2:12: warning: Division by zero [core.DivideZero]
  return a / b;
```

```
      ~~^~~
1 warning generated.
scan-build: Analysis run complete.
scan-build: 1 bug found.
scan-build: Run 'scan-view /tmp/scan-build-2021-03-01-023401-8721-1'
to examine bug reports.
```

The output on the screen already tells you that a problem was found – that is, the `core.DivideZero` checker was triggered. However, that is not all. You will find a complete report in HTML in the mentioned subdirectory of the `/tmp` directory. You can then use the `scan-view` command to view the report or open the `index.html` file found in the subdirectory in your browser.

The first page of the report shows you a summary of the bugs that were found:

sanitizers - scan-build results

User:	kai@freebsd
Working Directory:	/usr/home/kai/sanitizers
Command Line:	clang-12 -c div.c
Clang Version:	clang version 12.0.0
Date:	Sat Apr 3 22:47:20 2021

Bug Summary

Bug Type	Quantity	Display?
All Bugs	1	☑
Logic error		
Division by zero	1	☑

Reports

Bug Group	Bug Type ▾	File	Function/Method	Line	Path Length	
Logic error	Division by zero	div.c	divbyzero	2	3	View Report

Figure 10.3 – Summary page

For each error that was found, the summary page shows the type of the error, the location in the source, and the path length after which the analyzer found the error. A link to a detailed report for the error is also provided.

The following screenshot shows the detailed report for the error:

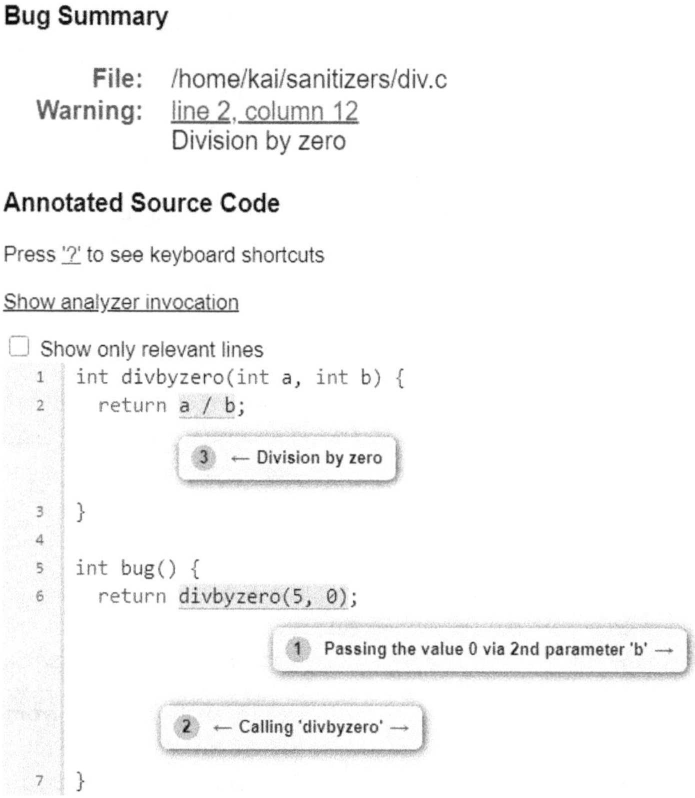

Bug Summary

> **File:** /home/kai/sanitizers/div.c
> **Warning:** line 2, column 12
> Division by zero

Annotated Source Code

Press '?' to see keyboard shortcuts

Show analyzer invocation

☐ Show only relevant lines

```
1    int divbyzero(int a, int b) {
2        return a / b;
```

       3 ← Division by zero

```
3    }
4
5    int bug() {
6        return divbyzero(5, 0);
```

      1 Passing the value 0 via 2nd parameter 'b' →

      2 ← Calling 'divbyzero' →

```
7    }
```

Figure 10.4 – Detailed report

With this detailed report, you can verify the error by following the numbered bubbles. Our simple example shows how passing 0 as a parameter value leads to a division by zero error.

Thus, human verification is required. If the derived constraints are not precise enough for a certain checker, then false positives are possible – that is, an error is reported for perfectly fine code. Based on the report, you can use them to identify false positives.

You are not limited to checkers that are provided with the tool – you can also add new checkers. The next section demonstrates how to do this.

Adding a new checker to the clang static analyzer

Many C libraries provide functions that must be used in pairs. For example, the C standard library provides the `malloc()` and `free()` functions. The memory that's allocated by `malloc()` function must be freed exactly one time by the `free()` function. Not calling the `free()` function, or calling it several times, is a programming error. There are many more instances of this coding pattern, and the static analyzer provides checkers for some of them.

The **iconv** library provides functions for converting text from one encoding into another – for example, from Latin-1 encoding into UTF-16 encoding. To perform the conversion, the implementation needs to allocate memory. To transparently manage the internal resources, the iconv library provides the `iconv_open()` and `iconv_close()` functions, which must be used in pairs, similar to the memory management functions. No checker is implemented for those functions, so let's implement one.

To add a new checker to the clang static analyzer, you must create a new subclass of the `Checker` class. The static analyzer tries all possible paths through the code. The analyzer engine generates events at certain points – for example, before or after a function call. Moreover, your class must provide callbacks for these events if you need to handle them. The `Checker` class and the registrations for the events are provided in the `clang/include/clang/StaticAnalyzer/Core/Checker.h` header file.

Usually, a checker needs to track some symbols. However, the checker cannot manage the state because it does not know which code path the analyzer engine currently tries. Therefore, the tracked state must be registered with the engine, and can only be changed using a `ProgramStateRef` instance.

To detect the errors, the checker needs to track the descriptor that's returned from the `iconv_open()` function. The analyzer engine returns a `SymbolRef` instance for the return value of the `iconv_open()` function. We associate this symbol with a state to reflect if `iconv_close()` was called or not. For the state, we create the `IconvState` class, which encapsulates a `bool` value.

The new `IconvChecker` class needs to handle four types of events:

- `PostCall`, which occurs after a function call. After the `iconv_open()` function was called, we retrieved the symbol for the return value and remembered it as being in an "opened" state.

- `PreCall`, which occurs before a function call. Before the `iconv_close()` function is called, we check if the symbol for the descriptor is in an "opened" state. If not, then the `iconv_close()` function was already called for the descriptor, and we have detected a double call to the function.

- `DeadSymbols`, which occurs when unused symbols are cleaned up. We check if an unused symbol for a descriptor is still in an "opened" state. If it is, then we have detected a missing call to `iconv_close()`, which is a resource leak.

- `PointerEscape`, which is called when the symbols can no longer be tracked by the analyzer. In this case, we remove the symbol from the state because we can no longer reason about whether the descriptor was closed or not.

We can create a new directory to implement the new checker as a clang plugin, and add the implementations within the `IconvChecker.cpp` file:

1. For the implementation, we need to include several header files. The `include` file, `BugType.h` is required for emitting reports. The header file, `Checker.h`, provides the declaration of the `Checker` class and the callbacks for the events, which are declared in the `CallEvent` file Moreover, the `CallDescription.h` file helps with matching functions and methods. Finally, the `CheckerContext.h` file is required for declaring the `CheckerContext` class, which is the central class that provides access to the state of the analyzer:

```
#include "clang/StaticAnalyzer/Core/BugReporter/BugType.h"
#include "clang/StaticAnalyzer/Core/Checker.h"
#include "clang/StaticAnalyzer/Core/PathSensitive/
CallDescription.h"
#include "clang/StaticAnalyzer/Core/PathSensitive/CallEvent.h"
#include "clang/StaticAnalyzer/Core/PathSensitive/
CheckerContext.h"
#include "clang/StaticAnalyzer/Frontend/CheckerRegistry.h"
#include <optional>
```

2. To avoid typing the namespace names, we can use the `clang` and `ento` namespaces:

```
using namespace clang;
using namespace ento;
```

3. We associate a state with each symbol representing an iconv descriptor. The state can be open or closed, and we use a `bool` typed variable, with `true` value for the open state. The state value is encapsulated in the `IconvState` struct. This struct is used with a `FoldingSet` data structure, which is a hash set that filters duplicate entries. To be usable with this data structure implementation, the `Profile()` method is added here, which sets the unique bits of this struct. We put the struct into an anonymous namespace to avoid polluting the global namespace. Instead of exposing the `bool` value, the class provides the `getOpened()` and `getClosed()` factory methods and the `isOpen()` query method:

```
namespace {
class IconvState {
  const bool IsOpen;

  IconvState(bool IsOpen) : IsOpen(IsOpen) {}
public:
  bool isOpen() const { return IsOpen; }

  static IconvState getOpened() {
    return IconvState(true);
  }
```

```
static IconvState getClosed() {
  return IconvState(false);
}

bool operator==(const IconvState &O) const {
  return IsOpen == O.IsOpen;
}

void Profile(llvm::FoldingSetNodeID &ID) const {
  ID.AddInteger(IsOpen);
}
};
} // namespace
```

4. The `IconvState` struct represents the state of an iconv descriptor, which is represented by a symbol of the `SymbolRef` class. This is best done with a map, which has the symbol as the key and the state as the value. As explained earlier, the checker cannot hold the state. Instead, the state must be registered with the global program state, which is done with the `REGISTER_MAP_WITH_PROGRAMSTATE` macro. This macro introduces the `IconvStateMap` name, which we will use later to access the map:

```
REGISTER_MAP_WITH_PROGRAMSTATE(IconvStateMap, SymbolRef,
                               IconvState)
```

5. We also implement the `IconvChecker` class in an anonymous namespace. The requested `PostCall`, `PreCall`, `DeadSymbols`, and `PointerEscape` events are template parameters to the `Checker` base class:

```
namespace {
class IconvChecker
    : public Checker<check::PostCall, check::PreCall,
                     check::DeadSymbols,
                     check::PointerEscape> {
```

6. The `IconvChecker` class has fields of the `CallDescription` type, which are used to identify function calls to `iconv_open()`, `iconv()`, and `iconv_close()` in the program:

```
CallDescription IconvOpenFn, IconvFn, IconvCloseFn;
```

7. The class also holds references to the detected bug types:

```
std::unique_ptr<BugType> DoubleCloseBugType;
std::unique_ptr<BugType> LeakBugType;
```

8. Finally, the class has a couple of methods. Besides the constructor and the methods for the call events, we also need a method to emit a bug report:

```cpp
    void report(ArrayRef<SymbolRef> Syms,
                const BugType &Bug, StringRef Desc,
                CheckerContext &C, ExplodedNode *ErrNode,
                std::optional<SourceRange> Range =
                    std::nullopt) const;

  public:
    IconvChecker();
    void checkPostCall(const CallEvent &Call,
                       CheckerContext &C) const;
    void checkPreCall(const CallEvent &Call,
                      CheckerContext &C) const;
    void checkDeadSymbols(SymbolReaper &SymReaper,
                          CheckerContext &C) const;
    ProgramStateRef
    checkPointerEscape(ProgramStateRef State,
                       const InvalidatedSymbols &Escaped,
                       const CallEvent *Call,
                       PointerEscapeKind Kind) const;
  };
} // namespace
```

9. The implementation of the constructor of the `IconvChecker` class initializes the `CallDescription` fields using the name of the functions, and creates the objects representing the bug types:

```cpp
IconvChecker::IconvChecker()
      : IconvOpenFn({"iconv_open"}), IconvFn({"iconv"}),
        IconvCloseFn({"iconv_close"}, 1) {
    DoubleCloseBugType.reset(new BugType(
        this, "Double iconv_close", "Iconv API Error"));
    LeakBugType.reset(new BugType(
        this, "Resource Leak", "Iconv API Error",
        /*SuppressOnSink=*/true));
  }
```

10. Now, we can implement the first call event method, `checkPostCall()`. This method is called after the analyzer has executed a function call. If the executed function is not a global C function and not named `iconv_open`, then there is nothing to do:

```
void IconvChecker::checkPostCall(
    const CallEvent &Call, CheckerContext &C) const {
  if (!Call.isGlobalCFunction())
    return;
  if (!IconvOpenFn.matches(Call))
    return;
```

11. Otherwise, we can try to get the return value of the function as a symbol. To store the symbol with the open state in the global program state, we need to get a `ProgramStateRef` instance from the `CheckerContext` instance. The state is immutable, so adding the symbol to the state results in a new state. Finally, the analyzer engine is informed about the new state with a call to the `addTransition()` method:

```
  if (SymbolRef Handle =
          Call.getReturnValue().getAsSymbol()) {
    ProgramStateRef State = C.getState();
    State = State->set<IconvStateMap>(
        Handle, IconvState::getOpened());
    C.addTransition(State);
  }
}
```

12. Likewise, the `checkPreCall()` method is called before the analyzer executes a function. Only a global C function called `iconv_close` is of interest to us:

```
void IconvChecker::checkPreCall(
    const CallEvent &Call, CheckerContext &C) const {
  if (!Call.isGlobalCFunction()) {
    return;
  }
  if (!IconvCloseFn.matches(Call)) {
    return;
  }
```

13. If the symbol for the first argument of the function, which is the iconv descriptor, is known, then we can retrieve the state of the symbol from the program state:

```
  if (SymbolRef Handle =
          Call.getArgSVal(0).getAsSymbol()) {
    ProgramStateRef State = C.getState();
```

```
if (const IconvState *St =
        State->get<IconvStateMap>(Handle)) {
```

14. If the state represents the closed state, then we have detected a double close error, and we can generate a bug report for it. The call to generateErrorNode() can return a nullptr value if an error report was already generated for this path, so we have to check for this situation:

```
if (!St->isOpen()) {
    if (ExplodedNode *N = C.generateErrorNode()) {
        report(Handle, *DoubleCloseBugType,
                "Closing a previous closed iconv "
                "descriptor",
                C, N, Call.getSourceRange());
    }
    return;
}
}
```

15. Otherwise, we must set the state for the symbol to the "closed" state:

```
State = State->set<IconvStateMap>(
    Handle, IconvState::getClosed());
C.addTransition(State);
}
}
```

16. The checkDeadSymbols() method is called to clean up unused symbols. We loop over all symbols we track and ask the SymbolReaper instance if the current symbol is dead:

```
void IconvChecker::checkDeadSymbols(
    SymbolReaper &SymReaper, CheckerContext &C) const {
    ProgramStateRef State = C.getState();
    SmallVector<SymbolRef, 8> LeakedSyms;
    for (auto [Sym, St] : State->get<IconvStateMap>()) {
        if (SymReaper.isDead(Sym)) {
```

17. If the symbol is dead, then we need to check the state. If the state is still open, then this is a potential resource leak. There is one exception: iconv_open() returns -1 in case of an error. If the analyzer is in a code path that handles this error, then it is wrong to assume a resource leak because the function call failed. We try to get the value of the symbol from the ConstraintManager instance, and we do not consider the symbol as a resource leak if this value is -1. We add a leaked symbol to a SmallVector instance to generate the error report later. Finally, we remove the dead symbol from the program state:

```
if (St.isOpen()) {
    bool IsLeaked = true;
```

```
    if (const llvm::APSInt *Val =
            State->getConstraintManager().getSymVal(
                State, Sym))
        IsLeaked = Val->getExtValue() != -1;
    if (IsLeaked)
        LeakedSyms.push_back(Sym);
    }
    State = State->remove<IconvStateMap>(Sym);
    }
}
```

18. After the loop, we call the `generateNonFatalErrorNode()` method. This method transitions to the new program state and returns an error node if there is not already an error node for this path. The `LeakedSyms` container holds the (possibly empty) list of leaked symbols, and we call the `report()` method to generate an error report:

```
    if (ExplodedNode *N =
            C.generateNonFatalErrorNode(State)) {
        report(LeakedSyms, *LeakBugType,
            "Opened iconv descriptor not closed", C, N);
    }
}
```

19. The `checkPointerEscape()` function is called when the analyzer detects a function call for which the parameters cannot be tracked. In such a case, we must assume that we do not know if the iconv descriptor will be closed inside the function or not. The exceptions are a call to `iconv()`, which does the conversion and is known to not call the `iconv_close()` function, and the `iconv_close()` function itself, which we handle in the `checkPreCall()` method. We also do not change the state if the call is inside a system header file, and if we know that the arguments do not escape in the called function. In all other cases, we remove the symbol from the state:

```
ProgramStateRef IconvChecker::checkPointerEscape(
    ProgramStateRef State,
    const InvalidatedSymbols &Escaped,
    const CallEvent *Call,
    PointerEscapeKind Kind) const {
    if (Kind == PSK_DirectEscapeOnCall) {
        if (IconvFn.matches(*Call) ||
            IconvCloseFn.matches(*Call))
            return State;
        if (Call->isInSystemHeader() ||
            !Call->argumentsMayEscape())
            return State;
```

```
    }
    for (SymbolRef Sym : Escaped)
      State = State->remove<IconvStateMap>(Sym);
    return State;
}
```

20. The `report()` method generates an error report. The important parameters of the method are an array of symbols, the type of the bug, and a bug description. Inside the method, a bug report is created for each symbol, and the symbol is marked as the interesting one for the bug. If a source range is provided as a parameter, then this is also added to the report. Finally, the report is emitted:

```
void IconvChecker::report(
    ArrayRef<SymbolRef> Syms, const BugType &Bug,
    StringRef Desc, CheckerContext &C,
    ExplodedNode *ErrNode,
    std::optional<SourceRange> Range) const {
  for (SymbolRef Sym : Syms) {
    auto R = std::make_unique<PathSensitiveBugReport>(
        Bug, Desc, ErrNode);
    R->markInteresting(Sym);
    if (Range)
      R->addRange(*Range);
    C.emitReport(std::move(R));
  }
}
```

21. Now, the new checker needs to be registered at a `CheckerRegistry` instance. When our plugin is loaded, the `clang_registerCheckers()` function is used, in which we perform the registration. Each checker has a name and belongs to a package. We call the `IconvChecker` checker and put it into the `unix` packager because the iconv library is a standard POSIX interface. This is the first parameter of the `addChecker()` method. The second parameter is a brief documentation of the functionality, and the third parameter can be a URI to a document that provides more information about the checker:

```
extern "C" void
clang_registerCheckers(CheckerRegistry &registry) {
  registry.addChecker<IconvChecker>(
      "unix.IconvChecker",
      "Check handling of iconv functions", "");
}
```

22. Finally, we need to declare the version of the static analyzer API we are using, which enables the system to determine if the plugin is compatible:

```
extern "C" const char clang_analyzerAPIVersionString[] =
    CLANG_ANALYZER_API_VERSION_STRING;
```

This finishes the implementation of the new checker. To build the plugin, we also need to create a build description in the CMakeLists.txt file which lives in the same directory as IconvChecker.cpp:

23. Begin by defining the required **CMake** version and the name of the project:

```
cmake_minimum_required(VERSION 3.20.0)
project(iconvchecker)
```

24. Next, include the LLVM files. If CMake can't find the files automatically, then you have to set the LLVM_DIR variable so that it points to the LLVM directory containing the CMake files:

```
find_package(LLVM REQUIRED CONFIG)
```

25. Append the LLVM directory with the CMake files to the search path, and include the required modules from LLVM:

```
list(APPEND CMAKE_MODULE_PATH ${LLVM_DIR})
include(AddLLVM)
include(HandleLLVMOptions)
```

26. Then, load the CMake definitions for clang. If CMake can't find the files automatically, then you have to set the Clang_DIR variable so that it points to the clang directory containing the CMake files:

```
find_package(Clang REQUIRED)
```

27. Next, append the Clang directory with the CMake files to the search path, and include the required modules from Clang:

```
list(APPEND CMAKE_MODULE_PATH ${Clang_DIR})
include(AddClang)
```

28. Then, define where the header files and the library files are located, and which definitions to use:

```
include_directories("${LLVM_INCLUDE_DIR}"
                    "${CLANG_INCLUDE_DIRS}")
add_definitions("${LLVM_DEFINITIONS}")
link_directories("${LLVM_LIBRARY_DIR}")
```

29. The previous definitions set up the build environment. Insert the following command, which defines the name of your plugin, the source file(s) of the plugin, and that it is a clang plugin:

```
add_llvm_library(IconvChecker MODULE IconvChecker.cpp
                 PLUGIN_TOOL clang)
```

30. On **Windows**, the plugin support is different from **Unix**, and the required LLVM and clang libraries must be linked in. The following code ensures this:

```
if(WIN32 OR CYGWIN)
  set(LLVM_LINK_COMPONENTS Support)
  clang_target_link_libraries(IconvChecker PRIVATE
    clangAnalysis
    clangAST
    clangStaticAnalyzerCore
    clangStaticAnalyzerFrontend)
endif()
```

Now, we can configure and build the plugin, assuming that the CMAKE_GENERATOR and CMAKE_ BUILD_TYPE environment variables are set:

```
$ cmake -DLLVM_DIR=~/LLVM/llvm-17/lib/cmake/llvm \
        -DClang_DIR=~/LLVM/llvm-17/lib/cmake/clang \
        -B build
$ cmake --build build
```

You can test the new checker with the following source saved in the conv.c file, which has two calls to the iconv_close() function:

```
#include <iconv.h>

void doconv() {
  iconv_t id = iconv_open("Latin1", "UTF-16");
  iconv_close(id);
  iconv_close(id);
}
```

To use the plugin with the scan-build script, you need to specify the path to the plugin via the -load-plugin option. A run with the conv.c file looks like:

```
$ scan-build -load-plugin build/IconvChecker.so clang-17 \
             -c conv.c
scan-build: Using '/home/kai/LLVM/llvm-17/bin/clang-17' for static
analysis
conv.c:6:3: warning: Closing a previous closed iconv descriptor [unix.
IconvChecker]
```

```
     6 |        iconv_close(id);
       |        ^~~~~~~~~~~~~~~
1 warning generated.
scan-build: Analysis run complete.
scan-build: 1 bug found.
scan-build: Run 'scan-view /tmp/scan-build-2023-08-08-114154-12451-1'
to examine bug reports.
```

With that, you've learned how to extend the clang static analyzer with your own checker. You can use this knowledge to either create new general checkers and contribute them to the community or create checkers specifically built for your needs, to raise the quality of your product.

The static analyzer is built by leveraging the clang infrastructure. The next section introduces you to how can build your own plugin extending clang.

Creating your own clang-based tool

The static analyzer is an impressive example of what you can do with the clang infrastructure. It is also possible to extend clang with plugins so that you can add your own functionality to clang. The technique is very similar to adding a pass plugin to LLVM.

Let's explore the functionality with a simple plugin. The LLVM coding standard requires function names to begin with a lowercase letter. However, the coding standard has evolved, and there are many instances in which a function begins with an uppercase letter. A plugin that warns about a violation of the naming rule can help fix this issue, so let's give it a try.

Because you want to run a user-defined action over the AST, you need to define a subclass of the PluginASTAction class. If you write your own tool using the clang libraries, then you can define subclasses of the ASTFrontendAction class for your actions. The PluginASTAction class is a subclass of the ASTFrontendAction class, with the additional ability to parse command-line options.

The other class you need is a subclass of the ASTConsumer class. An AST consumer is a class using which you can run an action over an AST, regardless of the origin of the AST. Nothing more is needed for our first plugin. You can create the implementation in the NamingPlugin.cpp file as follows:

1. Begin by including the required header files. Besides the mentioned ASTConsumer class, you also need an instance of the compiler and the plugin registry:

    ```
    #include "clang/AST/ASTConsumer.h"
    #include "clang/Frontend/CompilerInstance.h"
    #include "clang/Frontend/FrontendPluginRegistry.h"
    ```

2. Use the `clang` namespace and put your implementation into an anonymous `namespace` to avoid name clashes:

```
using namespace clang;
namespace {
```

3. Next, define your subclass of the `ASTConsumer` class. Later, you will want to emit warnings in case you detect a violation of the naming rule. To do so, you need a reference to a `DiagnosticsEngine` instance.

4. You'll need to store a `CompilerInstance` instance in the class, after which you can ask for a `DiagnosticsEngine` instance:

```
class NamingASTConsumer : public ASTConsumer {
  CompilerInstance &CI;

public:
  NamingASTConsumer(CompilerInstance &CI) : CI(CI) {}
```

5. An `ASTConsumer` instance has several entry methods. The `HandleTopLevelDecl()` method fits our purpose. The method is called for each declaration at the top level. This includes more than functions – for example, variables. So, you must use the LLVM RTTI `dyn_cast<>()` function to determine if the declaration is a function declaration. The `HandleTopLevelDecl()` method has a declaration group as a parameter, which can contain more than a single declaration. This requires a loop over the declarations. The following code shows the `HandleTopLevelDecl()` method:

```
bool HandleTopLevelDecl(DeclGroupRef DG) override {
  for (DeclGroupRef::iterator I = DG.begin(),
                              E = DG.end();
       I != E; ++I) {
    const Decl *D = *I;
    if (const FunctionDecl *FD =
            dyn_cast<FunctionDecl>(D)) {
```

6. After finding a function declaration, you'll need to retrieve the name of the function. You'll also need to make sure that the name is not empty:

```
std::string Name =
    FD->getNameInfo().getName().getAsString();
assert(Name.length() > 0 &&
        "Unexpected empty identifier");
```

If the function name does not start with a lowercase letter, then you'll have a violation of the naming rule that was found:

```
char &First = Name.at(0);
if (!(First >= 'a' && First <= 'z')) {
```

7. To emit a warning, you need a `DiagnosticsEngine` instance. Additionally, you need a message ID. Inside clang, the message ID is defined as an enumeration. Because your plugin is not part of clang, you need to create a custom ID, which you can then use to emit the warning:

```
DiagnosticsEngine &Diag = CI.getDiagnostics();
unsigned ID = Diag.getCustomDiagID(
    DiagnosticsEngine::Warning,
    "Function name should start with "
    "lowercase letter");
Diag.Report(FD->getLocation(), ID);
```

8. Except for closing all open braces, you need to return `true` from this function to indicate that processing can continue:

```
        }
      }
    }
    return true;
  }
};
```

9. Next, you need to create the `PluginASTAction` subclass, which implements the interface called by clang:

```
class PluginNamingAction : public PluginASTAction {
public:
```

The first method you must implement is the `CreateASTConsumer()` method, which returns an instance of your `NamingASTConsumer` class. This method is called by clang, and the passed `CompilerInstance` instance gives you access to all the important classes of the compiler. The following code demonstrates this:

```
std::unique_ptr<ASTConsumer>
CreateASTConsumer(CompilerInstance &CI,
                  StringRef file) override {
  return std::make_unique<NamingASTConsumer>(CI);
}
```

10. A plugin also has access to command-line options. Your plugin has no command-line parameters, and you will only return `true` to indicate success:

```
bool ParseArgs(const CompilerInstance &CI,
               const std::vector<std::string> &args)
                                            override {
    return true;
}
```

11. The action type of a plugin describes when the action is invoked. The default value is `Cmdline`, which means that the plugin must be named on the command line to be invoked. You'll need to override the method and change the value to `AddAfterMainAction`, which automatically runs the action:

```
PluginASTAction::ActionType getActionType() override {
    return AddAfterMainAction;
}
```

12. The implementation of your `PluginNamingAction` class is finished; only the closing braces for the class and the anonymous namespace are missing. Add them to the code, as follows:

```
};
}
```

13. Lastly, you need to register the plugin. The first parameter is the name of the plugin, while the second parameter is help text:

```
static FrontendPluginRegistry::Add<PluginNamingAction>
    X("naming-plugin", "naming plugin");
```

This finishes the implementation of the plugin. To compile the plugin, create a build description in the `CMakeLists.txt` file. The plugin lives outside the clang source tree, so you need to set up a complete project. You can do so by following these steps:

1. Begin by defining the required **CMake** version and the name of the project:

```
cmake_minimum_required(VERSION 3.20.0)
project(naminglugin)
```

2. Next, include the LLVM files. If CMake can't find the files automatically, then you have to set the `LLVM_DIR` variable so that it points to the LLVM directory containing the CMake files:

```
find_package(LLVM REQUIRED CONFIG)
```

3. Append the LLVM directory with the CMake files to the search path, and include some required modules:

    ```
    list(APPEND CMAKE_MODULE_PATH ${LLVM_DIR})
    include(AddLLVM)
    include(HandleLLVMOptions)
    ```

4. Then, load the CMake definitions for clang. If CMake can't find the files automatically, then you have to set the `Clang_DIR` variable so that it points to the clang directory containing the CMake files:

    ```
    find_package(Clang REQUIRED)
    ```

5. Next, define where the headers files and the library files are located, and which definitions to use:

    ```
    include_directories("${LLVM_INCLUDE_DIR}"
                        "${CLANG_INCLUDE_DIRS}")
    add_definitions("${LLVM_DEFINITIONS}")
    link_directories("${LLVM_LIBRARY_DIR}")
    ```

6. The previous definitions set up the build environment. Insert the following command, which defines the name of your plugin, the source file(s) of the plugin, and that it is a clang plugin:

    ```
    add_llvm_library(NamingPlugin MODULE NamingPlugin.cpp
                     PLUGIN_TOOL clang)
    ```

    On **Windows**, the plugin support is different from **Unix**, and the required LLVM and clang libraries must be linked in. The following code ensures this:

    ```
    if(WIN32 OR CYGWIN)
      set(LLVM_LINK_COMPONENTS Support)
      clang_target_link_libraries(NamingPlugin PRIVATE
        clangAST clangBasic clangFrontend clangLex)
    endif()
    ```

Now, we can configure and build the plugin, assuming that the CMAKE_GENERATOR and CMAKE_BUILD_TYPE environment variables are set:

```
$ cmake -DLLVM_DIR=~/LLVM/llvm-17/lib/cmake/llvm \
        -DClang_DIR=~/LLVM/llvm-17/lib/cmake/clang \
        -B build
$ cmake --build build
```

These steps create the `NamingPlugin.so` shared library in the `build` directory.

To test the plugin, save the following source as the `naming.c` file. The function name, `Func1`, violates the naming rule, but not the `main` name:

```
int Func1() { return 0; }
int main() { return Func1(); }
```

To invoke the plugin, you need to specify the `-fplugin=` option:

```
$ clang -fplugin=build/NamingPlugin.so naming.c
naming.c:1:5: warning: Function name should start with lowercase
letter
int Func1() { return 0; }
    ^
1 warning generated.
```

This kind of invocation requires that you override the `getActionType()` method of the `PluginASTAction` class and that you return a value different from the `Cmdline` default value.

If you did not do this – for example, because you want to have more control over the invocation of the plugin action – then you can run the plugin from the compiler command line:

```
$ clang -cc1 -load ./NamingPlugin.so -plugin naming-plugin\
    naming.c
```

Congratulations – you have built your first clang plugin!

The disadvantage of this approach is that it has certain limitations. The `ASTConsumer` class has different entry methods, but they are all coarse-grained. This can be solved by using the `RecursiveASTVisitor` class. This class traverses all AST nodes, and you can override the `VisitXXX()` methods you are interested in. You can rewrite the plugin so that it uses the visitor by following these steps:

1. You need an additional `include` for the definition of the `RecursiveASTVisitor` class. Insert it as follows:

    ```
    #include "clang/AST/RecursiveASTVisitor.h"
    ```

2. Then, define the visitor as the first class in the anonymous namespace. You will only store a reference to the AST context, which will give you access to all the important methods for AST manipulation, including the `DiagnosticsEngine` instance, which is required for emitting the warning:

    ```
    class NamingVisitor
        : public RecursiveASTVisitor<NamingVisitor> {
    private:
      ASTContext &ASTCtx;

    public:
    ```

```
    explicit NamingVisitor(CompilerInstance &CI)
        : ASTCtx(CI.getASTContext()) {}
```

3. During traversal, the `VisitFunctionDecl()` method is called whenever a function declaration is discovered. Copy the body of the inner loop inside the `HandleTopLevelDecl()` function here:

```
virtual bool VisitFunctionDecl(FunctionDecl *FD) {
  std::string Name =
      FD->getNameInfo().getName().getAsString();
  assert(Name.length() > 0 &&
         "Unexpected empty identifier");
  char &First = Name.at(0);
  if (!(First >= 'a' && First <= 'z')) {
    DiagnosticsEngine &Diag = ASTCtx.getDiagnostics();
    unsigned ID = Diag.getCustomDiagID(
        DiagnosticsEngine::Warning,
        "Function name should start with "
        "lowercase letter");
    Diag.Report(FD->getLocation(), ID);
  }
  return true;
}
};
```

4. This finishes the visitor's implementation. In your `NamingASTConsumer` class, you will now only store a visitor instance:

```
std::unique_ptr<NamingVisitor> Visitor;

public:
NamingASTConsumer(CompilerInstance &CI)
    : Visitor(std::make_unique<NamingVisitor>(CI)) {}
```

5. Remove the `HandleTopLevelDecl()` method – the functionality is now in the visitor class, so you'll need to override the `HandleTranslationUnit()` method instead. This class is called once for each translation unit. You will start the AST traversal here:

```
void
HandleTranslationUnit(ASTContext &ASTCtx) override {
  Visitor->TraverseDecl(
      ASTCtx.getTranslationUnitDecl());
}
```

This new implementation has the same functionality. The advantage is that it is easier to extend. For example, if you want to examine variable declarations, then you must implement the `VisitVarDecl()` method. Alternatively, if you want to work with a statement, then you must implement the `VisitStmt()` method. With this approach, you have a visitor method for each entity of the C, C++, and Objective C languages.

Having access to the AST allows you to build plugins that perform complex tasks. Enforcing naming conventions, as described in this section, is a useful addition to clang. Another useful addition you could implement as a plugin is the calculation of a software metric such as **cyclomatic complexity**. You can also add or replace AST nodes, allowing you, for example, to add runtime instrumentation. Adding plugins allows you to extend clang in the way you need it.

Summary

In this chapter, you learned how to apply various sanitizers. You detected pointer errors with the `address` sanitizer, uninitialized memory access with the `memory` sanitizer, and performed data races with the `thread` sanitizer. Application errors are often triggered by malformed input, and you implemented fuzz testing to test your application with random data.

You also instrumented your application with XRay to identify the performance bottlenecks, and you also learned about the various ways you can visualize the data. This chapter also taught you how to utilize the clang static analyzer for identifying potential errors by interpreting the source code, and how to create your own clang plugin.

These skills will help you raise the quality of the applications you build as it is certainly good to find runtime errors before your application users complain about them. Applying the knowledge you've gained in this chapter, you can not only find a wide range of common errors, but you can also extend clang with new functionality.

In the next chapter, you will learn how to add a new backend to LLVM.

Part 4:
Roll Your Own Backend

In this section, you will learn about adding a new backend target for a CPU architecture not supported by LLVM, utilizing the TableGen language. Additionally, you will explore the various instruction selection frameworks within LLVM and understand how to implement them. Finally, you will also delve into concepts beyond the instruction selection frameworks within LLVM, which are valuable for highly optimizing backends.

This section comprises the following chapters:

- *Chapter 11, The Target Description*
- *Chapter 12, Instruction Selection*
- *Chapter 13, Beyond Instruction Selection*

11

The Target Description

LLVM has a very flexible architecture. You can also add a new target backend to it. The core of a backend is the target description, from which most of the code is generated. In this chapter, you will learn how to add support for a historical CPU.

In this chapter, you will cover the following:

- *Setting the stage for a new backend* introduces you to the M88k CPU architecture and shows you where to find the required information

- *Adding the new architecture to the Triple class* teaches you how to make LLVM aware of a new CPU architecture

- *Extending the ELF file format definition in LLVM* shows you how to add support for the M88k-specific relocations to the libraries and tools that handle ELF object files

- *Creating the target description* applies your knowledge of the TableGen language to model the register file and instructions in the target description

- *Adding the M88k backend to LLVM* explains the minimal infrastructure required for an LLVM backend

- *Implementing an assembler parser* shows you how to develop the assembler

- *Creating the disassembler* teaches you how to create the disassembler

By the end of the chapter, you will know how to add a new backend to LLVM. You will acquire the knowledge to develop the register file definition and instruction definition in the target description, and you will know how to create the assembler and disassembler from that description.

Setting the stage for a new backend

Whether commercially needed to support a new CPU or only a hobby project to add support for some old architecture, adding a new backend to LLVM is a major task. This and the following two chapters outline what you need to develop for a new backend. We will add a backend for the Motorola M88k architecture, which is a RISC architecture from the 1980s.

> **References**
>
> You can read more about this Motorola architecture on Wikipedia at `https://en.wikipedia.org/wiki/Motorola_88000`. The most important information about this architecture is still available on the internet. You can find the CPU manuals with the instruction set and timing information at `http://www.bitsavers.org/components/motorola/88000/`, and the System V ABI M88k Processor supplement with the definitions of the ELF format and the calling convention at `https://archive.org/details/bitsavers_attunixSysa0138776555SystemVRelease488000ABI1990_8011463`.
>
> OpenBSD, available at `https://www.openbsd.org/`, still supports the LUNA-88k system. On the OpenBSD system, it is easy to create a GCC cross-compiler for M88k. And with GXemul, available at `http://gavare.se/gxemul/`, we get an emulator capable of running certain OpenBSD releases for the M88k architecture.

The M88k architecture is long out of production, but we found enough information and tools to make it an interesting goal to add an LLVM backend for it. We will begin with a very basic task of extending the `Triple` class.

Adding the new architecture to the Triple class

An instance of the `Triple` class represents the target platform LLVM is producing code for. To support a new architecture, the first task is to extend the `Triple` class. In the `llvm/include/llvm/TargetParser/Triple.h` file, add a member to the `ArchType` enumeration along with a new predicate:

```
class Triple {
public:
  enum ArchType {
      // Many more members
      m88k,      // M88000 (big endian): m88k
  };

  /// Tests whether the target is M88k.
  bool isM88k() const {
      return getArch() == Triple::m88k;
  }
```

```
// Many more methods
};
```

Inside the `llvm/lib/TargetParser/Triple.cpp` file, there are many methods that use the `ArchType` enumeration. You need to extend all of them; for example, in the `getArchTypeName()` method, you need to add a new `case` statement as follows:

```
switch (Kind) {
    // Many more cases
    case m88k:              return "m88k";
}
```

Most times, the compiler will warn you if you forget to handle the new m88k enumeration member in one of the functions. Next, we will expand the **Executable and Linkable Format** (ELF).

Extending the ELF file format definition in LLVM

The ELF file format is one of the binary object file formats LLVM supports. ELF itself is defined for many CPU architectures, and there is also a definition for the M88k architecture. All we need to do is to add the definition of the relocations and some flags. The relocations are given in *Chapter 4, Basics of IR Code Generation*, of the *System V ABI M88k Processor* supplement book (see link within the *Setting the stage for a new backend* section at the beginning of the chapter):

1. We need to type the following code into the `llvm/include/llvm/BinaryFormat/ELFRelocs/M88k.def` file:

    ```
    #ifndef ELF_RELOC
    #error "ELF_RELOC must be defined"
    #endif
    ELF_RELOC(R_88K_NONE, 0)
    ELF_RELOC(R_88K_COPY, 1)
    // Many more...
    ```

2. We also add the following flags into the `llvm/include/llvm/BinaryFormat/ELF.h` file, along with the definition of the relocations:

    ```
    // M88k Specific e_flags
    enum : unsigned {
        EF_88K_NABI = 0x80000000,    // Not ABI compliant
        EF_88K_M88110 = 0x00000004   // File uses 88110-specific
    features
    };

    // M88k relocations.
    enum {
    ```

```
    #include "ELFRelocs/M88k.def"
};
```

The code can be added anywhere in the file, but it is best to keep the file structured and insert it before the code for the MIPS architecture.

3. We also need to expand some other methods. In the `llvm/include/llvm/Object/ELFObjectFile.h` file are some methods that translate between enumeration members and strings. For example, we must add a new case statement to the `getFileFormatName()` method:

```
    switch (EF.getHeader()->e_ident[ELF::EI_CLASS]) {
// Many more cases
    case ELF::EM_88K:
        return "elf32-m88k";
}
```

4. Similarly, we extend the `getArch()` method:

```
    switch (EF.getHeader().e_machine) {
// Many more cases
    case ELF::EM_88K:
        return Triple::m88k;
```

5. Lastly, we use the relocation definitions in the `llvm/lib/Object/ELF.cpp` file, in the `getELFRelocationTypeName()` method:

```
    switch (Machine) {
// Many more cases
    case ELF::EM_88K:
        switch (Type) {
#include "llvm/BinaryFormat/ELFRelocs/M88k.def"
        default:
          break;
        }
        break;
    }
```

6. To complete the support, you can valso extend the `llvm/lib/ObjectYAML/ELFYAML.cpp` file. This file is used by the `yaml2obj` and `obj2yaml` tools, which create an ELF file based on a YAML description, and vice versa. The first addition needs to be done in the `ScalarEnumerationTraits<ELFYAML::ELF_EM>::enumeration()` method, which lists all the values for the ELF architectures:

```
    ECase(EM_88K);
```

7. Likewise, in the `ScalarEnumerationTraits<ELFYAML::ELF_REL>::enumeration()` method, you need to include the definitions of the relocations again:

```
   case ELF::EM_88K:
 #include "llvm/BinaryFormat/ELFRelocs/M88k.def"
     break;
```

At this point, we have completed the support of the m88k architecture in the ELF file format. You can use the `llvm-readobj` tool to inspect an ELF object file, for example, created by a cross-compiler on OpenBSD. Likewise, you can create an ELF object file for the m88k architecture with the `yaml2obj` tool.

> **Is adding support for an object file format mandatory?**
>
> Integrating support for an architecture into the ELF file format implementation requires only a couple of lines of code. If the architecture for which you are creating an LLVM backend uses the ELF format, then you should take this route. On the other hand, adding support for a completely new binary file format is a complicated task. If this is required, then an often-used approach is to only output assembler files and use an external assembler to create object files.

With these additions, the LLVM implementation of the ELF file format now supports the M88k architecture. In the next section, we create the target description for the M88k architecture, which describes the instructions, the registers, and many more details of the architecture.

Creating the target description

The **target description** is the heart of a backend implementation. It is written in the TableGen language and defines the basic properties of an architecture, such as the registers and the instruction formats and patterns for instruction selection. If you are not familiar with the TableGen language, then we recommend reading *Chapter 8, The TableGen Language*, first. The base definitions are in the `llvm/include/llvm/Target/Target.td` file, which can be found online at https://github.com/llvm/llvm-project/blob/main/llvm/include/llvm/Target/Target.td. This file is heavily commented on and is a useful source of information about the use of the definitions.

In an ideal world, we would generate the whole backend from the target description. This goal has not yet been reached, and therefore, we will need to extend the generated code later. Because of its size, the target description is split into several files. The top-level file will be `M88k.td`, inside the `llvm/lib/Target/M88k` directory, which also includes the other files. Let's have a look at some files, beginning with the register definition.

Adding the register definition

A CPU architecture usually defines a set of registers. The characteristics of these registers can vary. Some architectures allow access to sub-registers. For example, the x86 architecture has special register names to access only a part of a register value. Other architectures do not implement this. In addition to general-purpose, floating-point, and vector registers, an architecture may have special registers for status codes or configuration of floating-point operations. We need to define all this information for LLVM. The register definitions are stored in the M88kRegisterInfo.td file, also found within the llvm/lib/Target/M88k directory.

The M88k architecture defines general-purpose registers, extended registers for floating-point operations, and control registers. To keep the example small, we only define the general-purpose registers. We begin by defining a super-class for the registers. A register has a name and an encoding. The name is used in the textual representation of an instruction. Similarly, the encoding is used as part of the binary representation of an instruction. The architecture defines 32 registers and the encoding for registers therefore uses 5 bits, so we limit the field holding the encoding. We also define that all the generated C++ code should live in the M88k namespace:

```
class M88kReg<bits<5> Enc, string n> : Register<n> {
  let HWEncoding{15-5} = 0;
  let HWEncoding{4-0} = Enc;
  let Namespace = "M88k";
}
```

Next, we can define all 32 general-purpose registers. The r0 register is special because it always returns the constant 0 when read, so we set the isConstant flag to true for that register:

```
foreach I = 0-31 in {
  let isConstant = !eq(I, 0) in
    def R#I : M88kReg<I, "r"#I>;
}
```

For the register allocator, the single registers need to be grouped into register classes. The sequence order of the registers defines the allocation order. The register allocator also needs other information about the registers such as, for example, the value types, which can be stored in a register, the spill size of a register in bits, and the required alignment in memory. Instead of using the RegisterClass base class directly, we create a new M88kRegisterClass class. This allows us to change the parameter list to our needs. It also avoids the repetition of the C++ namespace name used for the generated code, which is the first argument for the RegisterClass class:

```
class M88kRegisterClass<list<ValueType> types, int size,
                        int alignment, dag regList,
                        int copycost = 1>
  : RegisterClass<"M88k", types, alignment, regList> {
    let Size = size;
```

```
      let CopyCost = copycost;
}
```

In addition, we define a class for register operands. Operands describe the input and output of an instruction. They are used during assembling and disassembling of an instruction, and also in the patterns used by the instruction selection phase. Using our own class, we can give the generated function used to decode a register operand a name that conforms to the LLVM coding guidelines:

```
class M88kRegisterOperand<RegisterClass RC>
    : RegisterOperand<RC> {
  let DecoderMethod = "decode"#RC#"RegisterClass";
}
```

Based on these definitions, we now define the general-purpose registers. Please note that a general-purpose register of the m88k architecture is 32-bits wide and can hold integer and floating-point values. To avoid writing all register names, we use the `sequence` generator, which generates a list of strings based on the template string:

```
def GPR : M88kRegisterClass<[i32, f32], 32, 32,
                           (add (sequence "R%u", 0, 31))>;
```

Likewise, we define the register operand. The `r0` register is special because it contains the constant `0`. This fact can be used by the global instruction selection framework, and therefore, we attach this information to the register operand:

```
def GPROpnd : M88kRegisterOperand<GPR> {
  let GIZeroRegister = R0;
}
```

There is an extension to the m88k architecture that defines an extended register file for floating-point values only. You would define those registers in the same way as the general-purpose registers.

The general-purpose registers are also used in pairs, mainly for 64-bit floating point operations, and we need to model them. We use the `sub_hi` and `sub_lo` sub-register indices to describe the high 32 bits and the low 32 bits. We also need to set the C++ namespace for the generated code:

```
let Namespace = "M88k" in {
  def sub_hi : SubRegIndex<32, 0>;
  def sub_lo : SubRegIndex<32, 32>;
}
```

The register pairs are then defined using the `RegisterTuples` class. The class takes a list of sub-register indices as the first argument and a list of registers as the second argument. We only need even/odd numbered pairs, and we achieve this with the optional fourth parameter of sequence, which is the stride to use when generating the sequence:

```
def GRPair : RegisterTuples<[sub_hi, sub_lo],
                            [(add (sequence "R%u", 0, 30, 2)),
                             (add (sequence "R%u", 1, 31, 2))]>;
```

To use the register pairs, we define a register class and a register operand:

```
def GPR64 : M88kRegisterClass<[i64, f64], 64, 32,
                              (add GRPair), /*copycost=*/ 2>;
def GPR64Opnd : M88kRegisterOperand<GPR64>;
```

Please note that we set the `copycost` parameter to 2 because we need two instructions instead of one to copy a register pair to another register pair.

This finishes our definition of the registers. In the next section, we will define the instruction formats.

Defining the instruction formats and the instruction information

An instruction is defined using the TableGen `Instruction` class. Defining an instruction is a complex task because we have to consider many details. An instruction has a textual representation used by the assembler and the disassembler. It has a name, for example, `and`, and it may have operands. The assembler transforms the textual representation into a binary format, therefore, we must define the layout of that format. For instruction selection, we need to attach a pattern to the instruction. To manage this complexity, we define a class hierarchy. The base classes will describe the various instruction formats and are stored in the `M88kIntrFormats.td` file. The instructions themselves and other definitions required for the instruction selection are stored in the `M88kInstrInfo.td` file.

Let's begin with defining a class for the instructions of the m88k architecture called `M88kInst`. We derive this class from the predefined `Instruction` class. Our new class has a couple of parameters. The `outs` and `ins` parameters describe the output and input operands as a list, using the special dag type. The textual representation of the instruction is split into the mnemonic given in the asm parameter, and the operands. Last, the `pattern` parameter can hold a pattern used for instruction selection.

We also need to define two new fields:

- The `Inst` field is used to hold the bit pattern of the instruction. Because the size of an instruction depends on the platform, this field cannot be predefined. All instructions of the m88k architecture are 32-bit wide, and so this field has the `bits<32>` type.

- The other field is called SoftFail and has the same type as Inst. It holds a bit mask used with an instruction for which the actual encoding can differ from the bits in the Inst field and still be valid. The only platform that requires this is ARM, so we can simply set this field to 0.

The other fields are defined in the superclass, and we only set the value. Simple computations are possible in the TableGen language, and we use this when we create the value for the AsmString field, which holds the full assembler representation. If the operands operand string is empty, then the AsmString field will just have the value of the asm parameter, otherwise, it will be the concatenation of both strings, with a space between them:

```
class InstM88k<dag outs, dag ins, string asm, string operands,
               list<dag> pattern = []>
  : Instruction {
  bits<32> Inst;
  bits<32> SoftFail = 0;
  let Namespace = "M88k";
  let Size = 4;
  dag OutOperandList = outs;
  dag InOperandList = ins;
  let AsmString = !if(!eq(operands, ""), asm,
                      !strconcat(asm, " ", operands));
  let Pattern = pattern;
  let DecoderNamespace = "M88k";
}
```

For the instruction encoding, the manufacturer usually groups instructions together, and the instructions of one group have a similar encoding. We can use those groups to systematically create classes defining the instruction formats. For example, all logical operations of the m88k architecture encode the destination register in the bits from 21 to 25 and the first source register in the bits from 16 to 20. Please note the implementation pattern here: we declare the rd and rs1 fields for the values, and we assign those values to the correct bit positions of the Inst field, which we defined previously in the superclass:

```
class F_L<dag outs, dag ins, string asm, string operands,
          list<dag> pattern = []>
  : InstM88k<outs, ins, asm, operands, pattern> {
  bits<5>  rd;
  bits<5>  rs1;
  let Inst{25-21} = rd;
  let Inst{20-16} = rs1;
}
```

There are several groups of logical operations based on this format. One of them is the group of instructions using three registers, which is called **triadic addressing mode** in the manual:

```
class F_LR<bits<5> func, bits<1> comp, string asm,
            list<dag> pattern = []>
   : F_L<(outs GPROpnd:$rd), (ins GPROpnd:$rs1, GPROpnd:$rs2),
          !if(comp, !strconcat(asm, ".c"), asm),
          "$rd, $rs1, $rs2", pattern> {
  bits<5>  rs2;
  let Inst{31-26} = 0b111101;
  let Inst{15-11} = func;
  let Inst{10}    = comp;
  let Inst{9-5}   = 0b00000;
  let Inst{4-0}   = rs2;
}
```

Let's examine the functionality provided by this class in more detail. The func parameter specifies the operation. As a special feature, the second operand can be complemented before the operation, which is indicated by setting the **flag** comp to 1. The mnemonic is given in the asm parameter, and an instruction selection pattern can be passed.

With initializing the superclass, we can give more information. The full assembler text template for the and instruction is and $rd, $rs1, $rs2. The operand string is fixed for all instructions of this group, so we can define it here. The mnemonic is given by the user of this class, but we can concatenate the .c suffix here, which denotes that the second operand should be complemented first. And last, we can define the output and input operands. These operands are expressed as **directed acyclic graphs** or **dag** for short. A dag has an operation and a list of arguments. An argument can also be a dag, which allows the construction of complex graphs. For example, the output operand is (outs GPROpnd:$rd).

The outs operation denotes this dag as the output operand list. The only argument, GPROpnd:$rd, consists of a type and a name. It connects several pieces we have already seen. The type is GPROnd, which is the name of the register operand we have defined in the previous section. The name $rd refers to the destination register. We used this name in the operand string earlier, and also as a field name in the F_L superclass. The input operands are defined similarly. The rest of the class initializes the other bits of the Inst field. Please take the time and check that all 32 bits are indeed now assigned.

We put the final instruction definition in the M88kInstrInfo.td file. Since we have two variants of each logical instruction, we use a multiclass to define both instructions at once. We also define here the pattern for the instruction selection as a directed acyclic graph. The operation in the pattern is set, and the first argument is the destination register. The second argument is a nested graph, which is the actual pattern. Once again, the name of the operation is the first OpNode element.

LLVM has many predefined operations, which you find in the `llvm/include/llvm/Target/TargetSelectionDAG.td` file (https://github.com/llvm/llvm-project/blob/main/llvm/include/llvm/Target/TargetSelectionDAG.td). For example, there is the `and` operation, which denotes a bitwise AND operation. The arguments are the two source registers, `$rs1` and `$rs2`. You read this pattern roughly as follows: if the input to the instruction selection contains an OpNode operation using two registers, then assign the result of this operation to the `$rd` register and generate this instruction. Utilizing the graph structure, you can define more complex patterns. For example, the second pattern integrates the complement into the pattern using the `not` operand.

A small detail to point out is that the logical operations are commutative. This can be helpful for the instruction selection, so we set the `isCommutable` flag to 1 for those instructions:

```
multiclass Logic<bits<5> Fun, string OpcStr, SDNode OpNode> {
  let isCommutable = 1 in
    def rr : F_LR<Fun, /*comp=*/0b0, OpcStr,
                  [(set i32:$rd,
                    (OpNode GPROpnd:$rs1, GPROpnd:$rs2))]>;
  def rrc : F_LR<Fun, /*comp=*/0b1, OpcStr,
                  [(set i32:$rd,
                    (OpNode GPROpnd:$rs1, (not GPROpnd:$rs2)))]>;
}
```

And finally, we define the records for the instructions:

```
defm AND : Logic<0b01000, "and", and>;
defm XOR : Logic<0b01010, "xor", xor>;
defm OR  : Logic<0b01011, "or", or>;
```

The first parameter is the bit pattern for the function, the second is the mnemonic, and the third parameter is the dag operation used in the pattern.

To fully understand the class hierarchy, revisit the class definitions. The guiding design principle is to avoid the repetition of information. For example, the `0b01000` function bit pattern is used exactly once. Without the `Logic` multiclass you would need to type this bit pattern twice and repeat the patterns several times, which is error-prone.

Please also note that it is good to establish a naming scheme for the instructions. For example, the record for the `and` instruction is named ANDrr, while the variant with the complemented register is named ANDrrc. Those names end up in the generated C++ source code, and using a naming scheme helps to understand to which assembler instruction the name refers.

Up to now, we modeled the register file of the m88k architecture and defined a couple of instructions. In the next section, we will create the top-level file.

Creating the top-level file for the target description

So far, we created the `M88kRegisterInfo.td`, `M88kInstrFormats.td`, and `M88kInstrInfo.td` files. The target description is a single file, called `M88k.td`. This file includes the LLVM definitions first, and the files that we have implemented follow afterwards.:

```
include "llvm/Target/Target.td"

include "M88kRegisterInfo.td"
include "M88kInstrFormats.td"
include "M88kInstrInfo.td"
```

We will extend this `include` section later when we add more backend functionality.

The top-level file also defines some global instances. The first record named `M88kInstrInfo` holds the information about all instructions:

```
def M88kInstrInfo : InstrInfo;
```

We call the assembler class `M88kAsmParser`. To enable TableGen to identify hardcoded registers, we specify that register names are prefixed with a percent sign, and we need to define an assembler parser variant to specify this:

```
def M88kAsmParser : AsmParser;
def M88kAsmParserVariant : AsmParserVariant {
  let RegisterPrefix = "%";
}
```

And last, we need to define the target:

```
def M88k : Target {
  let InstructionSet = M88kInstrInfo;
  let AssemblyParsers  = [M88kAsmParser];
  let AssemblyParserVariants = [M88kAsmParserVariant];
}
```

We now have defined enough of the target so that we can code the first utility. In the next section, we add the M88k backend to LLVM.

Adding the M88k backend to LLVM

We have not yet discussed where to place the target description files. Each backend in LLVM has a subdirectory in `llvm/lib/Target`. We create the `M88k` directory here and copy the target description files into it.

Of course, just adding the TableGen files is not enough. LLVM uses a registry to look up instances of a target implementation, and it expects certain global functions to register those instances. And since some parts are generated, we can already provide an implementation.

All information about a target, like the target triple and factory function for the target machine, assembler, disassembler, and so on, are stored in an instance of the `Target` class. Each target holds a static instance of this class, and this instance is registered in the central registry:

1. The implementation is in the `M88kTargetInfo.cpp` file in the `TargetInfo` subdirectory in our target. The single instance of the `Target` class is held inside the `getTheM88kTarget()` function:

```
using namespace llvm;
Target &llvm::getTheM88kTarget() {
    static Target TheM88kTarget;
    return TheM88kTarget;
}
```

2. LLVM requires that each target provides a `LLVMInitialize<Target Name>TargetInfo()` function to register the target instance. That function must have a C linkage because it is also used in the LLVM C API:

```
extern "C" LLVM_EXTERNAL_VISIBILITY void
LLVMInitializeM88kTargetInfo() {
    RegisterTarget<Triple::m88k, /*HasJIT=*/false> X(
        getTheM88kTarget(), "m88k", "M88k", "M88k");
}
```

3. We also need to create an `M88kTargetInfo.h` header file in the same directory, which just contains a single declaration:

```
namespace llvm {
class Target;
Target &getTheM88kTarget();
}
```

4. And last, we add a `CMakeLists.txt` file for building:

```
add_llvm_component_library(LLVMM88kInfo
  M88kTargetInfo.cpp
  LINK_COMPONENTS  Support
  ADD_TO_COMPONENT M88k)
```

Next, we partially populate the target instance with the information used at the **machine-code (MC)** level. Let's get started:

1. The implementation is in the `M88kMCTargetDesc.cpp` file in the `MCTargetDesc` subdirectory. TableGen turns the target description we created in the previous section into C++ source code fragments. Here, we include the parts for the register information, the instruction information, and the sub-target information:

   ```
   using namespace llvm;


   #define GET_INSTRINFO_MC_DESC
   #include "M88kGenInstrInfo.inc"


   #define GET_SUBTARGETINFO_MC_DESC
   #include "M88kGenSubtargetInfo.inc"


   #define GET_REGINFO_MC_DESC
   #include "M88kGenRegisterInfo.inc"
   ```

2. The target registry expects a factory method for each of the classes here. Let's begin with the instruction information. We allocate an instance of the `MCInstrInfo` class, and call the `InitM88kMCInstrInfo()` generated function to populate the object:

   ```
   static MCInstrInfo *createM88kMCInstrInfo() {
       MCInstrInfo *X = new MCInstrInfo();
       InitM88kMCInstrInfo(X);
       return X;
   }
   ```

3. Next, we allocate an object of the `MCRegisterInfo` class and call a generated function to populate it. The additional `M88k::R1` parameter value tells LLVM that the `r1` register holds the return address:

   ```
   static MCRegisterInfo *
   createM88kMCRegisterInfo(const Triple &TT) {
       MCRegisterInfo *X = new MCRegisterInfo();
       InitM88kMCRegisterInfo(X, M88k::R1);
       return X;
   }
   ```

4. And last, we need a factory method for the sub-target information. This method takes a target triple, a CPU name, and a feature string as parameters, and forwards them to the generated method:

   ```
   static MCSubtargetInfo *
   createM88kMCSubtargetInfo(const Triple &TT,
                             StringRef CPU, StringRef FS) {
   ```

```
                    return createM88kMCSubtargetInfoImpl(TT, CPU,
                                                         /*TuneCPU*/ CPU,
                                                         FS);
      }
```

5.   Having the factory methods defined, we can now register them. Similar to the target registration, LLVM expects a global function called LLVMInitialize<Target Name>TargetMC():

```
extern "C" LLVM_EXTERNAL_VISIBILITY void
LLVMInitializeM88kTargetMC() {
  TargetRegistry::RegisterMCInstrInfo(
    getTheM88kTarget(), createM88kMCInstrInfo);
  TargetRegistry::RegisterMCRegInfo(
    getTheM88kTarget(), createM88kMCRegisterInfo);
  TargetRegistry::RegisterMCSubtargetInfo(
    getTheM88kTarget(), createM88kMCSubtargetInfo);
}
```

6.   The M88kMCTargetDesc.h header file just makes some generated code available:

```
#define GET_REGINFO_ENUM
#include "M88kGenRegisterInfo.inc"

#define GET_INSTRINFO_ENUM
#include "M88kGenInstrInfo.inc"

#define GET_SUBTARGETINFO_ENUM
#include "M88kGenSubtargetInfo.inc"
```

The implementation is almost done. To prevent a linker error, we need to provide another function, which registers a factory method for an object of the TargetMachine class. This class is required for code generation, and we implement it in *Chapter 12, Instruction Selection*, up next. Here, we just define an empty function in the M88kTargetMachine.cpp file:

```
#include "TargetInfo/M88kTargetInfo.h"
#include "llvm/MC/TargetRegistry.h"

extern "C" LLVM_EXTERNAL_VISIBILITY void
LLVMInitializeM88kTarget() {
  // TODO Register the target machine. See chapter 12.
}
```

This concludes our first implementation. However, LLVM does not yet know about our new backend. To integrate it, open the `llvm/CMakeLists.txt` file, locate the section defining all the experimental targets, and add the M88k target to the list:

```
set(LLVM_ALL_EXPERIMENTAL_TARGETS ARC … M88k …)
```

Assuming the LLVM source code with our new backend is in the directory, you can configure the build by typing the following:

```
$ mkdir build
$ cd build
$ cmake -DLLVM_EXPERIMENTAL_TARGETS_TO_BUILD=M88k \
  ../llvm-m88k/llvm
…
-- Targeting M88k
…
```

After building LLVM, you can verify that the tools already know about our new target:

```
$ bin/llc –version
LLVM (http://llvm.org/):
  LLVM version 17.0.2

  Registered Targets:
    m88k    - M88k
```

The journey to get to this point was difficult, so take a moment to celebrate!

> **Fixing a possible compile error**
>
> There is a small oversight in LLVM 17.0.2, which causes a compile error. In one place in the code, the TableGen emitter for the sub-target information uses the removed value `llvm::None` instead of `std::nullopt`, causing an error while compiling `M88kMCTargetDesc.cpp`. The easiest way to fix this problem is to cherry-pick the fix from the LLVM 18 development branch: `git cherry-pick -x a587f429`.

In the next section, we implement the assembler parser, which will give us the first working LLVM tool.

Implementing the assembler parser

The assembler parser is easy to implement, since LLVM provides a framework for it, and large parts are generated from the target description.

The `ParseInstruction()` method in our class is called when the framework detects that an instruction needs to be parsed. That method parses in input via the provided lexer and constructs a so-called operand vector. An operand can be a token such as an instruction mnemonic, a register name, or an immediate, or it can be category-specific to the target. For example, two operands are constructed from the `jmp  %r2` input: a token operand for the mnemonic, and a register operand.

Then a generated matcher tries to match the operand vector against the instructions. If a match is found, then an instance of the `MCInst` class is created, which holds the parsed instruction. Otherwise, an error message is emitted. The advantage of this approach is that it automatically derives the matcher from the target description, without needing to handle all syntactical quirks.

However, we need to add a couple more support classes to make the assembler parser work. These additional classes are all stored in the `MCTargetDesc` directory.

Implementing the MCAsmInfo support class for the M88k Target

Within this section, we explore implementing the first required class for the configuration of the assembler parser: the `MCAsmInfo` class:

1. We need to set some customization parameters for the assembler parser. The `MCAsmInfo` base class (`https://github.com/llvm/llvm-project/blob/main/llvm/include/llvm/MC/MCAsmInfo.h`) contains the common parameters. In addition, a subclass is created for each supported object file format; for example, the `MCAsmInfoELF` class (`https://github.com/llvm/llvm-project/blob/main/llvm/include/llvm/MC/MCAsmInfoELF.h`). The reasoning behind it is that the system assemblers on systems using the same object file format share common characteristics because they must support similar features. Our target operating system is OpenBSD, and it uses the ELF file format, so we derive our own `M88kMCAsmInfo` class from the `MCAsmInfoELF` class. The declaration in the `M88kMCAsmInfo.h` file is as follows:

    ```
    namespace llvm {
    class Triple;

    class M88kMCAsmInfo : public MCAsmInfoELF {
    public:
      explicit M88kMCAsmInfo(const Triple &TT);
    };
    ```

2. The implementation in the `M88kMCAsmInfo.cpp` file only sets a couple of default values. Two crucial settings at present are the system using big-endian mode and employing the | symbol for comments. The other settings are for code generation later:

    ```
    using namespace llvm;

    M88kMCAsmInfo::M88kMCAsmInfo(const Triple &TT) {
    ```

```
    IsLittleEndian = false;
    UseDotAlignForAlignment = true;
    MinInstAlignment = 4;
    CommentString = "|"; // # as comment delimiter is only
                         // allowed at first column
    ZeroDirective = "\t.space\t";
    Data64bitsDirective = "\t.quad\t";
    UsesELFSectionDirectiveForBSS = true;
    SupportsDebugInformation = false;
    ExceptionsType = ExceptionHandling::SjLj;
}
```

Now we have completed the implementation for the MCAsmInfo class. The next class we will learn to implement helps us create a binary representation of the instructions within LLVM.

Implementing the MCCodeEmitter support class for the M88k Target

Internally in LLVM, an instruction is represented by an instance of the MCInst class. An instruction can be emitted as an assembler text or in binary into an object file. The M88kMCCodeEmitter class creates the binary representation of an instruction, while the M88kInstPrinter class emits the textual representation of it.

First, we will implement the M88kMCCodeEmitter class, which is stored in the M88kMCCodeEmitter.cpp file:

1. Most of the class is generated by TableGen. Therefore, we only need to add some boilerplate code. Note that there is no corresponding header file; the prototype of the factory function will be added to the M88kMCTargetDesc.h file. It begins with setting up a statistic counter for the number of emitted instructions:

    ```
    using namespace llvm;
    #define DEBUG_TYPE "mccodeemitter"
    STATISTIC(MCNumEmitted,
              "Number of MC instructions emitted");
    ```

2. The M88kMCCodeEmitter class lives in an anonymous namespace. We only need to implement the encodeInstruction() method, which is declared in the base class, and the getMachineOpValue() helper method. The other getBinaryCodeForInstr() method is generated by TableGen from the target description:

    ```
    namespace {
    class M88kMCCodeEmitter : public MCCodeEmitter {
      const MCInstrInfo &MCII;
      MCContext &Ctx;
    ```

```
public:
  M88kMCCodeEmitter(const MCInstrInfo &MCII,
                    MCContext &Ctx)
      : MCII(MCII), Ctx(Ctx) {}

  ~M88kMCCodeEmitter() override = default;

  void encodeInstruction(
      const MCInst &MI, raw_ostream &OS,
      SmallVectorImpl<MCFixup> &Fixups,
      const MCSubtargetInfo &STI) const override;

  uint64_t getBinaryCodeForInstr(
      const MCInst &MI,
      SmallVectorImpl<MCFixup> &Fixups,
      const MCSubtargetInfo &STI) const;

  unsigned
  getMachineOpValue(const MCInst &MI,
                    const MCOperand &MO,
                    SmallVectorImpl<MCFixup> &Fixups,
                    const MCSubtargetInfo &STI) const;
};
} // end anonymous namespace
```

3. The `encodeInstruction()` method just looks up the binary representation of the instruction, increments the statistic counter, and writes the bytes out in big-endian format. Remember that the instructions have a fixed size of 4 bytes, therefore we use the `uint32_t` type on the endian stream:

```
void M88kMCCodeEmitter::encodeInstruction(
    const MCInst &MI, raw_ostream &OS,
    SmallVectorImpl<MCFixup> &Fixups,
    const MCSubtargetInfo &STI) const {
  uint64_t Bits =
      getBinaryCodeForInstr(MI, Fixups, STI);
  ++MCNumEmitted;
  support::endian::write<uint32_t>(OS, Bits,
                    support::big);
}
```

4. The task of the `getMachineOpValue()` method is to return the binary representation of operands. In the target description, we defined the bit ranges where the registers used are stored in an instruction. Here, we compute the value, which is stored in these places. The method is called from the generated code. We only support two cases. For a register, the encoding of the register, which we defined in the target description, is returned. For an immediate, the immediate value is returned:

```
unsigned M88kMCCodeEmitter::getMachineOpValue(
    const MCInst &MI, const MCOperand &MO,
    SmallVectorImpl<MCFixup> &Fixups,
    const MCSubtargetInfo &STI) const {
  if (MO.isReg())
    return Ctx.getRegisterInfo()->getEncodingValue(
        MO.getReg());
  if (MO.isImm())
    return static_cast<uint64_t>(MO.getImm());
  return 0;
}
```

5. And last, we include the generated file and create a factory method for the class:

```
#include "M88kGenMCCodeEmitter.inc"

MCCodeEmitter *
llvm::createM88kMCCodeEmitter(const MCInstrInfo &MCII,
                             MCContext &Ctx) {
  return new M88kMCCodeEmitter(MCII, Ctx);
}
```

Implementing the instruction printer support class for the M88k Target

The `M88kInstPrinter` class has a similar structure to the `M88kMCCodeEmitter` class. As mentioned previously, the `InstPrinter` class is responsible for emitting the textual representation of LLVM instructions. Most of the class is generated by TableGen, but we have to add support for printing the operands. The class is declared in the `M88kInstPrinter.h` header file. The implementation is in the `M88kInstPrinter.cpp` file:

1. Let's begin with the header file. After including the required header files and declaring the `llvm` namespace, two forward references are declared to reduce the number of required includes:

```
namespace llvm {
class MCAsmInfo;
class MCOperand;
```

2. Besides the constructor, we only need to implement the `printOperand()` and `printInst()` methods. The other methods are generated by TableGen:

```
class M88kInstPrinter : public MCInstPrinter {
public:
  M88kInstPrinter(const MCAsmInfo &MAI,
                  const MCInstrInfo &MII,
                  const MCRegisterInfo &MRI)
      : MCInstPrinter(MAI, MII, MRI) {}

  std::pair<const char *, uint64_t>
  getMnemonic(const MCInst *MI) override;
  void printInstruction(const MCInst *MI,
                        uint64_t Address,
                        const MCSubtargetInfo &STI,
                        raw_ostream &O);

  static const char *getRegisterName(MCRegister RegNo);

  void printOperand(const MCInst *MI, int OpNum,
                    const MCSubtargetInfo &STI,
                    raw_ostream &O);

  void printInst(const MCInst *MI, uint64_t Address,
                 StringRef Annot,
                 const MCSubtargetInfo &STI,
                 raw_ostream &O) override;
};
} // end namespace llvm
```

3. The implementation lives in the `M88kInstPrint.cpp` file. After including the required header file and using the `llvm` namespace, the file with the generated C++ fragments is included:

```
using namespace llvm;

#define DEBUG_TYPE "asm-printer"

#include "M88kGenAsmWriter.inc"
```

4. The `printOperand()` method checks the type of the operand and emits either a register name or an immediate. The register name is looked up with the `getRegisterName()` generated method:

```
void M88kInstPrinter::printOperand(
    const MCInst *MI, int OpNum,
```

```
        const MCSubtargetInfo &STI, raw_ostream &O) {
      const MCOperand &MO = MI->getOperand(OpNum);
      if (MO.isReg()) {
        if (!MO.getReg())
          O << '0';
        else
          O << '%' << getRegisterName(MO.getReg());
      } else if (MO.isImm())
        O << MO.getImm();
      else
        llvm_unreachable("Invalid operand");
    }
```

5. The `printInst()` method only calls the `printInstruction()` generated method to print the instruction, and after that, the `printAnnotation()` method to print possible annotations:

```
void M88kInstPrinter::printInst(
    const MCInst *MI, uint64_t Address, StringRef Annot,
    const MCSubtargetInfo &STI, raw_ostream &O) {
  printInstruction(MI, Address, STI, O);
  printAnnotation(O, Annot);
}
```

Implementing M88k-specific target descriptions

In the `M88kMCTargetDesc.cpp` file, we need to make a couple of additions:

1. First, we need a new factory method for the `MCInstPrinter` class and the `MCAsmInfo` class:

```
static MCInstPrinter *createM88kMCInstPrinter(
    const Triple &T, unsigned SyntaxVariant,
    const MCAsmInfo &MAI, const MCInstrInfo &MII,
    const MCRegisterInfo &MRI) {
  return new M88kInstPrinter(MAI, MII, MRI);
}

static MCAsmInfo *
createM88kMCAsmInfo(const MCRegisterInfo &MRI,
                    const Triple &TT,
                    const MCTargetOptions &Options) {
  return new M88kMCAsmInfo(TT);
}
```

2. Finally, within the `LLVMInitializeM88kTargetMC()` function, we need to add the registration of the factory methods:

```
extern "C" LLVM_EXTERNAL_VISIBILITY void
LLVMInitializeM88kTargetMC() {
  // …
  TargetRegistry::RegisterMCAsmInfo(
      getTheM88kTarget(), createM88kMCAsmInfo);
  TargetRegistry::RegisterMCCodeEmitter(
      getTheM88kTarget(), createM88kMCCodeEmitter);
  TargetRegistry::RegisterMCInstPrinter(
      getTheM88kTarget(), createM88kMCInstPrinter);
}
```

Now we have implemented all required support classes, and we can finally add the assembler parser.

Creating the M88k assembler parser class

There is only an `M88kAsmParser.cpp` implementation file in the `AsmParser` directory. The `M88kOperand` class represents a parsed operand and is used by the generated source code and our assembler parser implementation in class `M88kAssembler`. Both classes are in an anonymous namespace, only the factory method is globally visible. Let's take a look at the `M88kOperand` class first:

1. An operand can be a token, a register, or an immediate. We define the `OperandKind` enumeration to distinguish between these cases. The current kind is stored in the `Kind` member. We also store the start and the end location of the operand, which is needed to print the error message:

```
class M88kOperand : public MCParsedAsmOperand {
  enum OperandKind { OpKind_Token, OpKind_Reg,
                     OpKind_Imm };
  OperandKind Kind;
  SMLoc StartLoc, EndLoc;
```

2. To store the value, we define a union. The token is stored as a `StringRef` and the register is identified by its number. The immediate is represented by the `MCExpr` class:

```
union {
  StringRef Token;
  unsigned RegNo;
  const MCExpr *Imm;
};
```

3. The constructor initializes all fields but the union. Furthermore, we define methods to return the value of the start and the end locations:

```cpp
public:
  M88kOperand(OperandKind Kind, SMLoc StartLoc,
              SMLoc EndLoc)
      : Kind(Kind), StartLoc(StartLoc), EndLoc(EndLoc) {}
  SMLoc getStartLoc() const override { return StartLoc; }
  SMLoc getEndLoc() const override { return EndLoc; }
```

4. For each operand type, we must define four methods. For a register, the methods are `isReg()` to check whether the operand is a register, `getReg()` to return the value, `createReg()` to create a register operand, and `addRegOperands()` to add an operant to an instruction. The latter function is called by the generated source code when an instruction is constructed. The methods for the token and the immediate are similar:

```cpp
bool isReg() const override {
  return Kind == OpKind_Reg;
}

unsigned getReg() const override { return RegNo; }

static std::unique_ptr<M88kOperand>
createReg(unsigned Num, SMLoc StartLoc,
          SMLoc EndLoc) {
  auto Op = std::make_unique<M88kOperand>(
      OpKind_Reg, StartLoc, EndLoc);
  Op->RegNo = Num;
  return Op;
}

void addRegOperands(MCInst &Inst, unsigned N) const {
  assert(N == 1 && "Invalid number of operands");
  Inst.addOperand(MCOperand::createReg(getReg()));
}
```

5. And last, the superclass defines an abstract `print()` virtual method that we need to implement. This is only used for debugging purposes:

```cpp
void print(raw_ostream &OS) const override {
  switch (Kind) {
  case OpKind_Imm:
    OS << "Imm: " << getImm() << "\n"; break;
  case OpKind_Token:
```

```
            OS << "Token: " << getToken() << "\n"; break;
        case OpKind_Reg:
            OS << "Reg: "
               << M88kInstPrinter::getRegisterName(getReg())
               << „\n"; break;
        }
    }
};
```

Next, we declare the M88kAsmParser class. The anonymous name space will end after the declaration:

1. At the beginning of the class we include the generated fragment:

```
class M88kAsmParser : public MCTargetAsmParser {
#define GET_ASSEMBLER_HEADER
#include "M88kGenAsmMatcher.inc"
```

2. Next, we define the required fields. We need a reference to the actual parser, which is of the MCAsmParser class, and a reference to the sub-target information:

```
MCAsmParser &Parser;
const MCSubtargetInfo &SubtargetInfo;
```

3. To implement the assembler, we override a couple of methods defined in the MCTargetAsmParser superclass. The MatchAndEmitInstruction() method tries to match an instruction and emits the instruction represented by an instance of the MCInst class. Parsing an instruction is done in the ParseInstruction() method, while the parseRegister() and tryParseRegister() methods are responsible for parsing the register. The other methods are required internally:

```
bool
ParseInstruction(ParseInstructionInfo &Info,
                 StringRef Name, SMLoc NameLoc,
                 OperandVector &Operands) override;
bool parseRegister(MCRegister &RegNo, SMLoc &StartLoc,
                   SMLoc &EndLoc) override;
OperandMatchResultTy
tryParseRegister(MCRegister &RegNo, SMLoc &StartLoc,
                 SMLoc &EndLoc) override;
bool parseRegister(MCRegister &RegNo, SMLoc &StartLoc,
                   SMLoc &EndLoc,
                   bool RestoreOnFailure);
bool parseOperand(OperandVector &Operands,
                  StringRef Mnemonic);
```

```
bool MatchAndEmitInstruction(
    SMLoc IdLoc, unsigned &Opcode,
    OperandVector &Operands, MCStreamer &Out,
    uint64_t &ErrorInfo,
    bool MatchingInlineAsm) override;
```

4. The constructor is defined inline. It mostly initializes all fields. This finishes the class declaration, after which the anonymous namespace ends:

```
public:
  M88kAsmParser(const MCSubtargetInfo &STI,
                MCAsmParser &Parser,
                const MCInstrInfo &MII,
                const MCTargetOptions &Options)
      : MCTargetAsmParser(Options, STI, MII),
        Parser(Parser), SubtargetInfo(STI) {
    setAvailableFeatures(ComputeAvailableFeatures(
        SubtargetInfo.getFeatureBits()));
  }
};
```

5. Now we include the generated parts of the assembler:

```
#define GET_REGISTER_MATCHER
#define GET_MATCHER_IMPLEMENTATION
#include "M88kGenAsmMatcher.inc"
```

6. The ParseInstruction() method is called whenever an instruction is expected. It must be able to parse all syntactical forms of an instruction. Currently, we only have instructions that take three operands, which are separated by a comma, so the parsing is simple. Be aware that the return value is true in case of an error!

```
bool M88kAsmParser::ParseInstruction(
    ParseInstructionInfo &Info, StringRef Name,
    SMLoc NameLoc, OperandVector &Operands) {
  Operands.push_back(
      M88kOperand::createToken(Name, NameLoc));
  if (getLexer().isNot(AsmToken::EndOfStatement)) {
    if (parseOperand(Operands, Name)) {
      return Error(getLexer().getLoc(),
                   "expected operand");
    }
    while (getLexer().is(AsmToken::Comma)) {
      Parser.Lex();
      if (parseOperand(Operands, Name)) {
```

```
            return Error(getLexer().getLoc(),
                            "expected operand");
        }
      }
      if (getLexer().isNot(AsmToken::EndOfStatement))
        return Error(getLexer().getLoc(),
                        "unexpected token in argument list");
    }
    Parser.Lex();
    return false;
  }
```

7. An operand can be a register or an immediate. We generalize a bit and parse an expression instead of just an integer. This helps later when adding address modes. When successful, the parsed operand is added to the Operands list:

```
bool M88kAsmParser::parseOperand(
    OperandVector &Operands, StringRef Mnemonic) {
  if (Parser.getTok().is(AsmToken::Percent)) {
    MCRegister RegNo;
    SMLoc StartLoc, EndLoc;
    if (parseRegister(RegNo, StartLoc, EndLoc,
                        /*RestoreOnFailure=*/false))
      return true;
    Operands.push_back(M88kOperand::createReg(
        RegNo, StartLoc, EndLoc));
    return false;
  }

  if (Parser.getTok().is(AsmToken::Integer)) {
    SMLoc StartLoc = Parser.getTok().getLoc();
    const MCExpr *Expr;
    if (Parser.parseExpression(Expr))
      return true;
    SMLoc EndLoc = Parser.getTok().getLoc();
    Operands.push_back(
        M88kOperand::createImm(Expr, StartLoc, EndLoc));
    return false;
  }
  return true;
}
```

8. The `parseRegister()` method tries to parse a register. First, it checks for a percent sign
    %. If this is followed by an identifier which matches a register name, then we successfully
    parsed a register, and return the register number in the RegNo parameter. However, if we
    cannot identify a register, then we may need to undo the lexing if the `RestoreOnFailure`
    parameter is `true`:

```cpp
bool M88kAsmParser::parseRegister(
    MCRegister &RegNo, SMLoc &StartLoc, SMLoc &EndLoc,
    bool RestoreOnFailure) {
  StartLoc = Parser.getTok().getLoc();

  if (Parser.getTok().isNot(AsmToken::Percent))
    return true;
  const AsmToken &PercentTok = Parser.getTok();
  Parser.Lex();

  if (Parser.getTok().isNot(AsmToken::Identifier) ||
      (RegNo = MatchRegisterName(
           Parser.getTok().getIdentifier())) == 0) {
    if (RestoreOnFailure)
      Parser.getLexer().UnLex(PercentTok);
    return Error(StartLoc, "invalid register");
  }
  Parser.Lex();
  EndLoc = Parser.getTok().getLoc();
  return false;
}
```

9. The `parseRegister()` and `tryparseRegister()` overridden methods are just
    wrappers around the previously defined method. The latter method also translates the boolean
    return value into an enumeration member of the `OperandMatchResultTy` enumeration:

```cpp
bool M88kAsmParser::parseRegister(MCRegister &RegNo,
                                  SMLoc &StartLoc,
                                  SMLoc &EndLoc) {
  return parseRegister(RegNo, StartLoc, EndLoc,
                       /*RestoreOnFailure=*/false);
}

OperandMatchResultTy M88kAsmParser::tryParseRegister(
    MCRegister &RegNo, SMLoc &StartLoc, SMLoc &EndLoc) {
  bool Result =
      parseRegister(RegNo, StartLoc, EndLoc,
                    /*RestoreOnFailure=*/true);
  bool PendingErrors = getParser().hasPendingError();
```

```
     getParser().clearPendingErrors();
     if (PendingErrors)
       return MatchOperand_ParseFail;
     if (Result)
       return MatchOperand_NoMatch;
     return MatchOperand_Success;
   }
```

10. Finally, the `MatchAndEmitInstruction()` method drives the parsing. Most of the method is dedicated to emitting error messages. To identify the instruction, the `MatchInstructionImpl()` generated method is called:

```
   bool M88kAsmParser::MatchAndEmitInstruction(
       SMLoc IdLoc, unsigned &Opcode,
       OperandVector &Operands, MCStreamer &Out,
       uint64_t &ErrorInfo, bool MatchingInlineAsm) {
     MCInst Inst;
     SMLoc ErrorLoc;

     switch (MatchInstructionImpl(
         Operands, Inst, ErrorInfo, MatchingInlineAsm)) {
     case Match_Success:
       Out.emitInstruction(Inst, SubtargetInfo);
       Opcode = Inst.getOpcode();
       return false;
     case Match_MissingFeature:
       return Error(IdLoc, "Instruction use requires "
                           "option to be enabled");
     case Match_MnemonicFail:
       return Error(IdLoc,
                   "Unrecognized instruction mnemonic");
     case Match_InvalidOperand: {
       ErrorLoc = IdLoc;
       if (ErrorInfo != ~0U) {
         if (ErrorInfo >= Operands.size())
           return Error(
               IdLoc, "Too few operands for instruction");

         ErrorLoc = ((M88kOperand &)*Operands[ErrorInfo])
                       .getStartLoc();
         if (ErrorLoc == SMLoc())
           ErrorLoc = IdLoc;
       }
       return Error(ErrorLoc,
                   "Invalid operand for instruction");
```

```
  }
  default:
    break;
  }
  llvm_unreachable("Unknown match type detected!");
}
```

11. And like some other classes, the assembler parser has its own factory method:

```
extern "C" LLVM_EXTERNAL_VISIBILITY void
LLVMInitializeM88kAsmParser() {
  RegisterMCAsmParser<M88kAsmParser> X(
      getTheM88kTarget());
}
```

This finishes the implementation of the assembler parser. After building LLVM, we can use the **llvm-mc** machine code playground tool to assemble an instruction:

```
$ echo 'and %r1,%r2,%r3' | \
  bin/llvm-mc --triple m88k-openbsd --show-encoding
        .text
        and %r1, %r2, %r3   | encoding: [0xf4,0x22,0x40,0x03]
```

Note the use of the vertical bar | as the comments sign. This is the value we configured in the M88kMCAsmInfo class.

> **Debugging the assembler matcher**
>
> To debug the assembler matcher, you specify the --debug-only=asm-matcher command-line option. This helps with understanding why a parsed instruction fails to match the instructions defined in the target description.

In the next section, we will add a disassembler feature to the llvm-mc tool.

Creating the disassembler

Implementing the disassembler is optional. However, the implementation does not require too much effort, and generating the disassembler table may catch encoding errors that are not checked by the other generators. The disassembler lives in the M88kDisassembler.cpp file, found in the Disassembler subdirectory:

1. We begin the implementation by defining a debug type and the DecodeStatus type. Both are required for the generated code:

```
using namespace llvm;
#define DEBUG_TYPE "m88k-disassembler"
using DecodeStatus = MCDisassembler::DecodeStatus;
```

2. The `M88kDisassmbler` class lives in an anonymous namespace. We only need to implement the `getInstruction()` method:

```cpp
namespace {
class M88kDisassembler : public MCDisassembler {
public:
  M88kDisassembler(const MCSubtargetInfo &STI,
                   MCContext &Ctx)
      : MCDisassembler(STI, Ctx) {}
  ~M88kDisassembler() override = default;

  DecodeStatus
  getInstruction(MCInst &instr, uint64_t &Size,
                 ArrayRef<uint8_t> Bytes,
                 uint64_t Address,
                 raw_ostream &CStream) const override;
};
} // end anonymous namespace
```

3. We also need to provide a factory method, which will be registered in the target registry:

```cpp
static MCDisassembler *
createM88kDisassembler(const Target &T,
                       const MCSubtargetInfo &STI,
                       MCContext &Ctx) {
  return new M88kDisassembler(STI, Ctx);
}

extern "C" LLVM_EXTERNAL_VISIBILITY void
LLVMInitializeM88kDisassembler() {
  TargetRegistry::RegisterMCDisassembler(
      getTheM88kTarget(), createM88kDisassembler);
}
```

4. The `decodeGPRRegisterClass()` function turns a register number into the register enum member generated by TableGen. This is the inverse operation of the `M88kInstPrinter::getMachineOpValue()` method. Note that we specified the name of this function in the `DecoderMethod` field in the `M88kRegisterOperand` class:

```cpp
static const uint16_t GPRDecoderTable[] = {
    M88k::R0,  M88k::R1,  M88k::R2,  M88k::R3,
    // …
};

static DecodeStatus
```

```
decodeGPRRegisterClass(MCInst &Inst, uint64_t RegNo,
                       uint64_t Address,
                       const void *Decoder) {
  if (RegNo > 31)
    return MCDisassembler::Fail;

  unsigned Register = GPRDecoderTable[RegNo];
  Inst.addOperand(MCOperand::createReg(Register));
  return MCDisassembler::Success;
}
```

5. Then we include the generated disassembler tables:

```
#include "M88kGenDisassemblerTables.inc"
```

6. And finally, we decode the instruction. For this, we need to take the next four bytes of the Bytes array, create the instruction encoding from them, and call the decodeInstruction() generated function:

```
DecodeStatus M88kDisassembler::getInstruction(
    MCInst &MI, uint64_t &Size, ArrayRef<uint8_t> Bytes,
    uint64_t Address, raw_ostream &CS) const {
  if (Bytes.size() < 4) {
    Size = 0;
    return MCDisassembler::Fail;
  }
  Size = 4;

  uint32_t Inst = 0;
  for (uint32_t I = 0; I < Size; ++I)
    Inst = (Inst << 8) | Bytes[I];

  if (decodeInstruction(DecoderTableM88k32, MI, Inst,
                        Address, this, STI) !=
      MCDisassembler::Success) {
    return MCDisassembler::Fail;
  }
  return MCDisassembler::Success;
}
```

That is all that needs to be done for the disassembler. After compiling LLVM, you can test the functionality again with the `llvm-mc` tool:

```
$ echo "0xf4,0x22,0x40,0x03" | \
  bin/llvm-mc --triple m88k-openbsd -disassemble
       .text
       and %r1, %r2, %r3
```

Moreover, we can now use the `llvm-objdump` tool to disassemble ELF files. However, for it to be really useful, we would need to add all instructions to the target description.

Summary

In this chapter, you learned how to create a LLVM target description, and you developed a simple backend target that supports the assembling and disassembling of instructions for LLVM. You first collected the required documentation and made LLVM aware of the new architecture by enhancing the `Triple` class. The documentation also includes the relocation definition for the ELF file format, and you added the support for them to LLVM.

You then learned about the register definition and the instruction definition in the target description and used the generated C++ source code to implement an instruction assembler and disassembler.

In the next chapter, we will add code generation to the backend.

12

Instruction Selection

The heart of any backend is instruction selection. LLVM implements several approaches; in this chapter, we will implement instruction selection via the selection **directed acyclic graph (DAG)** and with global instruction selection.

In this chapter, you will learn about the following topics:

- *Defining the rules of the calling convention*: This section shows you how to describe the rules of a calling convention in the target description

- *Instruction selection via the selection DAG*: This section teaches you how to implement instruction selection with a graph data structure

- *Adding register and instruction information*: This section explains how to access information in the target description, and what additional information you need to provide

- *Putting an empty frame lowering in place*: This section introduces you to the stack layout and the prologue of a function

- *Emitting machine instructions*: This section tells you how machine instructions are finally written into an object file or as assembly text

- *Creating the target machine and the sub-target*: This section shows you how a backend is configured

- *Global instruction selection*: This section demonstrates a different approach to instruction selection

- *How to further evolve the backend*: This section gives you some guidance about possible next steps

By the end of this chapter, you will know how to create an LLVM backend that can translate simple instructions. You will also acquire the knowledge to develop instruction selection via the selection DAG and with global instruction selection, and you will become familiar with all the important support classes you have to implement to get instruction selection working.

Defining the rules of the calling convention

Implementing the rules of the calling convention is an important part of lowering the LLVM **intermediate representation** (**IR**) to machine code. The basic rules can be defined in the target description. Let's have a look.

Most calling conventions follow a basic pattern: they define a subset of registers for parameter passing. If this subset is not exhausted, the next parameter is passed in the next free register. If there is no free register, then the value is passed on the stack. This can be realized by looping over the parameters and deciding how to pass each parameter to the called function while keeping track of the used registers. In LLVM, this loop is implemented inside the framework, and the state is held in a class called CCState. Furthermore, the rules are defined in the target description.

The rules are given as a sequence of conditions. If the condition holds, then an action is executed. Depending on the outcome of that action, either a place for the parameter is found, or the next condition is evaluated. For example, 32-bit integers are passed in a register. The condition is the type check, and the action is the assignment of a register to this parameter. In the target description, this is written as follows:

```
CCIfType<[i32],
         CCAssignToReg<[R2, R3, R4, R5, R6, R7, R8, R9]>>,
```

Of course, if the called function has more than eight parameters, then the register list will be exhausted, and the action will fail. The remaining parameters are passed on the stack, and we can specify this as the next action:

```
CCAssignToStack<4, 4>
```

The first parameter is the size of a stack slot in bytes, while the second is the alignment. Since it is a catch-all rule, no condition is used.

Implementing the rules of the calling convention

For a calling convention, there are also more predefined conditions and actions to note. For example, CCIfInReg checks if the argument is marked with the inreg attribute, and CCIfVarArg evaluates to true if the function has a variable argument list. The CCPromoteToType action promotes the type of the argument to a larger one, and the CCPassIndirect action indicates that the parameter value should be stored on the stack and that a pointer to that storage is passed as a normal argument. All of the predefined conditions and actions can be referenced within llvm/include/llvm/Target/TargetCallingConv.td.

Both the parameters and the return value are defined in this way. We will put the definition into the `M88kCallingConv.td` file:

1. First, we must define the rules for the parameters. To simplify the coding, we'll only consider 32-bit values:

    ```
    def CC_M88k : CallingConv<[
      CCIfType<[i8, i16], CCPromoteToType<i32>>,
      CCIfType<[i32,f32],
            CCAssignToReg<[R2, R3, R4, R5, R6, R7, R8, R9]>>,
      CCAssignToStack<4, 4>
    ]>;
    ```

2. After that, we must define the rules for return values:

    ```
    def RetCC_M88k : CallingConv<[
      CCIfType<[i32], CCAssignToReg<[R2]>>
    ]>;
    ```

3. Finally, the sequence of callee saved registers must be defined. Note that we use the `sequence` operator to generate a sequence of registers, instead of writing them down:

    ```
    def CSR_M88k :
        CalleeSavedRegs<(add R1, R30,
                             (sequence "R%d", 25, 14))>;
    ```

The benefit of defining the rules for the calling convention in the target description is that they can be reused for various instruction selection methods. We'll look at instruction selection via the selection DAG next.

Instruction selection via the selection DAG

Creating machine instructions from the IR is a very important task in the backend. One common way to implement it is to utilize a DAG:

1. First, we must create a DAG from the IR. A node of the DAG represents an operation and the edges model control and data flow dependencies.

2. Next, we must loop over the DAG and legalize the types and operations. Legalization means that we only use types and operations that are supported by the hardware. This requires us to create a configuration that tells the framework how to deal with non-legal types and operations. For instance, a 64-bit value could be split into two 32-bit values, the multiplication of two 64-bit values could be changed to a library call, and a complex operation such as count population could be expanded into a sequence of simpler operations for calculating this value.

3. After, pattern matching is utilized to match nodes in the DAG and replace them with machine instructions. We encountered such a pattern in the previous chapter.

4. Finally, an instruction scheduler reorders the machine instructions into a more performant order.

This is just a high-level description of the instruction selection process via the selection DAG. If you are interested in more details, you can find it in the *The LLVM Target-Independent Code Generator* user guide at `https://llvm.org/docs/CodeGenerator.html#selectiondag-instruction-selection-process`.

Furthermore, all backends in LLVM implement the selection DAG. The main advantage is that it generates performant code. However, this comes at a cost: creating the DAG is expensive, and it slows down compilation speed. Therefore, this has prompted LLVM developers to look for alternative and more desirable approaches. Some targets implement instruction selection via FastISel, which is only used for non-optimized code. It can quickly generate code, but the generated code is inferior to the one generated by the selection DAG method. In addition, it adds a whole new instruction selection method, which doubles the testing effort. Another method is also used for instruction selection called global instruction selection, which we'll examine later in the *Global instruction selection* section.

In this chapter, we aim to implement enough of the backend to lower a simple IR function, like this:

```
define i32 @f1(i32 %a, i32 %b) {
  %res = and i32 %a, %b
  ret i32 %res
}
```

Moreover, for a real backend, much more code is needed, and we must point out what needs to be added for greater functionality.

To implement instruction selection via the selection DAG, we need to create two new classes: `M88kISelLowering` and `M88kDAGToDAGISel`. The former class is used to customize the DAG, for example, by defining which types are legal. It also contains the code to support the lowering of functions and function calls. The latter class performs DAG transformations, and the implementation is mostly generated from the target description.

There are several classes within the backend that we will be adding implementation to, and *Figure 12.1* depicts the high-level relationship between the primary classes that we will be developing further:

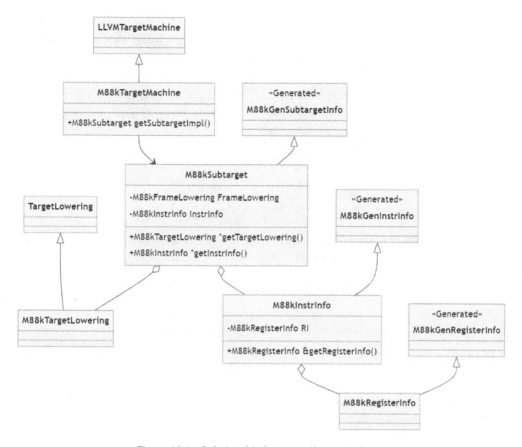

Figure 12.1 – Relationship between the main classes

Implementing DAG lowering – handling legal types and setting operations

Let's implement the `M88kISelLowering` class, which is stored in `M88kISelLowering.cpp` file, first. The constructor configures the legal types and operations:

1. The constructor takes references to the `TargetMachine` and `M88kSubtarget` classes as parameters. The `TargetMachine` class is responsible for the general configuration of the target, for example, which passes need to run. An LLVM backend usually targets a CPU family, and the `M88kSubtarget` class describes the characteristics of the chosen CPU. We'll look at both classes later in this chapter:

    ```
    M88kTargetLowering::M88kTargetLowering(
        const TargetMachine &TM, const M88kSubtarget &STI)
        : TargetLowering(TM), Subtarget(STI) {
    ```

2. The first action is to declare which machine value type uses which register class. Remember that the register classes are generated from the target description. Here, we only handle 32-bit values:

    ```
    addRegisterClass(MVT::i32, &M88k::GPRRegClass);
    ```

3. After adding all register classes, we must compute the derived properties of those register classes. We need to query the sub-target for register information, which is mostly generated from the target description:

    ```
    computeRegisterProperties(Subtarget.getRegisterInfo());
    ```

4. Next, we must declare which register contains the stack pointer:

    ```
    setStackPointerRegisterToSaveRestore(M88k::R31);
    ```

5. Boolean values are represented differently on different platforms. For our target, we will declare that a Boolean value is stored in bit 0; the other bits are cleared:

    ```
    setBooleanContents(ZeroOrOneBooleanContent);
    ```

6. After, we set the alignment of functions. The minimal function alignment is the alignment that is required for correct execution. In addition, we give the preferred alignment:

    ```
    setMinFunctionAlignment(Align(4));
    setPrefFunctionAlignment(Align(4));
    ```

7. Finally, we declare which operations are legal. In the previous chapter, we only defined three logical instructions, and they are legal for 32-bit values:

    ```
    setOperationAction(ISD::AND, MVT::i32, Legal);
    setOperationAction(ISD::OR, MVT::i32, Legal);
    setOperationAction(ISD::XOR, MVT::i32, Legal);
    ```

8. There are a couple of other actions we can use besides `Legal`. `Promote` widens the type, `Expand` replaces the operation with other operations, `LibCall` lowers the operation to a library call, and `Custom` calls the `LowerOperation()` hook method, which lets you implement your own custom handling. For example, in the M88k architecture, there is no count population instruction, so we request that this operation be expanded into other operations:

    ```
    setOperationAction(ISD::CTPOP, MVT::i32, Expand);
    }
    ```

Now, let's review some points to emphasize the connection between the definitions we made so far. In the target description mentioned in the `M88kInstrInfo.td` file, we defined a machine instruction with the and mnemonic, and we also attached a pattern to it. If we expand the AND multiclass record, and only look at the instruction using three registers, we get the TableGen definition:

```
let isCommutable = 1 in
  def ANDrr : F_LR<0b01000, Func, /*comp=*/0b0, "and",
                  [(set i32:$rd,
                      (and GPROpnd:$rs1, GPROpnd:$rs2))]>;
```

The `"and"` string is the mnemonic of the instruction. In C++ source code, we use `M88k::ANDrr` to refer to this instruction. Inside the pattern, the DAG and node type is used. In C++, it is named `ISD::AND`, and we used it in the call to the `setOperationAction()` method. During instruction selection, a DAG node of the and type is replaced by the `M88k::ANDrr` instruction if the pattern matches, which includes the input operands. Thus, when we develop instruction selection, the most important task is for us to define the correct legalization actions and attach the patterns to the instruction definitions.

Implementing DAG lowering – lowering formal arguments

Let's turn to another important task performed by the `M88kISelLowering` class. We defined the rules for the calling convention in the previous section, but we also need to map the physical registers and memory locations to virtual registers used in the DAG. For arguments, this is done in the `LowerFormalArguments()` method; return values are handled in the `LowerReturn()` method. First, we must handle the arguments:

1. We'll begin by including the generated source:

   ```
   #include "M88kGenCallingConv.inc"
   ```

2. The `LowerFormalArguments()` method takes several parameters. The `SDValue` class denotes a value associated with a DAG node and is often used when dealing with the DAG. The first parameter, `Chain`, represents the control flow, and the possible updated `Chain` is also the return value of the method. The `CallConv` parameter identifies the used calling convention, and `IsVarArg` is set to `true` if a variable argument list is part of the parameters. The arguments that need to be handled are passed in the `Ins` parameter, together with their location in the DL parameter. The DAG parameter gives us access to the `SelectionDAG` class. Lastly, the result of the mapping will be stored in the `InVals` vector argument:

   ```
   SDValue M88kTargetLowering::LowerFormalArguments(
       SDValue Chain, CallingConv::ID CallConv,
       bool IsVarArg,
       const SmallVectorImpl<ISD::InputArg> &Ins,
       const SDLoc &DL, SelectionDAG &DAG,
       SmallVectorImpl<SDValue> &InVals) const {
   ```

3. Our first action is to retrieve references to the machine function and the machine
    register information:

    ```
    MachineFunction &MF = DAG.getMachineFunction();
    MachineRegisterInfo &MRI = MF.getRegInfo();
    ```

4. Next, we must call the generated code. We need to instantiate an object of the CCState class.
    The CC_M88k parameter value that's used in the call to the AnalyzeFormalArguments()
    method is the name of the calling convention we used in the target description. The result is
    stored in the ArgLocs vector:

    ```
    SmallVector<CCValAssign, 16> ArgLocs;
    CCState CCInfo(CallConv, IsVarArg, MF, ArgLocs,
                   *DAG.getContext());
    CCInfo.AnalyzeFormalArguments(Ins, CC_M88k);
    ```

5. Once the locations of the arguments have been determined, we need to map them to the DAG.
    Therefore, we must loop over all locations:

    ```
    for (unsigned I = 0, E = ArgLocs.size(); I != E; ++I) {
      SDValue ArgValue;
      CCValAssign &VA = ArgLocs[I];
      EVT LocVT = VA.getLocVT();
    ```

6. The mapping depends on the determined location. First, we handle arguments assigned to
    registers. The goal is to copy the physical register to a virtual register. To do so, we need to
    determine the correct register class. Since we're only handling 32-bit values, it is easy to do this:

    ```
    if (VA.isRegLoc()) {
      const TargetRegisterClass *RC;
      switch (LocVT.getSimpleVT().SimpleTy) {
      default:
        llvm_unreachable("Unexpected argument type");
      case MVT::i32:
        RC = &M88k::GPRRegClass;
        break;
      }
    ```

7. With the register class stored in the RC variable, we can create the virtual register and copy the
    value. We also need to declare the physical register as a live-in:

    ```
    Register VReg = MRI.createVirtualRegister(RC);
    MRI.addLiveIn(VA.getLocReg(), VReg);
    ArgValue =
        DAG.getCopyFromReg(Chain, DL, VReg, LocVT);
    ```

8. In the definition of the calling convention, we added the rule that 8-bit and 16-bit values should be promoted to 32-bit, and we need to ensure the promotion here. To do so, a DAG node must be inserted, which makes sure that the value is promoted. After, the value is truncated to the right size. Note that we pass the value of ArgValue as an operand to the DAG node and store the result in the same variable:

```
if (VA.getLocInfo() == CCValAssign::SExt)
  ArgValue = DAG.getNode(
      ISD::AssertSext, DL, LocVT, ArgValue,
      DAG.getValueType(VA.getValVT()));
else if (VA.getLocInfo() == CCValAssign::ZExt)
  ArgValue = DAG.getNode(
      ISD::AssertZext, DL, LocVT, ArgValue,
      DAG.getValueType(VA.getValVT()));

if (VA.getLocInfo() != CCValAssign::Full)
  ArgValue = DAG.getNode(ISD::TRUNCATE, DL,
                         VA.getValVT(), ArgValue);
```

9. Lastly, we finish handling the register arguments by adding the DAG node to the result vector:

```
InVals.push_back(ArgValue);
}
```

10. The other possible location for a parameter is on the stack. However, we didn't define any load and store instructions, so we cannot handle this case yet. This ends the loop over all argument locations:

```
} else {
llvm_unreachable("Not implemented");
}
}
```

11. After that, we may need to add code to handle variable argument lists. Again, we have added some code to remind us that we have not implemented it:

```
assert(!IsVarArg && "Not implemented");
```

12. Finally, we must return the Chain argument:

```
return Chain;
}
```

Implementing DAG lowering – lowering return values

The return values are handled similarly. However, we must extend the target description for them. First, we need to define a new DAG node type called RET_GLUE. This DAG node type is used to glue the return values together, which prevents them from being rearranged, for example, by the instruction scheduler. The definition in M88kInstrInfo.td is as follows:

```
def retglue : SDNode<"M88kISD::RET_GLUE", SDTNone,
                     [SDNPHasChain, SDNPOptInGlue, SDNPVariadic]>;
```

In the same file, we also define a pseudo-instruction to represent a return from a function call, which will be selected for a RET_GLUE node:

```
let isReturn = 1, isTerminator = 1, isBarrier = 1,
    AsmString = "RET" in
  def RET : Pseudo<(outs), (ins), [(retglue)]>;
```

We will expand this pseudo-instruction when we generate the output.

With these definitions in place, we can implement the LowerReturn() method:

1. The parameters are the same as for LowerFormalArguments(), only the order is slightly different:

    ```
    SDValue M88kTargetLowering::LowerReturn(
        SDValue Chain, CallingConv::ID CallConv,
        bool IsVarArg,
        const SmallVectorImpl<ISD::OutputArg> &Outs,
        const SmallVectorImpl<SDValue> &OutVals,
        const SDLoc &DL, SelectionDAG &DAG) const {
    ```

2. First, we call the generated code, this time using the RetCC_M88k calling convention:

    ```
    SmallVector<CCValAssign, 16> RetLocs;
    CCState RetCCInfo(CallConv, IsVarArg,
                      DAG.getMachineFunction(), RetLocs,
                      *DAG.getContext());
    RetCCInfo.AnalyzeReturn(Outs, RetCC_M88k);
    ```

3. Then, we loop over the locations again. With the simple definition of the calling convention we currently have, this loop will be executed once at most. However, this would change if we would add support for returning 64-bit values, which need to be returned in two registers:

    ```
    SDValue Glue;
    SmallVector<SDValue, 4> RetOps(1, Chain);
    for (unsigned I = 0, E = RetLocs.size(); I != E; ++I) {
      CCValAssign &VA = RetLocs[I];
    ```

4. After, we copy the return values into the physical registers assigned to the return value. This is mostly similar to handling the arguments, with the exception that the values are glued together using the Glue variable:

```
Register Reg = VA.getLocReg();
Chain = DAG.getCopyToReg(Chain, DL, Reg, OutVals[I],
                         Glue);
Glue = Chain.getValue(1);
RetOps.push_back(
    DAG.getRegister(Reg, VA.getLocVT()));
}
```

5. The return value is the chain and the glued register copy operations. The latter is only returned if there is a value to return:

```
RetOps[0] = Chain;
if (Glue.getNode())
  RetOps.push_back(Glue);
```

6. Finally, we construct a DAG node of the RET_GLUE type, passing in the necessary values:

```
return DAG.getNode(M88kISD::RET_GLUE, DL, MVT::Other,
                   RetOps);
}
```

Congratulations! With these definitions, the foundation has been laid for instruction selection.

Implementing DAG-to-DAG transformations within instruction selection

One crucial part is still missing: we need to define the pass that performs the DAG transformations defined in the target descriptions. The class is called M88kDAGToDAGISel and is stored in the M88kISelDAGToDAG.cpp file. Most of the class is generated, but we still need to add some code:

1. We'll begin by defining the debug type and providing a descriptive name for the pass:

```
#define DEBUG_TYPE "m88k-isel"
#define PASS_NAME
            "M88k DAG->DAG Pattern Instruction Selection"
```

2. Then, we must declare the class inside an anonymous namespace. We will only override the Select() method; the other code is generated and included in the body of the class:

```
class M88kDAGToDAGISel : public SelectionDAGISel {
public:
```

```
static char ID;

M88kDAGToDAGISel(M88kTargetMachine &TM,
                CodeGenOpt::Level OptLevel)
    : SelectionDAGISel(ID, TM, OptLevel) {}

void Select(SDNode *Node) override;

#include "M88kGenDAGISel.inc"
};
} // end anonymous namespace
```

3. After, we must add the code to initialize the pass. The LLVM backends still use the legacy pass manager, and the setup differs from the pass manager used for IR transformations. The static member ID value is used to identify the pass. Initializing the pass can be implemented using the INITIALIZE_PASS macro, which expands to C++ code. We must also add a factory method to create an instance of the pass:

```
char M88kDAGToDAGISel::ID = 0;

INITIALIZE_PASS(M88kDAGToDAGISel, DEBUG_TYPE, PASS_NAME,
                false, false)

FunctionPass *
llvm::createM88kISelDag(M88kTargetMachine &TM,
                        CodeGenOpt::Level OptLevel) {
    return new M88kDAGToDAGISel(TM, OptLevel);
}
```

4. Finally, we must implement the Select() method. For now, we only call the generated code. However, if we encounter a complex transformation that we cannot express as a DAG pattern, then we can add our own code to perform the transformation before calling the generated code:

```
void M88kDAGToDAGISel::Select(SDNode *Node) {
    SelectCode(Node);
}
```

With that, we have implemented the instruction selection. However, we still need to add some support classes before we can do the first test. We'll look at those classes in the next few sections.

Adding register and instruction information

The target description captures most information about registers and instructions. To access that information, we must implement the `M88kRegisterInfo` and `M88kInstrInfo` classes. These classes also contain hooks that we can override to accomplish tasks that are too complex to express in the target description. Let's begin with the `M88kRegisterInfo` class, which is declared in the `M88kRegisterInfo.h` file:

1. The header file begins by including the code generated from the target description:

    ```
    #define GET_REGINFO_HEADER
    #include "M88kGenRegisterInfo.inc"
    ```

2. After that, we must declare the `M88kRegisterInfo` class in the `llvm` namespace. We only override a couple of methods:

    ```
    namespace llvm {
    struct M88kRegisterInfo : public M88kGenRegisterInfo {
      M88kRegisterInfo();

      const MCPhysReg *getCalleeSavedRegs(
          const MachineFunction *MF) const override;

      BitVector getReservedRegs(
          const MachineFunction &MF) const override;

      bool eliminateFrameIndex(
          MachineBasicBlock::iterator II, int SPAdj,
          unsigned FIOperandNum,
          RegScavenger *RS = nullptr) const override;

      Register getFrameRegister(
          const MachineFunction &MF) const override;
    };
    } // end namespace llvm
    ```

The definition of the class is stored in the `M88kRegisterInfo.cpp` file:

1. Again, the film begins with including the code generated from the target description:

    ```
    #define GET_REGINFO_TARGET_DESC
    #include "M88kGenRegisterInfo.inc"
    ```

2. The constructor initializes the superclass, passing the register holding the return address as a parameter:

```
M88kRegisterInfo::M88kRegisterInfo()
    : M88kGenRegisterInfo(M88k::R1) {}
```

3. Then, we implement the method that returns the list of callee-saved registers. We defined the list in the target description, and we only return that list:

```
const MCPhysReg *M88kRegisterInfo::getCalleeSavedRegs(
    const MachineFunction *MF) const {
  return CSR_M88k_SaveList;
}
```

4. After, we deal with the reserved registers. The reserved registers depend on the platform and the hardware. The r0 register contains a constant value of 0, so we treat it as a reserved register. The r28 and r29 registers are always reserved for use by a linker. Lastly, the r31 register is used as a stack pointer. This list may depend on the function, and it cannot be generated due to this dynamic behavior:

```
BitVector M88kRegisterInfo::getReservedRegs(
    const MachineFunction &MF) const {
  BitVector Reserved(getNumRegs());
  Reserved.set(M88k::R0);
  Reserved.set(M88k::R28);
  Reserved.set(M88k::R29);
  Reserved.set(M88k::R31);
  return Reserved;
}
```

5. If a frame register is required, then r30 is used. Please note that our code does not support creating a frame yet. If the function requires a frame, then r30 must also be marked as reserved in the getReservedRegs() method. However, we must implement this method because it is declared pure virtual in the superclass:

```
Register M88kRegisterInfo::getFrameRegister(
    const MachineFunction &MF) const {
  return M88k::R30;
}
```

6. Similarly, we need to implement the eliminateFrameIndex() method because it is declared *pure virtual*. It is called to replace a frame index in an operand with the correct value to use to address the value on the stack:

```
bool M88kRegisterInfo::eliminateFrameIndex(
    MachineBasicBlock::iterator MI, int SPAdj,
```

```
        unsigned FIOperandNum, RegScavenger *RS) const {
    return false;
  }
}
```

The `M88kInstrInfo` class has many hook methods we can override to accomplish special tasks, for example, for branch analysis and rematerialization. For now, we're only overriding the `expandPostRAPseudo()` method, in which we expand the pseudo-instruction RET. Let's begin with the header file, `M88kInstrInfo.h`:

1. The header file begins with including the generated code:

    ```
    #define GET_INSTRINFO_HEADER
    #include "M88kGenInstrInfo.inc"
    ```

2. The `M88kInstrInfo` class derives from the generated `M88kGenInstrInfo` class. Besides overriding the `expandPostRAPseudo()` method, the only other addition is that this class owns an instance of the previously defined class, `M88kRegisterInfo`:

    ```
    namespace llvm {
    class M88kInstrInfo : public M88kGenInstrInfo {
      const M88kRegisterInfo RI;
      [[maybe_unused]] M88kSubtarget &STI;

      virtual void anchor();

    public:
      explicit M88kInstrInfo(M88kSubtarget &STI);

      const M88kRegisterInfo &getRegisterInfo() const {
        return RI;
      }

      bool
      expandPostRAPseudo(MachineInstr &MI) const override;
    } // end namespace llvm
    ```

The implementation is stored in the `M88kInstrInfo.cpp` class:

1. Like the header file, the implementation begins with including the generated code:

    ```
    #define GET_INSTRINFO_CTOR_DTOR
    #define GET_INSTRMAP_INFO
    #include "M88kGenInstrInfo.inc"
    ```

2. Then, we define the anchor() method, which is used to pin the vtable to this file:

    ```
    void M88kInstrInfo::anchor() {}
    ```

3. Finally, we expand RET in the expandPostRAPseudo() method. As its name suggests, this method is invoked after the register allocator runs and is intended to expand the pseudo-instruction, which may still be mixed with the machine code. If the opcode of the machine instruction, MI, is the pseudo-instruction, RET, we must insert the jmp %r1 jump instruction, which is the instruction to exit a function. Then, we copy all implicit operands that represent the values to return and we delete the pseudo instruction. If we need other pseudo-instructions during code generation, then we can extend this function to expand them here too:

    ```
    bool M88kInstrInfo::expandPostRAPseudo(
        MachineInstr &MI) const {
      MachineBasicBlock &MBB = *MI.getParent();

      switch (MI.getOpcode()) {
      default:
        return false;
      case M88k::RET: {
        MachineInstrBuilder MIB =
            BuildMI(MBB, &MI, MI.getDebugLoc(),
                    get(M88k::JMP))
                .addReg(M88k::R1, RegState::Undef);

        for (auto &MO : MI.operands()) {
          if (MO.isImplicit())
            MIB.add(MO);
        }
        break;
      }
      }
      MBB.erase(MI);
      return true;
    }
    ```

Both classes have minimal implementations. If you continue to develop the target, then many more methods need to be overridden. It is worth reading the comments in the TargetInstrInfo and TargetRegisterInfo base classes, which you can find in the llvm/include/llvm/CodeGen directory.

We still need more classes to get the instruction selection running. Next, we'll look at frame lowering.

Putting an empty frame lowering in place

The binary interface of a platform not only defines how parameters are passed. It also includes how a stack frame is laid out: in which places are local variables stored, where registers are spilled to, and so on. Often, a special instruction sequence is required at the beginning and end of a function, called the **prolog** and the **epilog**. At the current development state, our target does not support the required machine instructions to create the prolog and the epilog. However, the framework code for instruction selection requires that a subclass of `TargetFrameLowering` is available. The easy solution is to provide the `M88kFrameLowering` class with an empty implementation.

The declaration of the class is in the `M88kFrameLowering.h` file. All we must do here is override the pure virtual functions:

```
namespace llvm {
class M88kFrameLowering : public TargetFrameLowering {
public:
  M88kFrameLowering();

  void
  emitPrologue(MachineFunction &MF,
               MachineBasicBlock &MBB) const override;
  void
  emitEpilogue(MachineFunction &MF,
               MachineBasicBlock &MBB) const override;
  bool hasFP(const MachineFunction &MF) const override;
};
}
```

The implementation, which is stored in the `M88kFrameLowering.cpp` file, provides some basic details about stack frames in the constructor. The stack grows downwards, to smaller addresses, and is aligned on 8-byte boundaries. When a function is called, the local variables are stored directly below the stack pointer of the calling function, so the offset of the local area is 0. Even during a function call, the stack should remain aligned at an 8-byte boundary. The last parameter implies that the stack cannot be realigned. The other functions just have an empty implementation:

```
M88kFrameLowering::M88kFrameLowering()
    : TargetFrameLowering(
          TargetFrameLowering::StackGrowsDown, Align(8),
          0, Align(8), false /* StackRealignable */) {}

void M88kFrameLowering::emitPrologue(
    MachineFunction &MF, MachineBasicBlock &MBB) const {}

void M88kFrameLowering::emitEpilogue(
```

```
        MachineFunction &MF, MachineBasicBlock &MBB) const {}

bool M88kFrameLowering::hasFP(
    const MachineFunction &MF) const { return false; }
```

Of course, as soon as our implementation grows, this class will be one of the first that needs to be fully implemented.

Before we can put all the pieces together, we need to implement the assembly printer, which is used to emit machine instructions.

Emitting machine instructions

The instruction selection creates machine instructions, represented by the MachineInstr class, from LLVM IR. But this is not the end. An instance of the MachineInstr class still carries additional information, such as labels or flags. To emit an instruction via the machine code component, we need to lower the instances of MachineInstr to instances of MCInst. By doing this, the machine code component provides the functionality to write instructions into object files or print them as assembler text. The M88kAsmPrinter class is responsible for emitting a whole compilation unit. Lowering an instruction is delegated to the M88kMCInstLower class.

The assembly printer is the last pass to run in a backend. Its implementation is stored in the M88kAsmPrinter.cpp file:

1. The declaration of the M88kAsmPrinter class is in an anonymous namespace. Besides the constructor, we only override the getPassName() function, which returns the name of the pass as a human-readable string, and the emitInstruction() function:

    ```
    namespace {
    class M88kAsmPrinter : public AsmPrinter {
    public:
        explicit M88kAsmPrinter(
            TargetMachine &TM,
            std::unique_ptr<MCStreamer> Streamer)
          : AsmPrinter(TM, std::move(Streamer)) {}

        StringRef getPassName() const override {
          return "M88k Assembly Printer";
        }

        void emitInstruction(const MachineInstr *MI) override;
    };
    } // end of anonymous namespace
    ```

2. Like many other classes, we have to register our assembly printer in the target registry:

```
extern "C" LLVM_EXTERNAL_VISIBILITY void
LLVMInitializeM88kAsmPrinter() {
  RegisterAsmPrinter<M88kAsmPrinter> X(
      getTheM88kTarget());
}
```

3. The `emitInstruction()` method is responsible for emitting the machine instruction, `MI`, to the output stream. In our implementation, we delegate the lowering of the instruction to the `M88kMCInstLower` class:

```
void M88kAsmPrinter::emitInstruction(
    const MachineInstr *MI) {
  MCInst LoweredMI;
  M88kMCInstLower Lower(MF->getContext(), *this);
  Lower.lower(MI, LoweredMI);
  EmitToStreamer(*OutStreamer, LoweredMI);
}
```

This is already the full implementation. The base class, `AsmPrinter`, provides many useful hooks you can override. For example, the `emitStartOfAsmFile()` method is called before anything is emitted, and `emitEndOfAsmFile()` is called after everything is emitted. These methods can emit target-specific data or code at the beginning and the end of a file. Similarly, the `emitFunctionBodyStart()` and `emitFunctionBodyEnd()` methods are called before and after a function body is emitted. Read the comments in the `llvm/include/llvm/CodeGen/AsmPrinter.h` file to understand what can be customized.

The `M88kMCInstLower` class lowers operands and instructions, and our implementation contains two methods for that purpose. The declaration is in the `M88kMCInstLower.h` file:

```
class LLVM_LIBRARY_VISIBILITY M88kMCInstLower {
public:
  void lower(const MachineInstr *MI, MCInst &OutMI) const;
  MCOperand lowerOperand(const MachineOperand &MO) const;
};
```

The definition goes into the `M88kMCInstLower.cpp` file:

1. To lower `MachineOperand` to `MCOperand`, we need to check the operand type. Here, we only handle registers and immediates by creating `MCOperand`-equivalent register and immediate values by supplying the original `MachineOperand` values. As soon as expressions are introduced as operands, this method needs to be enhanced:

```
MCOperand M88kMCInstLower::lowerOperand(
    const MachineOperand &MO) const {
```

```
switch (MO.getType()) {
case MachineOperand::MO_Register:
  return MCOperand::createReg(MO.getReg());

case MachineOperand::MO_Immediate:
  return MCOperand::createImm(MO.getImm());

default:
  llvm_unreachable("Operand type not handled");
}
}
```

2. The lowering of an instruction is similar. First, the opcode is copied, and then the operands are handled. An instance of `MachineInstr` can have implicit operands attached, which are not lowered, and we need to filter them:

```
void M88kMCInstLower::lower(const MachineInstr *MI,
                            MCInst &OutMI) const {
  OutMI.setOpcode(MI->getOpcode());
  for (auto &MO : MI->operands()) {
    if (!MO.isReg() || !MO.isImplicit())
      OutMI.addOperand(lowerOperand(MO));
  }
}
```

With that, we've implemented the assembly printer. Now, we need to bring all the pieces together. We'll do this in the next section.

Creating the target machine and the sub-target

So far, we've implemented the instruction selection classes and a couple of other classes. Now, we need to set up how our backend will work. Like the optimization pipeline, a backend is divided into passes. Configuring those passes is the main task of the `M88kTargetMachine` class. In addition, we need to specify which features are available for instruction selection. Usually, a platform is a family of CPUs, which all have a common set of instructions but differ by specific extensions. For example, some CPUs have vector instructions, while others do not. In LLVM IR, a function can have attributes attached that specify for which CPU this function should be compiled, or what features are available. In other words, each function could have a different configuration, which is captured in the `M88kSubTarget` class.

Implementing M88kSubtarget

Let's implement the `M88kSubtarget` class first. The declaration is stored in the `M88kSubtarget.h` class:

1. Parts of the sub-target are generated from the target description, and we include those codes first:

```
#define GET_SUBTARGETINFO_HEADER
#include "M88kGenSubtargetInfo.inc"
```

2. Then, we declare the class, deriving it from the generated `M88kGenSubtargetInfo` class. The class owns a couple of previously defined classes – the instruction information, the target lowering class, and the frame lowering class:

```
namespace llvm {
class StringRef;
class TargetMachine;

class M88kSubtarget : public M88kGenSubtargetInfo {
  virtual void anchor();

  Triple TargetTriple;
  M88kInstrInfo InstrInfo;
  M88kTargetLowering TLInfo;
  M88kFrameLowering FrameLowering;
```

3. The sub-target is initialized with the target triple, the name of the CPU, and a feature string, as well as with the target machine. All these parameters describe the hardware for which our backend will generate code:

```
public:
  M88kSubtarget(const Triple &TT,
                const std::string &CPU,
                const std::string &FS,
                const TargetMachine &TM);
```

4. Next, we include the generated file again, this time for automatically defining getter methods for features defined in the target description:

```
#define GET_SUBTARGETINFO_MACRO(ATTRIBUTE, DEFAULT,     \
                                GETTER)                 \
  bool GETTER() const { return ATTRIBUTE; }
#include "M88kGenSubtargetInfo.inc"
```

5. In addition, we need to declare the `ParseSubtargetFeatures()` method. The method itself is generated from the target description:

```
void ParseSubtargetFeatures(StringRef CPU,
                            StringRef TuneCPU,
                            StringRef FS);
```

6. Next, we must add getter methods for the member variables:

```
const TargetFrameLowering *
getFrameLowering() const override {
  return &FrameLowering;
}
const M88kInstrInfo *getInstrInfo() const override {
  return &InstrInfo;
}
const M88kTargetLowering *
getTargetLowering() const override {
  return &TLInfo;
}
```

7. Finally, we must add a getter method for the register information, which is owned by the instruction information class. This finishes the declaration:

```
const M88kRegisterInfo *
getRegisterInfo() const override {
  return &InstrInfo.getRegisterInfo();
}
};
} // end namespace llvm
```

Next, we must implement the actual subtarget class. The implementation is stored in the `M88kSubtarget.cpp` file:

1. Again, we begin the file by including the generated source:

```
#define GET_SUBTARGETINFO_TARGET_DESC
#define GET_SUBTARGETINFO_CTOR
#include "M88kGenSubtargetInfo.inc"
```

2. Then, we define the anchor method, which pins the vtable to this file:

```
void M88kSubtarget::anchor() {}
```

3. Finally, we define the constructor. Note that the generated class expected two CPU parameters: the first one for the instruction set, and the second one for scheduling. The use case here is

that you want to optimize the code for the latest CPU but still be able to run the code on an older CPU. We do not support this feature and use the same CPU name for both parameters:

```
M88kSubtarget::M88kSubtarget(const Triple &TT,
                             const std::string &CPU,
                             const std::string &FS,
                             const TargetMachine &TM)
    : M88kGenSubtargetInfo(TT, CPU, /*TuneCPU*/ CPU, FS),
      TargetTriple(TT), InstrInfo(*this),
      TLInfo(TM, *this), FrameLowering() {}
```

Implementing M88kTargetMachine – defining the definitions

Finally, we can implement the M88kTargetMachine class. This class holds all used sub-target instances. It also owns a subclass of TargetLoweringObjectFile, which provides details such as section names to the lowering process. Lastly, it creates the configuration of the passes that runs in this backend.

The declaration in the M88kTargetMachine.h file is as follows:

1. The M88kTargetMachine class derives from the LLVMTargetMachine class. The only members are an instance of TargetLoweringObjectFile and the sub-target map:

    ```
    namespace llvm {
    class M88kTargetMachine : public LLVMTargetMachine {
      std::unique_ptr<TargetLoweringObjectFile> TLOF;
      mutable StringMap<std::unique_ptr<M88kSubtarget>>
          SubtargetMap;
    ```

2. The parameters of the constructor completely describe the target configuration for which we will generate code. With the TargetOptions class, many details of the code generations can be controlled – for example, if floating-point multiply-and-add instructions can be used or not. Also, the relocation model, the code model, and the optimization level are passed to the constructor. Notably, the JIT parameter is set to true if the target machine is used for just-in-time compilation.

    ```
    public:
      M88kTargetMachine(const Target &T, const Triple &TT,
                        StringRef CPU, StringRef FS,
                        const TargetOptions &Options,
                        std::optional<Reloc::Model> RM,
                        std::optional<CodeModel::Model> CM,
                        CodeGenOpt::Level OL, bool JIT);
    ```

3. We also need to override some methods. The `getSubtargetImpl()` method returns the sub-target instance to use for the given function, and the `getObjFileLowering()` method just returns the member variable. In addition, we override the `createPassConfig()` method, which returns our configuration for the backend passes:

```
~M88kTargetMachine() override;

const M88kSubtarget *
getSubtargetImpl(const Function &) const override;

TargetPassConfig *
createPassConfig(PassManagerBase &PM) override;

TargetLoweringObjectFile *
getObjFileLowering() const override {
  return TLOF.get();
}
};
} // end namespace llvm
```

Implementing M88kTargetMachine – adding the implementation

The implementation of the class is stored in the `M88kTargetMachine.cpp` file. Please note that we created this file in *Chapter 11*. Now, we will replace this file with a complete implementation:

1. First, we must register the target machine. In addition, we must initialize the DAG-to-DAG pass via the initialization function we defined earlier:

```
extern "C" LLVM_EXTERNAL_VISIBILITY void
LLVMInitializeM88kTarget() {
  RegisterTargetMachine<M88kTargetMachine> X(
      getTheM88kTarget());
  auto &PR = *PassRegistry::getPassRegistry();
  initializeM88kDAGToDAGISelPass(PR);
}
```

2. Next, we must define the support function, `computeDataLayout()`. We talked about the data layout string in *Chapter 4, Basics of IR Code Generation*. In this function, the data layout, as the backend, expects it to be defined. Since the data layout depends on hardware features, the triple, the name of the CPU, and the feature set string are passed to this function. We create the data layout string with the following components. The target is big-endian (E) and uses the ELF symbol mangling.

Pointers are 32-bit wide and 32-bit aligned. All scalar types are naturally aligned. The MC88110 CPU has an extended register set and supports 80-bit wide floating points. If we were to support this special feature, then we'd need to add a check of the CPU name here and extend the string with the floating-point values accordingly. Next, we must state that all globals have a preferred alignment of 16-bit and that the hardware has only 32-bit registers:

```
namespace {
std::string computeDataLayout(const Triple &TT,
                              StringRef CPU,
                              StringRef FS) {
  std::string Ret;
  Ret += "E";
  Ret += DataLayout::getManglingComponent(TT);
  Ret += "-p:32:32:32";
  Ret += "-i1:8:8-i8:8:8-i16:16:16-i32:32:32-i64:64:64";
  Ret += "-f32:32:32-f64:64:64";
  Ret += "-a:8:16";
  Ret += "-n32";
  return Ret;
}
} // namespace
```

3. Now, we can define the constructor and destructor. Many of the parameters are just passed to the superclass constructor. Note that our computeDataLayout() function is called here. In addition, the TLOF member is initialized with an instance of TargetLoweringObjectFileELF, since we are using the ELF file format. In the body of the constructor, we must call the initAsmInfo() method, which initializes many data members of the superclass:

```
M88kTargetMachine::M88kTargetMachine(
    const Target &T, const Triple &TT, StringRef CPU,
    StringRef FS, const TargetOptions &Options,
    std::optional<Reloc::Model> RM,
    std::optional<CodeModel::Model> CM,
    CodeGenOpt::Level OL, bool JIT)
    : LLVMTargetMachine(
          T, computeDataLayout(TT, CPU, FS), TT, CPU,
          FS, Options, !RM ? Reloc::Static : *RM,
          getEffectiveCodeModel(CM, CodeModel::Medium),
          OL),
      TLOF(std::make_unique<
          TargetLoweringObjectFileELF>()) {
  initAsmInfo();
}

M88kTargetMachine::~M88kTargetMachine() {}
```

4. After, we define the `getSubtargetImpl()` method. The sub-target instance to use depends on the `target-cpu` and `target-features` function attributes. For example, the `target-cpu` attribute could be set to `MC88110`, thus targeting the second-generation CPU. However, the attribute target feature could describe that we should not use the graphics instructions of that CPU. We have not defined the CPUs and their features in the target description yet, so we are doing a bit more than what's necessary here. However, the implementation is simple enough: we query the function attributes and use either the returned strings or the default values. With this information, we can query the `SubtargetMap` member, and if it's not found, we create the sub-target:

```
const M88kSubtarget *
M88kTargetMachine::getSubtargetImpl(
    const Function &F) const {
  Attribute CPUAttr = F.getFnAttribute("target-cpu");
  Attribute FSAttr =
      F.getFnAttribute("target-features");

  std::string CPU =
      !CPUAttr.hasAttribute(Attribute::None)
          ? CPUAttr.getValueAsString().str()
          : TargetCPU;
  std::string FS = !FSAttr.hasAttribute(Attribute::None)
                       ? FSAttr.getValueAsString().str()
                       : TargetFS;

  auto &I = SubtargetMap[CPU + FS];
  if (!I) {
    resetTargetOptions(F);
    I = std::make_unique<M88kSubtarget>(TargetTriple,
                                        CPU, FS, *this);

  }
  return I.get();
}
```

5. Finally, we create the pass configuration. For this, we need our own class, `M88kPassConfig`, which derives from the `TargetPassConfig` class. We only override the `addInstSelector` method:

```
namespace {
class M88kPassConfig : public TargetPassConfig {
public:
  M88kPassConfig(M88kTargetMachine &TM,
                 PassManagerBase &PM)
      : TargetPassConfig(TM, PM) {}
```

```
    bool addInstSelector() override;
};
} // namespace
```

6. With this definition, we can implement the `createPassConfig` factory method:

```
TargetPassConfig *M88kTargetMachine::createPassConfig(
    PassManagerBase &PM) {
  return new M88kPassConfig(*this, PM);
}
```

7. Lastly, we must add our instruction selection class to the pass pipeline in the `addInstSelector()` method. The return value, `false`, indicates that we have added a pass that converts LLVM IR into machine instructions:

```
bool M88kPassConfig::addInstSelector() {
  addPass(createM88kISelDag(getTM<M88kTargetMachine>(),
                           getOptLevel()));
  return false;
}
```

That was a long journey to finish the implementation! Now that we've built the `llc` tool, we can run an example. Save the following simple IR in the `and.ll` file:

```
define i32 @f1(i32 %a, i32 %b) {
  %res = and i32 %a, %b
  ret i32 %res
}
```

Now, we can run `llc` and verify that the generated assembly looks reasonable:

```
$ llc -mtriple m88k-openbsd < and.ll
        .text
        .file   "<stdin>"
        .globl  f1                              | -- Begin function f1
        .align  2
        .type   f1,@function
f1:                                             | @f1
| %bb.0:
        and %r2, %r2, %r3
        jmp %r1
.Lfunc_end0:
        .size   f1, .Lfunc_end0-f1
                                                | -- End function
        .section        ".note.GNU-stack","",@progbits
```

To compile for the m88k target, we must specify the triple on the command line, as in this example, or in the IR file.

Enjoy your success for a bit before we look at global instruction selection.

Global instruction selection

Instruction selection via the selection DAG produces fast code, but it takes time to do so. The speed of the compiler is often critical for developers, who want to quickly try out the changes they've made. Usually, the compiler should be very fast at optimization level 0, but it can take more time with increased optimization levels. However, constructing the selection DAG costs so much time that this approach does not scale as required. The first solution was to create another instruction selection algorithm called FastISel, which is fast but does not generate good code. It also does not share code with the selection DAG implementation, which is an obvious problem. Because of this, not all targets support FastISel.

The selection DAG approach does not scale because it is a large, monolithic algorithm. If we can avoid creating a new data structure such as the selection DAG, then we should be able to perform the instruction selection using small components. The backend already has a pass pipeline, so using passes is a natural choice. Based on these thoughts, GlobalISel performs the following steps:

1. First, the LLVM IR is lowered into generic machine instructions. Generic machine instructions represent the most common operation found in real hardware. Note that this translation uses the machine functions and machine basic blocks, which means it directly translates into the data structures used by the other parts of the backend.

2. The generic machine instructions are then legalized.

3. After, the operands of generic machine instructions are mapped to register banks.

4. Finally, the generic instructions are replaced with real machine instructions, using the patterns defined in the target description.

Since these are all passes, we can insert as many passes as we want in between. For example, a combiner pass could be used to replace a sequence of generic machine instructions with another generic machine instruction, or with a real machine instruction. Turning these additional passes off increases the compilation speed while turning them on improves the quality of the generated code. Hence, we can scale as we need.

There is another advantage in this approach. The selection DAG translates basic block by basic block, but a machine pass works on a machine function, which enables us to consider all basic blocks of a function during instruction selections. Therefore, this instruction selection method is called global instruction selection (GlobalISel). Let's have a look at how this approach works, starting with the transformation of calls.

Lowering arguments and return values

For translating the LLVM IR to generic machine instructions, we only need to implement how arguments and return values are handled. Again, the implementation can be simplified by using the generated code from the target description. The class we'll create is called M88kCallLowering, and the declaration is in the GISel/M88kCallLowering.h header file:

```
class M88kCallLowering : public CallLowering {

public:
  M88kCallLowering(const M88kTargetLowering &TLI);

  bool
  lowerReturn(MachineIRBuilder &MIRBuilder,
              const Value *Val,
              ArrayRef<Register> VRegs,
              FunctionLoweringInfo &FLI,
              Register SwiftErrorVReg) const override;

  bool lowerFormalArguments(
      MachineIRBuilder &MIRBuilder, const Function &F,
      ArrayRef<ArrayRef<Register>> VRegs,
      FunctionLoweringInfo &FLI) const override;
  bool enableBigEndian() const override { return true; }
};
```

The GlobalISel framework will call the lowerReturn() and lowerFormalArguments() methods when a function is translated. To translate a function call, you would need to override and implement the lowerCall() method as well. Please note that we also need to override enableBigEndian(). Without it, the wrong machine code would be generated.

For the implementation in the GISel/M88kCallLowering.cpp file, we need to define to support classes. The generated code from the target description tells us how a parameter is passed – for example, in a register. We need to create a subclass of ValueHandler to generate the machine instructions for it. For incoming parameters, we need to derive our class from IncomingValueHandler, as well as for the return value from OutgoingValueHandler. Both are very similar, so we'll only look at the handler for incoming arguments:

```
namespace {
struct FormalArgHandler
    : public CallLowering::IncomingValueHandler {
  FormalArgHandler(MachineIRBuilder &MIRBuilder,
                   MachineRegisterInfo &MRI)
      : CallLowering::IncomingValueHandler(MIRBuilder,
```

```
                                          MRI) {}

    void assignValueToReg(Register ValVReg,
                          Register PhysReg,
                          CCValAssign VA) override;

    void assignValueToAddress(Register ValVReg,
                              Register Addr, LLT MemTy,
                              MachinePointerInfo &MPO,
                              CCValAssign &VA) override{};

    Register
    getStackAddress(uint64_t Size, int64_t Offset,
                    MachinePointerInfo &MPO,
                    ISD::ArgFlagsTy Flags) override {
      return Register();
    };
  };
} // namespace
```

So far, we can only handle parameters passed in registers, so we must provide a dummy implementation for the other methods. The `assignValueToReg()` method copies the value of the incoming physical register to a virtual register, performing a truncation if necessary. All we have to do here is mark the physical register as live-in to the function, and call the superclass implementation:

```
void FormalArgHandler::assignValueToReg(
    Register ValVReg, Register PhysReg,
    CCValAssign VA) {
  MIRBuilder.getMRI()->addLiveIn(PhysReg);
  MIRBuilder.getMBB().addLiveIn(PhysReg);
  CallLowering::IncomingValueHandler::assignValueToReg(
      ValVReg, PhysReg, VA);
}
```

Now, we can implement the `lowerFormalArgument()` method:

1. First, the parameters of the IR function are translated into instances of the `ArgInfo` class. The `setArgFlags()` and `splitToValueTypes()` framework methods help with copying the parameter attributes and splitting the value type in case an incoming argument needs more than one virtual register:

    ```
    bool M88kCallLowering::lowerFormalArguments(
        MachineIRBuilder &MIRBuilder, const Function &F,
        ArrayRef<ArrayRef<Register>> VRegs,
    ```

```
        FunctionLoweringInfo &FLI) const {
    MachineFunction &MF = MIRBuilder.getMF();
    MachineRegisterInfo &MRI = MF.getRegInfo();
    const auto &DL = F.getParent()->getDataLayout();

    SmallVector<ArgInfo, 8> SplitArgs;
    for (const auto &[I, Arg] :
        llvm::enumerate(F.args())) {
      ArgInfo OrigArg{VRegs[I], Arg.getType(),
                      static_cast<unsigned>(I)};
      setArgFlags(OrigArg,
                  I + AttributeList::FirstArgIndex, DL,
                  F);
      splitToValueTypes(OrigArg, SplitArgs, DL,
                        F.getCallingConv());
    }
```

2. With the arguments prepared in the `SplitArgs` variable, we are ready to generate the machine code. This is all done by the framework code, with the help of the generated calling convention, `CC_M88k`, and our helper class, `FormalArghandler`:

```
    IncomingValueAssigner ArgAssigner(CC_M88k);
    FormalArgHandler ArgHandler(MIRBuilder, MRI);
    return determineAndHandleAssignments(
        ArgHandler, ArgAssigner, SplitArgs, MIRBuilder,
        F.getCallingConv(), F.isVarArg());
  }
```

Return values are handled similarly, with the main difference being that one value is returned at most. The next task is to legalize the generic machine instructions.

Legalizing the generic machine instructions

The translation from LLVM IR to generic machine code is mostly fixed. As a result, instructions can be generated that use unsupported data types, among other challenges. The task of the legalizer pass is to define which operations and instructions are legal. With this information, the GlobalISel framework tries to transform the instructions into a legal form. For example, the m88k architecture only has 32-bit registers, so a bitwise and operation with 64-bit values is not legal. However, if we split the 64-bit value into two 32-bit values, and use two bitwise and operations instead, then we have legal code. This can be translated into a legalization rule:

```
getActionDefinitionsBuilder({G_AND, G_OR, G_XOR})
    .legalFor({S32})
    .clampScalar(0, S32, S32);
```

Whenever the legalizer pass processes a G_AND instruction, then it is legal if all operands are 32-bit wide. Otherwise, the operands are clamped to 32-bit, effectively splitting larger values into multiple 32-bit values, and the rule is applied again. If an instruction cannot be legalized, then the backend terminates with an error message.

All legalization rules are defined in the constructor of the M88kLegalizerInfo class, which makes the class very simple.

> **What does legal mean?**
>
> In GlobalISel, a generic instruction is legal if it can be translated by the instruction selector. This gives us more freedom in the implementation. For example, we can state that an instruction works on a bit value, even if the hardware only operates on 32-bit values, so long as the instruction selector can handle the type correctly.

The next pass we need to look at is the register bank selector.

Selecting a register bank for operands

Many architectures define several register banks. A register bank is a set of registers. Typical register banks are general-purpose register banks and floating-point register banks. Why is this information important? Moving a value from one register to another is usually cheap inside a register bank, but copying the value to another register bank can be costly or impossible. Thus, we must select a good register bank for each operand.

The implementation of this class involves an addition to the target description. In the GISel/ M88lRegisterbanks.td file, we define our single register bank, referencing the register classes we have defined:

```
def GRRegBank : RegisterBank<"GRRB", [GPR, GPR64]>;
```

From this line, some support code is generated. However, we still need to add some code that could be potentially generated. First, we need to define partial mappings. This tells the framework at which bit index a value begins, how wide it is, and to which register bank it maps. We have two entries, one for each register class:

```
RegisterBankInfo::PartialMapping
    M88kGenRegisterBankInfo::PartMappings[]{
        {0, 32, M88k::GRRegBank},
        {0, 64, M88k::GRRegBank},
    };
```

To index this array, we must define an enumeration:

```
enum PartialMappingIdx { PMI_GR32 = 0, PMI_GR64, };
```

Since we have only three address instructions, we need three partial mappings, one for each operand. We must create an array with all those pointers, with the first entry denoting an invalid mapping:

```
RegisterBankInfo::ValueMapping
    M88kGenRegisterBankInfo::ValMappings[]{
        {nullptr, 0},
        {&M88kGenRegisterBankInfo::PartMappings[PMI_GR32], 1},
        {&M88kGenRegisterBankInfo::PartMappings[PMI_GR32], 1},
        {&M88kGenRegisterBankInfo::PartMappings[PMI_GR32], 1},
        {&M88kGenRegisterBankInfo::PartMappings[PMI_GR64], 1},
        {&M88kGenRegisterBankInfo::PartMappings[PMI_GR64], 1},
        {&M88kGenRegisterBankInfo::PartMappings[PMI_GR64], 1},
    };
```

To access that array, we must define a function:

```
const RegisterBankInfo::ValueMapping *
M88kGenRegisterBankInfo::getValueMapping(
    PartialMappingIdx RBIdx) {
  return &ValMappings[1 + 3*RBIdx];
}
```

When creating these tables, it is easy to make errors. At first glance, all this information can be derived from the target description, and a comment in the source states that this code should be generated by TableGen! However, this hasn't been implemented yet, so we have to create the code manually.

The most important function we have to implement in the `M88kRegisterBankInfo` class is `getInstrMapping()`, which returns the mapped register banks for each operand of the instruction. This now becomes easy because we can look up the array of partial mappings, which we can then pass to the `getInstructionMapping()` method, which constructs the full instruction mapping:

```
const RegisterBankInfo::InstructionMapping &
M88kRegisterBankInfo::getInstrMapping(
    const MachineInstr &MI) const {
  const ValueMapping *OperandsMapping = nullptr;
  switch (MI.getOpcode()) {
  case TargetOpcode::G_AND:
  case TargetOpcode::G_OR:
  case TargetOpcode::G_XOR:
    OperandsMapping = getValueMapping(PMI_GR32);
    break;

  default:
#if !defined(NDEBUG) || defined(LLVM_ENABLE_DUMP)
    MI.dump();
```

```
#endif
    return getInvalidInstructionMapping();
  }

  return getInstructionMapping(DefaultMappingID, /*Cost=*/1,
                               OperandsMapping,
                               MI.getNumOperands());
}
```

During development, it is common to forget the register bank mapping for a generic instruction. Unfortunately, the error message that's generated at runtime does not mention for which instruction the mapping failed. The easy fix is to dump the instruction before returning the invalid mapping. However, we need to be careful here because the dump() method is not available in all build types.

After mapping the register banks, we must translate the generic machine instructions into real machine instructions.

Translating generic machine instructions

For instruction selection via the selection DAG, we added patterns to the target description, which use DAG operations and operands. To reuse those patterns, a mapping from DAG node types to generic machine instructions was introduced. For example, the DAG and operation maps to the generic G_AND machine instruction. Not all DAG operations have an equivalent generic machine instruction; however, the most common cases are covered. Therefore, it is beneficial to define all code selection patterns in the target description.

Most of the implementation of the M88kInstructionSelector class, which can be found in the GISel/M88kInstructionSelector.cpp file, is generated from the target description. However, we need to override the select() method, which allows us to translate generic machine instructions that are not covered by the patterns in the target description. Since we only support a very small subset of generic instructions, we can simply call the generated pattern matcher:

```
bool M88kInstructionSelector::select(MachineInstr &I) {
  if (selectImpl(I, *CoverageInfo))
    return true;
  return false;
}
```

With the instruction selection implemented, we can translate LLVM IR using GlobalISel!

Running an example

To translate LLVM IR using GlobalISel, we need to add the `-global-isel` option to the command line of `llc`. For example, you can use the previously defined IR file, `and.ll`:

```
$ llc -mtriple m88k-openbsd -global-isel < and.ll
```

The printed assembly text is the same. To convince ourselves that the translation uses GlobalISel, we must take advantage of the fact that we can stop the translation after a specified pass is run with the `-stop-after=` option. For example, to see the generic instructions after legalization, you would run the following command:

```
$ llc -mtriple m88k-openbsd -global-isel < and.ll \
   -stop-after=legalizer
```

The ability to stop after (or before) a pass is run is another advantage of GlobalISel because it makes it easy to debug and test the implementation.

At this point, we have a working backend that can translate some LLVM IR into machine code for the m88k architecture. Let's think about how to move from here to a more complete backend.

How to further evolve the backend

With the code from this and the previous chapter, we have created a backend that can translate some LLVM IR into machine code. It is very satisfying to see the backend working, but it is far from being usable for serious tasks. Much more coding is needed. Here is a recipe for how you can further evolve the backend:

- The first decision you should make is if you want to use GlobalISel or the selection DAG. In our experience, GlobalISel is easier to understand and develop, but all targets in the LLVM source tree implement the selection DAG, and you may already have experience in using it.

- Next, you should define the instructions for adding and subtracting integer values, which can be done similarly to the bitwise and instruction.

- After, you should implement the load and store instructions. This is more involved since you need to translate the different addressing modes. Most likely, you will deal with indexing, for example, to address an element of an array, which most likely requires the previously defined instruction for addition.

- Finally, you can fully implement frame lowering and call lowering. At this point, you can translate a simple "Hello, world!" style application into a running program.

- The next logical step is to implement branch instructions, which enable the translation of loops. To generate optimal code, you need to implement the branch analyzing methods in the instruction information class.

When you reach this point, your backend can already translate simple algorithms. You should also have gained enough experience to develop the missing parts based on your priorities.

Summary

In this chapter, you added two different instruction selections to your backend: instruction selection via the selection DAG, and global instruction selection. For this, you had to define the calling convention in the target description. In addition, you needed to implement register and instruction information classes, which give you access to information generated from the target description but which you also needed to enhance with additional information. You learned that the stack frame layout and prolog generation are needed later. To translate an example, you added a class to emit machine instructions, and you created the configuration of the backend. You also learned how global instruction selection works. Finally, you gained some guidance on how you can develop the backend on your own.

In the next chapter, we will look at some tasks that can be done after instruction selection – we will add a new pass in the pipeline of the backend, look at how to integrate the backend into the clang compiler, and how to cross-compile to a different architecture.

13

Beyond Instruction Selection

Now that we've learned about instruction selection using the SelectionDAG and GlobalISel LLVM-based frameworks in the previous chapters, we can explore other interesting concepts beyond instruction selection. This chapter encapsulates some more advanced topics outside of the backend that can be interesting for a highly optimizing compiler. For instance, some passes run beyond instruction selection and can perform different optimizations on various instructions, which can mean that developers have the luxury to introduce their own passes to perform meaningful target-specific tasks at this point in the compiler.

Ultimately, within this chapter, we will dive into the following concepts:

- Adding a new machine function pass to LLVM
- Integrating a new target into the clang frontend
- How to target a different CPU architecture

Adding a new machine function pass to LLVM

In this section, we will explore how to implement a new machine function pass within LLVM that runs after instruction selection. Specifically, a `MachineFunctionPass` class will be created, which is a subset of the original `FunctionPass` class within LLVM that can be run with `opt`. This class adapts the original infrastructure to allow for the implementation of passes that operate on the `MachineFunction` representation in the backend through `llc`.

It is important to note that the implementation of passes within the backend utilizes the interfaces of the legacy pass manager, rather than the new pass manager. This is because LLVM currently does not have a complete working implementation of the new pass manager within the backend. Due to this, this chapter will follow the method of adding a new pass within the legacy pass manager pipeline.

In terms of the actual implementation, such as function passes, machine function passes optimize a single (machine) function at a time, but instead of overriding the runOnFunction() method, machine function passes override the runOnMachineFunction() method. The machine function pass that will be implemented in this section is a pass that checks for when a division by zero occurs, specifically, to insert code that traps in the backend. This type of pass is important for the M88k target due to hardware limitations on MC88100 as this CPU does not reliably detect division by zero situations.

Continuing from the previous chapter's implementation of the backend, let's examine how a backend machine function pass is implemented!

Implementing the top-level interface for the M88k target

Firstly, within `llvm/lib/Target/M88k/M88k.h`, let's add two prototypes inside the `llvm` namespace declaration that will be used later:

1. The machine function pass that will be implemented will be called `M88kDivInstrPass`. We will add a function declaration that initializes this pass and takes in the pass registry, which is a class that manages the registration and initialization of all passes:

   ```
   void initializeM88kDivInstrPass(PassRegistry &);
   ```

2. Next, the actual function that creates the `M88kDivInstr` pass is declared, with the M88k target machine information as its argument:

   ```
   FunctionPass *createM88kDivInstr(const M88kTargetMachine &);
   ```

Adding the TargetMachine implementation for machine function passes

Next, we will analyze some of the changes that are required in `llvm/lib/Target/M88k/M88kTargetMachine.cpp`:

1. Within LLVM, it's common to give the user the option to toggle passes on or off. So, let's provide the user the same flexibility with our machine function pass. We'll start by declaring a command-line option called `m88k-no-check-zero-division` and initializing it to `false`, which implies that there will always be a check for zero division unless the user explicitly turns this off. We'll add this under the `llvm` namespace declaration and is an option for `llc`:

   ```
   using namespace llvm;
   static cl::opt<bool>
       NoZeroDivCheck("m88k-no-check-zero-division", cl::Hidden,
                   cl::desc("M88k: Don't trap on integer
   division by zero."),
                       cl::init(false));
   ```

2. It is also customary to create a formal method that returns the command-line value so that we can query it to determine whether the pass will be run. Our original command-line option will be wrapped in the `noZeroDivCheck()` method so that we can utilize the command-line result later:

```
M88kTargetMachine::~M88kTargetMachine() {}
bool M88kTargetMachine::noZeroDivCheck() const { return
NoZeroDivCheck; }
```

3. Next, inside `LLVMInitializeM88kTarget()`, where the M88k target and passes are registered and initialized, we will insert a call to the `initializeM88kDivInstrPass()` method that was declared earlier in `llvm/lib/Target/M88k/M88k.h`:

```
extern "C" LLVM_EXTERNAL_VISIBILITY void
LLVMInitializeM88kTarget() {
    RegisterTargetMachine<M88kTargetMachine>
X(getTheM88kTarget());
    auto &PR = *PassRegistry::getPassRegistry();
    initializeM88kDAGToDAGISelPass(PR);
    initializeM88kDivInstrPass(PR);
}
```

4. The M88k target also needs to override `addMachineSSAOptimization()`, which is a method that adds passes to optimize machine instructions when they are in SSA form. Essentially, our machine function pass is added as a type of machine SSA optimization. This method is declared as a function that is to be overridden. We will add the full implementation at the end of `M88kTargetMachine.cpp`:

```
bool addInstSelector() override;
void addPreEmitPass() override;
void addMachineSSAOptimization() override;

. . .

void M88kPassConfig::addMachineSSAOptimization() {
    addPass(createM88kDivInstr(getTM<M88kTargetMachine>()));
    TargetPassConfig::addMachineSSAOptimization();
}
```

5. Our method that returns the command-line option to toggle the machine function pass on and off (the `noZeroDivCheck()` method) is also declared in `M88kTargetMachine.h`:

```
~M88kTargetMachine() override;
bool noZeroDivCheck() const;
```

Developing the specifics of the machine function pass

Now that the implementation in the M88k target machine is completed, the next step will be to develop the machine function pass itself. The implementation is contained within the new file, `llvm/lib/Target/M88k/M88kDivInstr.cpp`:

1. The necessary headers for our machine function pass are added first. This includes headers that give us access to the M88k target information and headers that allow us to operate on machine functions and machine instructions:

    ```
    #include "M88k.h"
    #include "M88kInstrInfo.h"
    #include "M88kTargetMachine.h"
    #include "MCTargetDesc/M88kMCTargetDesc.h"
    #include "llvm/ADT/Statistic.h"
    #include "llvm/CodeGen/MachineFunction.h"
    #include "llvm/CodeGen/MachineFunctionPass.h"
    #include "llvm/CodeGen/MachineInstrBuilder.h"
    #include "llvm/CodeGen/MachineRegisterInfo.h"
    #include "llvm/IR/Instructions.h"
    #include "llvm/Support/Debug.h"
    ```

2. After that, we will add some code to prepare for our machine function pass. The first is a `DEBUG_TYPE` definition that is named `m88k-div-instr`, which is used for fine-grained control when debugging. Specifically, defining this `DEBUG_TYPE` allows users to specify the machine function pass name and view any debugging information that is pertinent to the pass when debug information is enabled:

    ```
    #define DEBUG_TYPE "m88k-div-instr"
    ```

3. We also specify that the `llvm` namespace is being used, and a `STATISTIC` value for our machine function is also declared. This statistic, called `InsertedChecks`, keeps track of how many division-by-zero checks are inserted by the compiler. Finally, an anonymous namespace is declared to encapsulate the subsequent machine function pass implementation:

    ```
    using namespace llvm;
    STATISTIC(InsertedChecks, "Number of inserted checks for
    division by zero");
    namespace {
    ```

4. As mentioned previously, this machine function pass aims to check for division by zero cases and inserts instructions that will cause the CPU to trap. These instructions require condition codes, so an enum value that we call `CC0` is defined with condition codes that are valid for the M88k target, along with their encodings:

```
enum class CC0 : unsigned {
    EQ0 = 0x2,
    NE0 = 0xd,
    GT0 = 0x1,
    LT0 = 0xc,
    GE0 = 0x3,
    LE0 = 0xe
};
```

5. Let's create the actual class for our machine function pass next, called `M88kDivInstr`. Firstly, we create it as an instance that inherits and is of the `MachineFunctionPass` type. Next, we declare various necessary instances that our `M88kDivInstr` pass will require. This includes `M88kBuilder`, which we will create and elaborate on later, and `M88kTargetMachine`, which contains target instruction and register information. Furthermore, we also require the register bank information and the machine register information when emitting instructions. An `AddZeroDivCheck` Boolean is also added to represent the previous command-line option, which turns our pass on or off:

```
class M88kDivInstr : public MachineFunctionPass {
    friend class M88kBuilder;
    const M88kTargetMachine *TM;
    const TargetInstrInfo *TII;
    const TargetRegisterInfo *TRI;
    const RegisterBankInfo *RBI;
    MachineRegisterInfo *MRI;
    bool AddZeroDivCheck;
```

6. For the public variables and methods of the `M88kDivInstr` class, we declare an identification number that LLVM will use to identify our pass, as well as the `M88kDivInstr` constructor, which takes in `M88kTargetMachine`. Next, we override the `getRequiredProperties()` method, which represents the properties that `MachineFunction` may have at any time during the optimization, and we also override the `runOnMachineFunction()` method, which will be one of the primary methods that our pass will run when checking for any division by zero. The second important function that is publicly declared is the `runOnMachineBasicBlock()` function, which will be executed from inside `runOnMachineFunction()`:

```
public:
    static char ID;
    M88kDivInstr(const M88kTargetMachine *TM = nullptr);
    MachineFunctionProperties getRequiredProperties() const
```

```
override;
  bool runOnMachineFunction(MachineFunction &MF) override;
  bool runOnMachineBasicBlock(MachineBasicBlock &MBB);
```

7. Finally, the last part is declaring the private methods and closing off the class. The only private method we declare within the `M88kDivInstr` class is the `addZeroDivCheck()` method, which inserts the checks for division by zero after any divide instruction. As we will see later, `MachineInstr` will need to point to specific divide instructions on the M88k target:

```
private:
  void addZeroDivCheck(MachineBasicBlock &MBB, MachineInstr
*DivInst);
};
```

8. An `M88kBuilder` class is created next, which is a specialized builder instance that creates M88k-specific instructions. This class keeps an instance of `MachineBasicBlock` (and a corresponding iterator), and `DebugLoc` to keep track of the debug location of this builder class. Other necessary instances include the target instruction information, the target register information, and the register bank information of the M88k target:

```
class M88kBuilder {
  MachineBasicBlock *MBB;
  MachineBasicBlock::iterator I;
  const DebugLoc &DL;
  const TargetInstrInfo &TII;
  const TargetRegisterInfo &TRI;
  const RegisterBankInfo &RBI;
```

9. For public methods of the `M88kBuilder` class, we must implement the constructor for this builder. Upon initialization, our specialized builder requires an instance of the `M88kDivInstr` pass to initialize the target instruction, register information, and the register bank information, as well as `MachineBasicBlock` and a debug location:

```
public:
  M88kBuilder(M88kDivInstr &Pass, MachineBasicBlock *MBB, const
DebugLoc &DL)
        : MBB(MBB), I(MBB->end()), DL(DL), TII(*Pass.TII),
TRI(*Pass.TRI),
          RBI(*Pass.RBI) {}
```

10. Next, a method to set `MachineBasicBlock` inside the M88k builder is created, and the `MachineBasicBlock` iterator is also set accordingly:

```
void setMBB(MachineBasicBlock *NewMBB) {
  MBB = NewMBB;
```

```
    I = MBB->end();
  }
```

11. The `constrainInst()` function needs to be implemented next and is needed for when `MachineInstr` instances are processed. For a given `MachineInstr`, we check if the register class of the `MachineInstr` instance's operands can be constrained through the pre-existing function, `constrainSelectedInstRegOperands()`. As shown here, this machine function pass requires that the register operands of the machine instruction can be constrained:

```
void constrainInst(MachineInstr *MI) {
  if (!constrainSelectedInstRegOperands(*MI, TII, TRI, RBI))
    llvm_unreachable("Could not constrain register operands");
}
```

12. One of the instructions that this pass inserts is a `BCND` instruction, as defined in `M88kInstrInfo.td`, which is a conditional branch on the M88k target. To create this instruction, we require a condition code, which are the `CC0` enums that were implemented at the beginning of `M88kDivInstr.cpp` – that is, a register and `MachineBasicBlock`. The BCND instruction is simply returned upon creation and after checking if the newly created instruction can be constrained:

```
MachineInstr *bcnd(CC0 Cc, Register Reg, MachineBasicBlock
*TargetMBB) {
  MachineInstr *MI = BuildMI(*MBB, I, DL, TII.get(M88k::BCND))
                       .addImm(static_cast<int64_t>(Cc))
                       .addReg(Reg)
                       .addMBB(TargetMBB);
  constrainInst(MI);
  return MI;
}
```

13. Similarly, we also require a trap instruction for our machine function pass, which is a `TRAP503` instruction. This instruction requires a register and raises a trap with vector 503 if the 0-th bit of the register is not set, which will be raised after a zero division. Upon creating the `TRAP503` instruction, `TRAP503` is checked for constraints before being returned. Moreover, this concludes the class implementation of the `M88kBuilder` class and completes the previously declared anonymous namespace:

```
MachineInstr *trap503(Register Reg) {
  MachineInstr *MI = BuildMI(*MBB, I, DL, TII.
get(M88k::TRAP503)).addReg(Reg);
  constrainInst(MI);
  return MI;
}
```

```
    };
} // end anonymous namespace
```

14. We can now start implementing the functions that perform the actual checks in the machine function pass. First, let's explore how `addZeroDivCheck()` is implemented. This function simply inserts a check for division by zero between the current machine instruction, which is expected to point to either `DIVSrr` or `DIVUrr`; these are mnemonics for signed and unsigned divisions, respectively. The `BCND` and `TRAP503` instructions are inserted, and the `InsertedChecks` statistic is incremented to indicate the addition of the two instructions:

```
void M88kDivInstr::addZeroDivCheck(MachineBasicBlock &MBB,
                                   MachineInstr *DivInst) {
  assert(DivInst->getOpcode() == M88k::DIVSrr ||
         DivInst->getOpcode() == M88k::DIVUrr && "Unexpected
           opcode");
  MachineBasicBlock *TailBB = MBB.splitAt(*DivInst);
  M88kBuilder B(*this, &MBB, DivInst->getDebugLoc());
  B.bcnd(CC0::NE0, DivInst->getOperand(2).getReg(), TailBB);
  B.trap503(DivInst->getOperand(2).getReg());
  ++InsertedChecks;
}
```

15. `runOnMachineFunction()` is implemented next and is one of the important functions to override when creating a type of function pass within LLVM. This function returns true or false, depending on if any changes have been made during the duration of the machine function pass. Furthermore, for a given machine function, we gather all the relevant M88k subtarget information, including the target instruction, target register, register bank, and machine register information. Details regarding whether or not the user turns the `M88kDivInstr` machine function pass on or off are also queried and stored in the `AddZeroDivCheck` variable. Additionally, all machine basic blocks in the machine function are analyzed for the division by zero. The function that performs the machine basic block analysis is `runOnMachineBasicBlock()`; we will implement this next. Finally, if the machine function has changed, this is indicated by the `Changed` variable that is returned:

```
bool M88kDivInstr::runOnMachineFunction(MachineFunction &MF) {
  const M88kSubtarget &Subtarget =
  MF.getSubtarget<M88kSubtarget>();
  TII = Subtarget.getInstrInfo();
  TRI = Subtarget.getRegisterInfo();
  RBI = Subtarget.getRegBankInfo();
  MRI = &MF.getRegInfo();
  AddZeroDivCheck = !TM->noZeroDivCheck();
  bool Changed = false;
  for (MachineBasicBlock &MBB : reverse(MF))
    Changed |= runOnMachineBasicBlock(MBB);
```

```
    return Changed;
}
```

16. For the `runOnMachineBasicBlock()` function, a Changed Boolean flag is also returned to indicate if the machine basic block has been changed; however, it is initially set to `false`. Furthermore, within a machine basic block, we need to analyze all the machine instructions and check if the instructions are the `DIVUrr` or `DIVSrr` opcodes, respectively. In addition to checking if the opcodes are divide instructions, we need to check if the user has turned our machine function pass on or off. If all of these conditions are satisfied, the division by zero checks with the conditional branch and the trap instructions are added accordingly through the `addZeroDivCheck()` function, which was implemented previously.

```
bool M88kDivInstr::runOnMachineBasicBlock(MachineBasicBlock
&MBB) {
    bool Changed = false;
    for (MachineBasicBlock::reverse_instr_iterator I =
    MBB.instr_rbegin();
        I != MBB.instr_rend(); ++I) {
      unsigned Opc = I->getOpcode();
      if ((Opc == M88k::DIVUrr || Opc == M88k::DIVSrr) &&
      AddZeroDivCheck) {
        addZeroDivCheck(MBB, &*I);
        Changed = true;
      }
    }
    return Changed;
}
```

17. After, we need to implement the constructor to initialize our function pass and set the appropriate machine function properties. This can be achieved by calling the `initializeM88kDivInstrPass()` function with the `PassRegistry` instance inside the constructor of the `M88kDivInstr` class, and also by setting the machine function properties to indicate that our pass requires machine functions to be in SSA form:

```
M88kDivInstr::M88kDivInstr(const M88kTargetMachine *TM)
      : MachineFunctionPass(ID), TM(TM) {
    initializeM88kDivInstrPass(*PassRegistry::getPassRegistry());
}

MachineFunctionProperties M88kDivInstr::getRequiredProperties()
const {
    return MachineFunctionProperties().set(
      MachineFunctionProperties::Property::IsSSA);
}
```

18. The next step is to initialize the ID for our machine function pass and to instantiate the `INITIALIZE_PASS` macro with the details of our machine function pass. This requires the pass instance, naming information, and two Boolean arguments that indicate if the pass only examines the CFG and if the pass is an analysis pass. Since `M88kDivInstr` performs neither of those, two `false` arguments are specified to the pass initialization macro:

```
char M88kDivInstr::ID = 0;
INITIALIZE_PASS(M88kDivInstr, DEBUG_TYPE, "Handle div
instructions", false, false)
```

19. Finally, the `createM88kDivInstr()` function creates a new instance of the `M88kDivInstr` pass, with a `M88kTargetMachine` instance. This is encapsulated into an `llvm` namespace, and the namespace is ended after finishing this function:

```
namespace llvm {
FunctionPass *createM88kDivInstr(const M88kTargetMachine &TM) {
  return new M88kDivInstr(&TM);
}
} // end namespace llvm
```

Building newly implemented machine function passes

We're almost done with implementing our new machine function pass! Now, we need to ensure CMake is aware of the new machine function pass within `M88kDivinstr.cpp`. This file is then added to `llvm/lib/Target/M88k/CMakeLists.txt`:

```
add_llvm_target(M88kCodeGen
    M88kAsmPrinter.cpp
    M88kDivInstr.cpp
    M88kFrameLowering.cpp
    M88kInstrInfo.cpp
    M88kISelDAGToDAG.cpp
```

The last step is to build LLVM with our new machine function pass implementation with the following commands. We require the `-DLLVM_EXPERIMENTAL_TARGETS_TO_BUILD=M88k` CMake option to build the M88k target:

```
$ cmake -G Ninja ../llvm-project/llvm -DLLVM_EXPERIMENTAL_TARGETS_TO_
BUILD=M88k -DCMAKE_BUILD_TYPE=Release -DLLVM_ENABLE_PROJECTS="llvm"
$ ninja
```

With that, we've implemented the machine function pass, but wouldn't it be interesting to see how it works? We can demonstrate the result of this pass by passing LLVM IR through `llc`.

A glimpse of running a machine function pass with llc

We have the following IR, which contains a division by zero:

```
$ cat m88k-divzero.ll
target datalayout = "E-m:e-p:32:32:32-i1:8:8-i8:8:8-i16:16:16-
i32:32:32-i64:64:64-f32:32:32-f64:64:64-a:8:16-n32"
target triple = "m88k-unknown-openbsd"

@dividend = dso_local global i32 5, align 4
define dso_local i32 @testDivZero() #0 {
  %1 = load i32, ptr @dividend, align 4
  %2 = sdiv i32 %1, 0
  ret i32 %2
}
```

Let's feed it into llc:

```
$ llc m88k-divzero.ll
```

By doing this, we'll see that, in the resulting assembly, by default, the division by zero checks, which are represented by bcnd.n (BCND) and tb0 (TRAP503), are inserted by our new machine function pass:

```
| %bb.1:
      subu %r2, %r0, %r2
      bcnd.n ne0, %r0, .LBB0_2
      divu %r2, %r2, 0
      tb0 0, %r3, 503
  . . .
.LBB0_3:
      bcnd.n ne0, %r0, .LBB0_4
      divu %r2, %r2, 0
      tb0 0, %r3, 503
```

However, let's see what happens when we specify --m88k-no-check-zero-division to llc:

```
$ llc m88k-divzero.ll —m88k-no-check-zero-division
```

This option to the backend instructs llc not to run the pass that checks for the division by zero. The resulting assembly will not contain any BCND or TRAP503 instructions. Here's an example:

```
| %bb.1:
      subu %r2, %r0, %r2
      divu %r2, %r2, 0
      jmp.n %r1
      subu %r2, %r0, %r2
```

As we can see, implementing a machine function pass requires several steps, but these procedures can be used as a guideline for you to implement any type of machine function pass that fits your needs. Since we have extensively explored the backend within this section, let's switch gears and see how we can teach the frontend about the M88k target.

Integrating a new target into the clang frontend

In the previous chapters, we developed the M88k target's backend implementation within LLVM. To complete the compiler implementation for the M88k target, we will investigate connecting our new target to the frontend by adding a clang implementation for our M88k target.

Implementing the driver integration within clang

Let's start by adding driver integration into clang for M88k:

1. The first change we will be making is inside the clang/include/clang/Basic/TargetInfo.h file. The BuiltinVaListKind enum lists the different kinds of __builtin_va_list types for each target, which is used for variadic functions support, so a corresponding type for M88k is added:

   ```
   enum BuiltinVaListKind {
   . . .
       // typedef struct __va_list_tag {
       //     int __va_arg;
       //     int *__va_stk;
       //     int *__va_reg;
       //} va_list;
       M88kBuiltinVaList
   };
   ```

2. Next, we must add a new header file, clang/lib/Basic/Targets/M88k.h. This file is a header for the M88k target feature support within the frontend. The first step is to define a new macro, to prevent multiple inclusive of the same header files, types, variables, and more. We must also include various headers that we require for the implementation to follow:

   ```
   #ifndef LLVM_CLANG_LIB_BASIC_TARGETS_M88K_H
   #define LLVM_CLANG_LIB_BASIC_TARGETS_M88K_H
   #include "OSTargets.h"
   #include "clang/Basic/TargetInfo.h"
   #include "clang/Basic/TargetOptions.h"
   #include "llvm/Support/Compiler.h"
   #include "llvm/TargetParser/Triple.h"
   ```

3. The methods we will declare will be added to the `clang` and `targets` namespaces accordingly, much like the other targets within `llvm-project`:

```
namespace clang {
namespace targets {
```

4. Let's declare the actual `M88kTargetInfo` class now, and have it extend the original `TargetInfo` class. This class is marked with LLVM_LIBRARY_VISIBILITY because if this class is linked to a shared library, this attribute allows the `M88kTargetInfo` class to only be visible from within the library, and inaccessible externally:

```
class LLVM_LIBRARY_VISIBILITY M88kTargetInfo: public TargetInfo
{
```

5. Additionally, we must declare two variables – an array of characters to represent the register names and an enum value containing the type of CPUs available in the M88k target that can be selected. The default CPU that we set is the CK_Unknown CPU. Later, we will see that this can be overwritten by user options:

```
static const char *const GCCRegNames[];
enum CPUKind { CK_Unknown, CK_88000, CK_88100, CK_88110 } CPU
= CK_Unknown;
```

6. After, we begin by declaring the public methods that will be needed in our class implementation. Aside from the constructor of our class, we define various getter methods. This includes methods that get target-specific #define values, ones that get a list of built-ins supported by the target, methods that return the GCC register names along with their aliases, and finally, a method that returns our M88k `BuiltinVaListKind` that we previously added to `clang/include/clang/Basic/TargetInfo.h`:

```
public:
  M88kTargetInfo(const llvm::Triple &Triple, const TargetOptions
&);

  void getTargetDefines(const LangOptions &Opts,
                        MacroBuilder &Builder) const override;
  ArrayRef<Builtin::Info> getTargetBuiltins() const override;
  ArrayRef<const char *> getGCCRegNames() const override;
  ArrayRef<TargetInfo::GCCRegAlias> getGCCRegAliases() const
override;
  BuiltinVaListKind getBuiltinVaListKind() const override {
    return TargetInfo::M88kBuiltinVaList;
  }
```

7. Following the getter methods, we must also define methods that perform various checks on the M88k target. The first one checks if the M88k target has a particular target feature, supplied in the form of a string. Secondly, we add a function to validate the constraints when inline assembly is used. Finally, we have a function that checks if a specific CPU is valid for the M88k target, also supplied in the form of a string:

```
bool hasFeature(StringRef Feature) const override;
bool validateAsmConstraint(const char *&Name,
                           TargetInfo::ConstraintInfo &info)
                           const override;
bool isValidCPUName(StringRef Name) const override;
```

8. Next, let's declare setter methods for our M88kTargetInfo class. The first one simply sets the specific M88k CPU that we want to target, while the second method sets a vector to contain all of the valid supported CPUs for M88k:

```
bool setCPU(const std::string &Name) override;
void fillValidCPUList(SmallVectorImpl<StringRef> &Values)
const override;
};
```

9. To complete our header for the driver implementation, let's conclude our namespaces and macro definition that we added in the beginning:

```
} // namespace targets
} // namespace clang
#endif // LLVM_CLANG_LIB_BASIC_TARGETS_M88K_H
```

10. Now that we've completed the M88k header file within clang/lib/Basic/Targets, we must add the corresponding TargetInfo C++ implementation within clang/lib/Basic/Targets/M88k.cpp. We'll start by including the required header files, especially the new M88k.h header we have just created:

```
#include "M88k.h"
#include "clang/Basic/Builtins.h"
#include "clang/Basic/Diagnostic.h"
#include "clang/Basic/TargetBuiltins.h"
#include "llvm/ADT/StringExtras.h"
#include "llvm/ADT/StringRef.h"
#include "llvm/ADT/StringSwitch.h"
#include "llvm/TargetParser/TargetParser.h"
#include <cstring>
```

11. As we did previously in the header, we start with the `clang` and `targets` namespaces, and then also begin implementing the constructor for the `M88kTargetInfo` class:

```
namespace clang {
namespace targets {
M88kTargetInfo::M88kTargetInfo(const llvm::Triple &Triple,
                               const TargetOptions &)
    : TargetInfo(Triple) {
```

12. Within the constructor, we set the data layout string for the M88k target. As you may have seen before, this data layout string is seen at the top of the emitted LLVM IR files. An explanation of each section of the data layout string is described here:

```
std::string Layout = "";
Layout += "E"; // M68k is Big Endian
Layout += "-m:e";
Layout += "-p:32:32:32"; // Pointers are 32 bit.
// All scalar types are naturally aligned.
Layout += "-i1:8:8-i8:8:8-i16:16:16-i32:32:32-i64:64:64";

// Floats and doubles are also naturally aligned.
Layout += "-f32:32:32-f64:64:64";
// We prefer 16 bits of aligned for all globals; see above.
Layout += "-a:8:16";

Layout += "-n32"; // Integer registers are 32bits.
resetDataLayout(Layout);
```

13. The constructor for the `M88kTargetInfo` class concludes by setting the various variable types as `signed long long`, `unsigned long`, or `signed int`:

```
IntMaxType = SignedLongLong;
Int64Type = SignedLongLong;
SizeType = UnsignedLong;
PtrDiffType = SignedInt;
IntPtrType = SignedInt;
}
```

14. After that, the function to set the CPU for the target is implemented. This function takes a string and sets the CPU to be the particular CPU string that is supplied by the user within `llvm::StringSwitch`, which is essentially just a regular switch but specifically for strings with LLVM. We can see that there are three supported CPU types on the M88k target, and there is a CK_Unknown type for if the supplied string does not match any of the expected types:

```
bool M88kTargetInfo::setCPU(const std::string &Name) {
  StringRef N = Name;
```

```
CPU = llvm::StringSwitch<CPUKind>(N)
           .Case("generic", CK_88000)
           .Case("mc88000", CK_88000)
           .Case("mc88100", CK_88100)
           .Case("mc88110", CK_88110)
           .Default(CK_Unknown);
    return CPU != CK_Unknown;
}
```

15. It was previously stated that there are three supported and valid CPU types on the M88k target: mc88000, mc88100, and mc88110, with the generic type simply being the mc88000 CPU. We must implement the following functions to enforce these valid CPUs within clang. First, we must declare an array of strings, ValidCPUNames[], to denote the valid CPU names on M88k. Secondly, the fillValidCPUList() method populates the array of valid CPU names into a vector. This vector is then used in the isValidCPUName() method, to check whether a particular CPU name supplied is indeed valid for our M88k target:

```
static constexpr llvm::StringLiteral ValidCPUNames[] = {
    {"generic"}, {"mc88000"}, {"mc88100"}, {"mc88110"}};

void M88kTargetInfo::fillValidCPUList(
    SmallVectorImpl<StringRef> &Values) const {
  Values.append(std::begin(ValidCPUNames),
    std::end(ValidCPUNames));
}
bool M88kTargetInfo::isValidCPUName(StringRef Name) const {
  return llvm::is_contained(ValidCPUNames, Name);
}
```

16. Next, the getTargetDefines() method is implemented. This function defines the macros that are necessary for the frontend, such as the valid CPU types. Aside from the __m88k__ and __m88k macros, we must also define corresponding CPU macros for the valid CPUs:

```
void M88kTargetInfo::getTargetDefines(const LangOptions &Opts,
                                      MacroBuilder &Builder)
const {
  using llvm::Twine;
  Builder.defineMacro("__m88k__");
  Builder.defineMacro("__m88k");
  switch (CPU) { // For sub-architecture
  case CK_88000:
    Builder.defineMacro("__mc88000__");
    break;
  case CK_88100:
```

```
        Builder.defineMacro("__mc88100__");
        break;
    case CK_88110:
        Builder.defineMacro("__mc88110__");
        break;
    default:
        break;
    }
}
```

17. The next few functions are stub functions but are required for the frontend for basic support. This includes the functions to get builtins from a target and a function to query the target if a specific feature of the target is supported. For now, we'll leave them unimplemented and set default return values for these functions so that they can be implemented later:

```
ArrayRef<Builtin::Info> M88kTargetInfo::getTargetBuiltins()
const {
    return std::nullopt;
}
bool M88kTargetInfo::hasFeature(StringRef Feature) const {
    return Feature == "M88000";
}
```

18. Following these functions, we will add an implementation for the register names on M88k. Usually, the list of supported register names and their purposes can be found on the ABI of the specific platform of interest. Within this implementation, we'll implement the main general-purpose registers from 0-31 and also create an array to store this information in. In terms of register aliases, note that there are no aliases for the registers that we implement currently:

```
const char *const M88kTargetInfo::GCCRegNames[] = {
    "r0",  "r1",  "r2",  "r3",  "r4",  "r5",  "r6",  "r7",
    "r8",  "r9",  "r10", "r11", "r12", "r13", "r14", "r15",
    "r16", "r17", "r18", "r19", "r20", "r21", "r22", "r23",
    "r24", "r25", "r26", "r27", "r28", "r29", "r39", "r31"};

ArrayRef<const char *> M88kTargetInfo::getGCCRegNames() const {
    return llvm::makeArrayRef(GCCRegNames);
}
ArrayRef<TargetInfo::GCCRegAlias>
M88kTargetInfo::getGCCRegAliases() const {
    return std::nullopt; // No aliases.
}
```

19. The last function we'll implement is a function that validates the inline assembly constraints on our target. This function simply takes a character, which represents the inline assembly constraint, and handles the constraint accordingly. A few inline assembly register constraints are implemented, such as for the address, data, and floating-point registers, and a select few constraints for constants are also accounted for:

```
bool M88kTargetInfo::validateAsmConstraint(
    const char *&Name, TargetInfo::ConstraintInfo &info) const {
  switch (*Name) {
  case 'a': // address register
  case 'd': // data register
  case 'f': // floating point register
    info.setAllowsRegister();
    return true;
  case 'K': // the constant 1
  case 'L': // constant -1^20 .. 1^19
  case 'M': // constant 1-4:
    return true;
  }
  return false;
}
```

20. We conclude the file by closing off the `clang` and `targets` namespaces that we initiated at the beginning of the file:

```
} // namespace targets
} // namespace clang
```

After completing the implementation for `clang/lib/Basic/Targets/M88k.cpp`, the following implementation of adding the M88k features group and valid CPU types within `clang/include/clang/Driver/Options.td`. is required.

Recall from earlier that we previously defined three valid CPU types for our M88k target: `mc88000`, `mc88100`, and `mc88110`. These CPU types also need to be defined in `Options.td` since this file is the central place that defines all options and flags that will be accepted by clang:

1. First, we must add `m_m88k_Features_Group`, which represents a group of features that will be available to the M88k target:

```
def m_m88k_Features_Group: OptionGroup<"<m88k features group>">,
                           Group<m_Group>, DocName<"M88k">;
```

2. Next, we must define the three valid M88k CPU types as a feature in the M88k features group:

```
def m88000 : Flag<["-"], "m88000">, Group<m_m88k_Features_
Group>;
def m88100 : Flag<["-"], "m88100">, Group<m_m88k_Features_
```

```
Group>;
def m88110 : Flag<["-"], "m88110">, Group<m_m88k_Features_
Group>;
```

With that, we have implemented the driver integration portion for connecting the M88k target with clang.

Implementing ABI support for M88k within clang

Now, we need to add ABI support within the frontend for clang, which allows us to produce code specific to the M88k target from the frontend:

1. Let's start by adding the following `clang/lib/CodeGen/TargetInfo.h`. This is a prototype that creates the code generation information for the M88k target:

    ```
    std::unique_ptr<TargetCodeGenInfo>
    createM88kTargetCodeGenInfo(CodeGenModule &CGM);
    ```

2. We also need to add the following code to `clang/lib/Basic/Targets.cpp`, which will help teach clang the acceptable target triples for M88k. As we can see, for the M88k target, the acceptable operating system is OpenBSD. This means that clang accepts `m88k-openbsd` as a target triple:

    ```
    #include "Targets/M88k.h"
    #include "Targets/MSP430.h"
    . . .
        case llvm::Triple::m88k:
          switch (os) {
          case llvm::Triple::OpenBSD:
            return std::make_
    unique<OpenBSDTargetInfo<M88kTargetInfo>>(Triple, Opts);
          default:
            return std::make_unique<M88kTargetInfo>(Triple, Opts);
          }
        case llvm::Triple::le32:
    . . .
    ```

    Now, we need to create a file called `clang/lib/CodeGen/Targets/M88k.cpp` so that we can continue the code generation information and ABI implementation for M88k.

3. Within `clang/lib/CodeGen/Targets/M88k.cpp`, we must add the following necessary headers, one of which is the `TargetInfo.h` header that we have just modified. Then, we must specify that we're using the `clang` and `clang::codegen` namespaces:

    ```
    #include "ABIInfoImpl.h"
    #include "TargetInfo.h"
    using namespace clang;
    using namespace clang::CodeGen;
    ```

4. Next, we must declare a new anonymous namespace and place our `M88kABIInfo` inside of it. `M88kABIInfo` inherits from the existing `ABIInfo` from clang and contains `DefaultABIInfo` inside it. For our target, we rely heavily on the existing `ABIInfo` and `DefaultABIInfo`, which simplifies the `M88kABIInfo` class significantly:

```
namespace {
class M88kABIInfo final : public ABIInfo {
  DefaultABIInfo defaultInfo;
```

5. Furthermore, aside from adding the constructor for the `M88kABIInfo` class, a couple of methods are also added. `computeInfo()` implements the default `clang::CodeGen::ABIInfo` class. There's also the `EmitVAArg()` function, which generates code that retrieves an argument from a pointer that is passed in; this is updated after. This is primarily used for variadic function support:

```
public:
  explicit M88kABIInfo(CodeGen::CodeGenTypes &CGT)
      : ABIInfo(CGT), defaultInfo(CGT) {}
  void computeInfo(CodeGen::CGFunctionInfo &FI) const override
{}
  CodeGen::Address EmitVAArg(CodeGen::CodeGenFunction &CGF,
                             CodeGen::Address VAListAddr,
                             QualType Ty) const override {
    return VAListAddr;
  }
};
```

6. The class constructor for the `M88kTargetCodeGenInfo` class is added next, which extends from the original `TargetCodeGenInfo`. After, we must close off the anonymous namespace that was created initially:

```
class M88kTargetCodeGenInfo final : public TargetCodeGenInfo {
public:
  explicit M88kTargetCodeGenInfo(CodeGen::CodeGenTypes &CGT)
      : TargetCodeGenInfo(std::make_unique<DefaultABIInfo>(CGT))
{} };
}
```

7. Finally, we must add the implementation to create the actual `M88kTargetCodeGenInfo` class as `std::unique_ptr`, which takes in a single `CodeGenModule` that generates LLVM IR code. This directly corresponds to what was originally added to `TargetInfo.h`:

```
std::unique_ptr<TargetCodeGenInfo>
CodeGen::createM88kTargetCodeGenInfo(CodeGenModule &CGM) {
  return std::make_unique<M88kTargetCodeGenInfo>(CGM.
```

```
getTypes());
}
```

That concludes the ABI support for the M88k in the frontend.

Implementing the toolchain support for M88k within clang

The final portion of the M88k target integration within clang will be to implement toolchain support for our target. Like before, we'll need to create a header file for toolchain support. We call this header `clang/lib/Driver/ToolChains/Arch/M88k.h`:

1. First, we must define `LLVM_CLANG_LIB_DRIVER_TOOLCHAINS_ARCH_M88K_H` to prevent multiple inclusion later, and also add any necessary headers for later use. Following this, we must declare the `clang`, `driver`, `tools`, and `m88k` namespaces, with each nesting inside the other:

```
#ifndef LLVM_CLANG_LIB_DRIVER_TOOLCHAINS_ARCH_M88K_H
#define LLVM_CLANG_LIB_DRIVER_TOOLCHAINS_ARCH_M88K_H
#include "clang/Driver/Driver.h"
#include "llvm/ADT/StringRef.h"
#include "llvm/Option/Option.h"
#include <string>
#include <vector>
namespace clang {
namespace driver {
namespace tools {
namespace m88k {
```

2. Next, we must declare an `enum` value that depicts the floating-point ABI, which is for soft and hard floating points. This means that floating-point computations can either be done by the floating-point hardware itself, which is fast, or through software emulation, which would be slower:

```
enum class FloatABI { Invalid, Soft, Hard, };
```

3. Following this, we must add definitions to get the float ABI through the driver, and the CPU through clang's `-mcpu=` and `-mtune=` options. We must also declare a function that retrieves the target features from the driver:

```
FloatABI getM88kFloatABI(const Driver &D, const
llvm::opt::ArgList &Args);
StringRef getM88kTargetCPU(const llvm::opt::ArgList &Args);
StringRef getM88kTuneCPU(const llvm::opt::ArgList &Args);
void getM88kTargetFeatures(const Driver &D, const
llvm::Triple &Triple, const llvm::opt::ArgList &Args,
std::vector<llvm::StringRef> &Features);
```

4. Finally, we conclude the header file by ending the namespaces and the macro that we originally defined:

```
} // end namespace m88k
} // end namespace tools
} // end namespace driver
} // end namespace clang
#endif // LLVM_CLANG_LIB_DRIVER_TOOLCHAINS_ARCH_M88K_H
```

The last file we will implement is the C++ implementation for the toolchain support, within `clang/lib/Driver/ToolChains/Arch/M88k.cpp`:

1. Once again, we'll begin the implementation by including the necessary headers and namespaces that we will use later. We must also include the `M88k.h` header that we created earlier:

```
#include "M88k.h"
#include "ToolChains/CommonArgs.h"
#include "clang/Driver/Driver.h"
#include "clang/Driver/DriverDiagnostic.h"
#include "clang/Driver/Options.h"
#include "llvm/ADT/SmallVector.h"
#include "llvm/ADT/StringSwitch.h"
#include "llvm/Option/ArgList.h"
#include "llvm/Support/Host.h"
#include "llvm/Support/Regex.h"
#include <sstream>
using namespace clang::driver;
using namespace clang::driver::tools;
using namespace clang;
using namespace llvm::opt;
```

2. The `normalizeCPU()` function is implemented next, which processes the CPU name into the `-mcpu=` option in clang. As we can see, each CPU name has several accepted variations. Furthermore, when a user specifies `-mcpu=native`, it allows them to compile for the current host's CPU type:

```
static StringRef normalizeCPU(StringRef CPUName) {
  if (CPUName == "native") {
    StringRef CPU = std::string(llvm::sys::getHostCPUName());
    if (!CPU.empty() && CPU != "generic")
      return CPU;
  }
  return llvm::StringSwitch<StringRef>(CPUName)
```

```
      .Cases("mc88000", "m88000", "88000", "generic", "mc88000")
      .Cases("mc88100", "m88100", "88100", "mc88100")
      .Cases("mc88110", "m88110", "88110", "mc88110")
      .Default(CPUName);
   }
```

3. Next up, we must implement the getM88kTargetCPU() function, in which, given the clang CPU name that we implemented earlier in clang/include/clang/Driver/Options. td, we get the corresponding LLVM name for the M88k CPU we are targeting:

```
StringRef m88k::getM88kTargetCPU(const ArgList &Args) {
  Arg *A = Args.getLastArg(options::OPT_m88000, options::OPT_
m88100, options::OPT_m88110, options::OPT_mcpu_EQ);
  if (!A)
    return StringRef();
  switch (A->getOption().getID()) {
  case options::OPT_m88000:
    return "mc88000";
  case options::OPT_m88100:
    return "mc88100";
  case options::OPT_m88110:
    return "mc88110";
  case options::OPT_mcpu_EQ:
    return normalizeCPU(A->getValue());
  default:
    llvm_unreachable("Impossible option ID");
  }
}
```

4. The getM88kTuneCPU() function is implemented after. This is the behavior of the clang -mtune= option, which changes the instruction scheduling model to use data from a given CPU for M88k. We simply tune for whatever CPU that we are currently targeting:

```
StringRef m88k::getM88kTuneCPU(const ArgList &Args) {
  if (const Arg *A = Args.getLastArg(options::OPT_mtune_EQ))
    return normalizeCPU(A->getValue());
  return StringRef();
}
```

5. We'll also implement the getM88kFloatABI() method, which gets the floating-point ABI. Initially, we'll set the ABI to be m88k::FloatABI::Invalid as a default value. Next, we must check if any of the -msoft-float or -mhard-float options are passed to the command line. If -msoft-float is specified, then we set the ABI to m88k::FloatABI::Soft accordingly. Likewise, we set m88k::FloatABI::Hard when -mhard-float is specified

to clang. Finally, if none of these options are specified, we choose the default on the current platform, which would be a hard floating-point value for M88k:

```
m88k::FloatABI m88k::getM88kFloatABI(const Driver &D, const
ArgList &Args) {
  m88k::FloatABI ABI = m88k::FloatABI::Invalid;

  if (Arg *A =
          Args.getLastArg(options::OPT_msoft_float,
options::OPT_mhard_float)) {
    if (A->getOption().matches(options::OPT_msoft_float))
      ABI = m88k::FloatABI::Soft;
    else if (A->getOption().matches(options::OPT_mhard_float))
      ABI = m88k::FloatABI::Hard;
  }
  if (ABI == m88k::FloatABI::Invalid)
    ABI = m88k::FloatABI::Hard;
  return ABI;
}
```

6. We'll add the implementation for getM88kTargetFeatures() next. The important part of this function is the vector of Features that are passed as a parameter. As we can see, the only target features that are handled are the floating-point ABI. From the driver and arguments passed to it, we'll get the appropriate floating-point ABI from what we implemented in the previous step. Note that we add the -hard-float target features to the Features vector for soft float ABI as well, which means that currently, M88k only supports hard float:

```
void m88k::getM88kTargetFeatures(const Driver &D, const
llvm::Triple &Triple,
                                 const ArgList &Args,
                                 std::vector<StringRef>
&Features) {
  m88k::FloatABI FloatABI = m88k::getM88kFloatABI(D, Args);
  if (FloatABI == m88k::FloatABI::Soft)
    Features.push_back("-hard-float");
}
```

Building the M88k target with clang integration

We're almost done with the implementation for integrating M88k into clang. The last step is to add the new clang files that we have added into their corresponding CMakeLists.txt file, which allows us to build the clang project with our M88k target implementation:

1. First, add the Targets/M88k.cpp line to clang/lib/Basic/CMakeLists.txt.

2. Next, add Targets/M88k.cpp to clang/lib/CodeGen/CMakeLists.txt.

3. Finally, add `ToolChains/Arch/M88k.cpp` to `clang/lib/Driver/CMakeLists.txt`.

There we have it! That concludes our toolchain implementation for the toolchain support for the M88k target, which subsequently means we've completed the integration into clang for M88k!

The last step we need to do is build clang with the M88k target. The following commands will build the clang and LLVM project. For clang, be aware of the M88k target. Here, the `-DLLVM_EXPERIMENTAL_TARGETS_TO_BUILD=M88k` CMake option must be added, as in the previous section:

```
$ cmake -G Ninja ../llvm-project/llvm -DLLVM_EXPERIMENTAL_
TARGETS_TO_BUILD=M88k -DCMAKE_BUILD_TYPE=Release -DLLVM_ENABLE_
PROJECTS="clang;llvm"
$ ninja
```

We should now have a version of clang that recognizes the M88k target! We can confirm this by checking the list of targets that clang supports, through the `–print-targets` option:

```
$ clang --print-targets | grep M88k
    m88k          - M88k
```

In this section, we delved into the technical details of integrating a new backend target into clang and having it recognized. In the next section, we'll explore the concept of cross-compiling, where we detail the procedure of targeting a different CPU architecture from the current host.

Targeting a different CPU architecture

Today, many small computers, such as the Raspberry Pi, are in use despite having only limited resources. Running a compiler on such a computer is often not possible or it takes too much time. Thus, a common requirement for a compiler is to generate code for a different CPU architecture. The whole process of having a host compile an executable for a different target is called cross-compiling.

In cross-compiling, two systems are involved: the host system and the target system. The compiler runs on the host system and produces code for the target system. To denote the systems, the so-called triple is used. This is a configuration string that usually consists of the CPU architecture, the vendor, and the operating system. Furthermore, additional information about the environment is often added to the configuration string. For example, the `x86_64-pc-win32` triple is used for a Windows system running on a 64-bit X86 CPU. The CPU architecture is `x86_64`, `pc` is a generic vendor, and `win32` is the operating system, and all of these pieces are connected by a hyphen. A Linux system running on an ARMv8 CPU uses `aarch64-unknown-linux-gnu` as the triple, with `aarch64` as the CPU architecture. Moreover, the operating system is `linux`, running a `gnu` environment. There is no real vendor for a Linux-based system, so this part is `unknown`. Additionally, parts that are not known or unimportant for a specific purpose are often omitted: the `aarch64-linux-gnu` triple describes the same Linux system.

Let's assume your development machine runs Linux on an X86 64-bit CPU and you want to cross-compile to an ARMv8 CPU system running Linux. The host triple is x86_64-linux-gnu and the target triple is aarch64-linux-gnu. Different systems have different characteristics. Thus, your application must be written in a portable fashion; otherwise, complications may arise. Some common pitfalls are as follows:

- **Endianness**: The order in which multi-byte values are stored in memory can be different.

- **Pointer size**: The size of a pointer varies with the CPU architecture (usually 16, 32, or 64-bit). The C int type may not be large enough to hold a pointer.

- **Type differences**: Data types are often closely related to the hardware. The long double type can use 64-bit (ARM), 80-bit (X86), or 128-bit (ARMv8). PowerPC systems may use double-double arithmetic for long double, which gives more precision by using a combination of two 64-bit double values.

If you do not pay attention to these points, then your application can act surprisingly or crash on the target platform, even if it runs perfectly on your host system. The LLVM libraries are tested on different platforms and also contain portable solutions to the aforementioned issues.

For cross-compiling, the following tools are required:

- A compiler that generates code for the target

- A linker capable of generating binaries for the target

- Header files and libraries for the target

Fortunately, the Ubuntu and Debian distributions have packages that support cross-compiling. We're taking advantage of this in the following setup. The gcc and g++ compilers, the linker, ld, and the libraries are available as precompiled binaries that produce ARMv8 code and executables. The following command installs all of these packages:

```
$ sudo apt -y install gcc-12-aarch64-linux-gnu \
  g++-12-aarch64-linux-gnu binutils-aarch64-linux-gnu \
  libstdc++-12-dev-arm64-cross
```

The new files are installed under the /usr/aarch64-linux-gnu directory. This directory is the (logical) root directory of the target system. It contains the usual bin, lib, and include directories. The cross-compilers (aarch64-linux-gnu-gcc-8 and aarch64-linux-gnu-g++-8) are aware of this directory.

> **Cross-compiling on other systems**
>
> Some distributions, such as Fedora, only provide cross-compiling support for bare-metal targets such as the Linux kernel, but the header and library files needed for user land applications are not provided. In such a case, you can simply copy the missing files from your target system.
>
> If your distribution does not come with the required toolchain, then you can build it from source. For the compiler, you can use clang or gcc/g++. The gcc and g++ compilers must be configured to produce code for the target system and the binutils tools need to handle files for the target system. Moreover, the C and C++ libraries need to be compiled with this toolchain. The steps vary by operating system and host and target architecture. On the web, you can find instructions if you search for `gcc cross-compile <architecture>`.

With this preparation, you are almost ready to cross-compile the sample application (including the LLVM libraries) except for one little detail. LLVM uses the **TableGen** tool during the build. During cross-compilation, everything is compiled for the target architecture, including this tool. You can use `llvm-tblgen` from the build in *Chapter 1* or you can compile only this tool. Assuming you are in the directory that contains the clone of this book's GitHub repository, type the following:

```
$ mkdir build-host
$ cd build-host
$ cmake -G Ninja \
   -DLLVM_TARGETS_TO_BUILD="X86" \
   -DLLVM_ENABLE_ASSERTIONS=ON \
   -DCMAKE_BUILD_TYPE=Release \
   ../llvm-project/llvm
$ ninja llvm-tblgen
$ cd ..
```

These steps should be familiar by now. A build directory is created and entered. The cmake command creates the build files for LLVM only for the X86 target. To save space and time, a release build is done but assertions are enabled to catch possible errors. Only the `llvm-tblgen` tool is compiled with `ninja`.

With the `llvm-tblgen` tool at hand, you can now start the cross-compilation process. The CMake command line is very long, so you may want to store the command in a script file. The difference from previous builds is that more information must be provided:

```
$ mkdir build-target
$ cd build-target
$ cmake -G Ninja \
   -DCMAKE_CROSSCOMPILING=True \
   -DLLVM_TABLEGEN=../build-host/bin/llvm-tblgen \
   -DLLVM_DEFAULT_TARGET_TRIPLE=aarch64-linux-gnu \
   -DLLVM_TARGET_ARCH=AArch64 \
```

```
     -DLLVM_TARGETS_TO_BUILD=AArch64 \
     -DLLVM_ENABLE_ASSERTIONS=ON \
     -DLLVM_EXTERNAL_PROJECTS=tinylang \
     -DLLVM_EXTERNAL_TINYLANG_SOURCE_DIR=../tinylang \
     -DCMAKE_INSTALL_PREFIX=../target-tinylang \
     -DCMAKE_BUILD_TYPE=Release \
     -DCMAKE_C_COMPILER=aarch64-linux-gnu-gcc-12 \
     -DCMAKE_CXX_COMPILER=aarch64-linux-gnu-g++-12 \
     ../llvm-project/llvm
  $ ninja
```

Again, you create a build directory and enter it before running the CMake command. Some of these CMake parameters have not been used before and require some explanation:

- CMAKE_CROSSCOMPILING set to ON tells CMake that we are cross-compiling.

- LLVM_TABLEGEN specifies the path to the llvm-tblgen tool to use. This is the one from the previous build.

- LLVM_DEFAULT_TARGET_TRIPLE is the triple of the target architecture.

- LLVM_TARGET_ARCH is used for **JIT** code generation. It defaults to the architecture of the host. For cross-compiling, this must be set to the target architecture.

- LLVM_TARGETS_TO_BUILD is the list of targets for which LLVM should include code generators. The list should at least include the target architecture.

- CMAKE_C_COMPILER and CMAKE_CXX_COMPILER specify the C and C++ compilers used for the build, respectively. The binaries of the cross-compilers are prefixed with the target triple and are not found automatically by CMake.

With the other parameters, a release build with assertions enabled is requested and our tinylang application is built as part of LLVM. Once the compilation process has finished, the file command can demonstrate that we have created a binary for ARMv8. Specifically, we can run $ file bin/ tinylang and check that the output says ELF 64-bit object for the ARM aarch64 architecture.

> **Cross-compiling with clang**
>
> As LLVM generates code for different architectures, it seems obvious to use clang to cross-compile. The obstacle here is that LLVM does not provide all the required parts – for example, the C library is missing. Because of this, you must use a mix of LLVM and GNU tools and as a result, you need to tell CMake even more about the environment you are using. As a minimum, you need to specify the following options for clang and clang++: `--target=<target-triple>` (enables code generation for a different target), `--sysroot=<path>` (path to the root directory for the target), `I` (search path for header files), and `−L` (search path for libraries). During the CMake run, a small application is compiled and CMake complains if something is wrong with your setup. This step is sufficient to check if you have a working environment. Common problems are picking the wrong header files or link failures due to different library names or wrong search paths.

Cross-compiling is surprisingly complex. With the instructions from this section, you will be able to cross-compile your application for a target architecture of your choice.

Summary

In this chapter, you learned about creating passes that run beyond instruction selection, specifically exploring the creation behind machine function passes in the backend! You also discovered how to add a new experimental target into clang, and some of the driver, ABI, and toolchain changes that are required. Finally, while considering the supreme discipline of compiler construction, you learned how to cross-compile your application for another target architecture.

Now that we're at the end of *Learn LLVM 17*, you are equipped with the knowledge to use LLVM in creative ways in your projects and have explored many interesting topics. The LLVM ecosystem is very active, and new features are added all the time, so be sure to follow its development!

As compiler developers ourselves, it was a pleasure for us to write about LLVM and discover some new features along the way. Have fun with LLVM!

Index

X

Z

Packtpub.com

Subscribe to our online digital library for full access to over 7,000 books and videos, as well as industry leading tools to help you plan your personal development and advance your career. For more information, please visit our website.

Why subscribe?

- Spend less time learning and more time coding with practical eBooks and Videos from over 4,000 industry professionals

- Improve your learning with Skill Plans built especially for you

- Get a free eBook or video every month

- Fully searchable for easy access to vital information

- Copy and paste, print, and bookmark content

Did you know that Packt offers eBook versions of every book published, with PDF and ePub files available? You can upgrade to the eBook version at packtpub.com and as a print book customer, you are entitled to a discount on the eBook copy. Get in touch with us at customercare@packtpub.com for more details.

At www.packt.com, you can also read a collection of free technical articles, sign up for a range of free newsletters, and receive exclusive discounts and offers on Packt books and eBooks.

Other Books You May Enjoy

If you enjoyed this book, you may be interested in these other books by Packt:

LLVM Techniques, Tips, and Best Practices Clang and Middle-End Libraries

Min-Yih Hsu

ISBN: 978-1-83882-495-2

- Find out how LLVM's build system works and how to reduce the building resource
- Get to grips with running custom testing with LLVM's LIT framework
- Build different types of plugins and extensions for Clang
- Customize Clang's toolchain and compiler flags
- Write LLVM passes for the new PassManager
- Discover how to inspect and modify LLVM IR
- Understand how to use LLVM's profile-guided optimizations (PGO) framework
- Create custom compiler sanitizers

C++20 STL Cookbook

Bill Weinman

ISBN: 978-1-80324-871-4

- Understand the new language features and the problems they can solve
- Implement generic features of the STL with practical examples
- Understand standard support classes for concurrency and synchronization
- Perform efficient memory management using the STL
- Implement seamless formatting using std::format
- Work with strings the STL way instead of handcrafting C-style code

Packt is searching for authors like you

If you're interested in becoming an author for Packt, please visit `authors.packtpub.com` and apply today. We have worked with thousands of developers and tech professionals, just like you, to help them share their insight with the global tech community. You can make a general application, apply for a specific hot topic that we are recruiting an author for, or submit your own idea.

Share Your Thoughts

Now you've finished *Learn LLVM 17*, we'd love to hear your thoughts! Scan the QR code below to go straight to the Amazon review page for this book and share your feedback or leave a review on the site that you purchased it from.

`https://packt.link/r/1837631344`

Your review is important to us and the tech community and will help us make sure we're delivering excellent quality content.

Download a free PDF copy of this book

Thanks for purchasing this book!

Do you like to read on the go but are unable to carry your print books everywhere?

Is your eBook purchase not compatible with the device of your choice?

Don't worry, now with every Packt book you get a DRM-free PDF version of that book at no cost.

Read anywhere, any place, on any device. Search, copy, and paste code from your favorite technical books directly into your application.

The perks don't stop there, you can get exclusive access to discounts, newsletters, and great free content in your inbox daily

Follow these simple steps to get the benefits:

1. Scan the QR code or visit the link below

https://packt.link/free-ebook/9781837631346

2. Submit your proof of purchase
3. That's it! We'll send your free PDF and other benefits to your email directly